T0234010

# Polymer Processing Instabilities
## Control and Understanding

Edited by
### Savvas G. Hatzikiriakos
### Kalman B. Migler

CRC Press
Taylor & Francis Group
Boca Raton London New York

CRC Press is an imprint of the
Taylor & Francis Group, an **informa** business

First published 2005 by Marcel Dekker

Published 2020 by CRC Press
Taylor & Francis Group
6000 Broken Sound Parkway NW, Suite 300
Boca Raton, FL 33487-2742

First issued in paperback 2020

© 2005 by Taylor & Francis Group, LLC
CRC Press is an imprint of Taylor & Francis Group, an Informa business

No claim to original U.S. Government works

ISBN 13: 978-0-367-57818-3 (pbk)
ISBN 13: 978-0-8247-5386-3 (hbk)

**Visit the Taylor & Francis Web site at**
**http://www.taylorandfrancis.com**

**and the CRC Press Web site at**
**http://www.crcpress.com**

**Library of Congress Cataloging-in-Publication Data**
A catalog record for this book is available from the Library of Congress.

# Preface

Polymer processing has grown in the past 50 years into a multi-billion dollar industry with production facilities and development labs all over the world. The primary reason for this phenomenal growth compared to other materials is the relative ease of manufacture and processing. Numerous methods have been developed to process polymeric materials at high volume and at relatively low temperatures.

But despite this success, manufacturing is limited by the occurrence of polymer processing instabilities. These limitations manifest themselves in two ways; first as the rate limiting step in the optimization of existing operations, and second in the introduction of new materials to the marketplace. For example, it is natural to ask what is the rate limiting step for processing operations; why not run a given operation 20% faster? Quite frequently, the answer to this question concerns flow instabilities. As the polymeric material is processed in the molten state; it retains characteristics of both liquids and solids. The faster one processes it, the more solid like its response becomes; seemingly simple processing operations become intractably difficult and the polymer flow becomes "chaotic" and uncontrollable.

A second problem concerns the development of new materials; this is particularly pressing because new materials with enhanced properties offer relief from the commodities nature of the polymer processing industry. But in order to gain market acceptance, they must also enjoy ease of processability and there are numerous examples of new materials which processed poorly and suffered in the market.

The past decade has seen great progress in our efforts to understand and control polymer processing instabilities; however much of the success is scattered throughout the scientific and technical literature. The intention of this book is to coherently collect these recent triumphs and present them to a wide audience. This book is intended for polymer rheologists, scientists, engineers, technologists and graduate students who are engaged in the field of polymer processing operations and need to understand the impact of flow instabilities. It is also intended for those who are already active in fields such as instabilities in polymer rheology and processing and wish to widen their knowledge and understanding further.

The chapters in this book seek to impart both fundamental and practical understanding on various flows that occur during processing. Processes where instabilities pose serious limitations in the rate of production include extrusion and co-extrusion, blow molding, film blowing, film casting, and injection molding. Methods to cure and eliminate such instabilities is also of concern in this book. For example, conventional polymer processing additives that eliminate flow instabilities such as sharkskin melt fracture, and non-conventional polymer processing additives that eliminate flow instabilities such as gross melt fracture are also integral parts of this book. Materials of interest that are covered in this book include most of the commodity polymers that are processed as melts at temperatures above their melting point (polyethylenes, polypropylenes, fluoropolymers, and others) or as concentrated solutions at lower temperatures.

Greater emphasis has been given to the flow instability of melt fracture since such phenomena have drawn the attention of many researchers in recent years. Moreover, these phenomena take place in a variety of processes such as film blowing, film casting, blow molding, extrusion and various coating flows. Equally important, however, instabilities take place in other processing operations such as draw resonance and the "dog-bone effect" in film casting, tiger-skin instabilities in injection molding, interfacial instabilities in co-extrusion and several secondary flows in various contraction flows. An overall overview of these instabilities can be found in the introduction (Chapter 1).

It is hoped that this book fills the gap in the polymer processing literature where polymer flow instabilities are not treated in-depth in any book. Research in this field, in particular over the last ten years, has produced a significant amount of data. An attempt is made to distil these data and to define the state-of-the art in the field. It is hoped that this will be useful to researchers active in the field as a starting point as well as a guide to obtain helpful direction in their research. It would be almost impossible to include all the knowledge generated over the past fifty-to-sixty years in a single book. As such, we would like to apologize for not citing several important reports and contributions to the field.

*Savvas G. Hatzikiriakos*
*Kalman B. Migler*

# Contents

Contents

# Contributors

**Lynden A. Archer**   Cornell University, Ithaca, New York, U.S.A.

**F.P.T. Baaijens**   Eindhoven University of Technology, Eindhoven, The Netherlands

**A.C.B. Bogaerds**   Eindhoven University of Technology, Eindhoven, and DSM Research, Geleen, The Netherlands

**Albert Co**   University of Maine, Orono, Maine, U.S.A.

**John M. Dealy**   McGill University, Montreal, Quebec, Canada

**Joseph Dooley**   The Dow Chemical Company, Midland, Michigan, U.S.A.

**Georgios Georgiou**   University of Cyprus, Nicosia, Cyprus

**Savvas G. Hatzikiriakos**   The University of British Columbia, Vancouver, British Columbia, Canada

**Jae Chun Hyun**   Korea University, Seoul, South Korea

**Hyun Wook Jung**   Korea University, Seoul, South Korea

**Semen B. Kharchenko**     National Institute of Standards and Technology, Gaithersburg, Maryland, U.S.A.

**Seungoh Kim**     Verdun, Quebec, Canada

**Kalman B. Migler**     National Institute of Standards and Technology, Gaithersburg, Maryland, U.S.A.

**Evan Mitsoulis**     National Technical University of Athens, Athens, Greece

**G.W.M. Peters**     Eindhoven University of Technology, Eindhoven, The Netherlands

# 1

# Overview of Processing Instabilities

**Savvas G. Hatzikiriakos**
The University of British Columbia, Vancouver,
British Columbia, Canada

**Kalman B. Migler**
National Institute of Standards and Technology
Gaithersburg, Maryland, U.S.A.

## 1.1 POLYMER FLOW INSTABILITIES

Hydrodynamic stability is one of the central problems of fluid dynamics. It is concerned with the breakdown of laminar flow and its subsequent development and transition to turbulent flow (1). The flow of polymeric liquids differs significantly from that of their low-viscosity counterparts in several ways and, consequently, the nature of flow instabilities is completely different. Most notably, whereas for low-viscosity fluids it is the inertial forces on the fluid that cause turbulence (as measured by the *Reynolds number*), for high-viscosity polymers, it is the elasticity of the fluid that causes a breakdown in laminar fluid flow (as measured by the *Weissenberg number*). Additionally, the high viscosity and the propensity for the molten fluid to slip against solid surfaces contribute to a rich and diverse set of phenomena, which this book aims to review.

In this introductory chapter, an overview of most flow instabilities that are discussed and examined in detail in this book is presented. Both experimental observations as well as modeling of flows with the purpose of predicting flow instabilities are of concern in subsequent chapters. In addition, ways of overcoming these instabilities with the aim of increasing the rate of production of polymer processes are also of central importance [e.g., use of processing aids to eliminate surface defects (*melt fracture*), adjustment of molecular parameters, and rational adjustment of operating procedures and geometries to obtain better flow properties].

## 1.2 PART A: THE NATURE OF POLYMERIC FLOW

Polymeric liquids exhibit many idiosyncracies that their Newtonian counterparts lack. Most notably are: 1) the normal stress and elasticity effects; 2) strong extensional viscosity effects; and 3) wall slip effects. These properties of polymeric liquids change dramatically the nature and structure of the flow compared to the corresponding flow of Newtonian liquids. Chapter 2 of Ref. 2 presents a comprehensive overview of such striking differences in Newtonian flow compared to their non-Newtonian counterparts. Part A of this book (Chapters 2–4) is thus devoted to the fundamental aspects of polymeric flow from the perspective of processing regime, where the above effects are most manifest.

Chapter 2 by Dealy introduces the reader to the concepts of polymer rheology that are most important to polymer processing and defines a large number of the terms needed to read the rest of the book. It describes how to relate the stress on a fluid to an imposed strain, first for the simple linear case and then for the cases more relevant to processing, such as nonlinear flows (such as shear thinning) and also extensional flows (which are prevalent in most processing operations.) It carefully defines the *Weissenberg* number, which characterizes the degree of nonlinearity or anisotropy exhibited by the fluid in a particular deformation, as well as the *Deborah* number, which is a measure of the degree to which a material exhibits elastic behavior. It cautions the reader that our knowledge and measurement ability in this area are still in their infancy.

Chapter 3 by Mitsoulis describes secondary flows in processing operations, in particular those in which vortices or helices appear. This is a classic case in which a flow phenomenon (vortices), which occurs for Newtonian flows, also occurs for polymeric flows, but the nature of the flow is quite distinct. For polymeric fluids, the transition to vortices is governed largely by the Deborah number. Depending on the response of the fluid to extensional flows, the occurrence and magnitude of the vortices can be quite different. This chapter describes the occurrence of these flows in a number of processing flows including extrusion, calendaring, roll/wire coating, and coextrusion.

Chapter 4 by Archer describes the phenomenon of wall slip, where the fluid velocity does not go to zero at a solid wall. Again, although this phenomenon has been reported for Newtonian fluids, it is relatively weak and difficult to observe. However, for polymeric liquids, wall slip can be quite large; it significantly impacts on the flow behavior (and necessary modeling) of processing operations. This chapter describes the significant advances made recently on the measurement of wall slip, on the molecular theory describing it, and on the relationship at the interface between a polymer and a solid substrate. Although wall slip has been studied extensively in regard to its effects on extrusion instabilities, it plays a critical role in other processing operations as well.

Unlike Newtonian fluids, polymer melts slip over solid surfaces when the wall shear stress exceeds a critical value. For example, Fig. 1.1 illustrates the well-known Haagen–Poiseuille steady-state flow of a Newtonian fluid (Fig. 1.1a) and two typical corresponding profiles for the case of a molten polymer (Fig. 1.1b and c), indicating a small deviation from the no-slip boundary condition and plug flow. Although the classical no-slip boundary condition applied in Newtonian fluid mechanics (perhaps with the exception of rarefied gas dynamics where the continuum hypothesis does not apply), wall slip is typical in the case of molten polymers.

For the case of a passive polymer–wall interface where there is no interaction between the polymer and the solid surface, de Gennes (4) proposed an interfacial rheological law suggesting that a melt would slip at all shear rates. This theory was extended by Brochard-Wyart and de Gennes (5) to distinguish the case of a passive interface (no polymer adsorption) from that of an adsorbing one. It has been predicted that there exists a critical wall

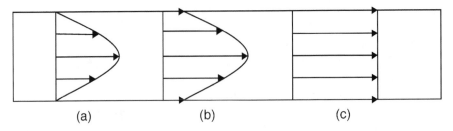

      (a)           (b)           (c)

**FIGURE 1.1** Typical velocity profiles of polymer melt flow in capillaries: (a) no-slip; (b) partial slip; and (c) plug flow.

shear stress value at which a transition from a weak to a strong slip takes place. These predictions have been suggested to be true through experimentation (6,7), and Chapter 4 contains a further enhancement of the theory by Archer. The various mechanisms of wall slip in the case of flow of molten polymers are still under debate and many observations need better and complete explanation.

## 1.3  PART B: MELT FRACTURE IN EXTRUSION

Part B is devoted to describing extrusion—one of the simpler processing operations but is critically important because it is ubiquitous in polymer manufacturing and exhibits a full range of instabilities (Chapters 5–9). Here the instabilities revolve around the phenomena of *melt fracture*, wall slip, and polymer elasticity. The phenomena of melt fracture and wall slip have been studied for the past 50 years but have not yet been explained (3). Not only are these phenomena of academic interest, but they are also industrially relevant as they may limit the rate of production in processing operations. For example, in the continuous extrusion of a typical linear polyethylene at some specific output rate, the extrudate starts losing its glossy appearance; instead, a matte surface finish is evident and, at slightly higher output rates, small-amplitude periodic distortions appear on its surface (see Fig. 1.2). This phenomenon, known as *sharkskin* or *surface melt fracture*, is described in Chapter 5 by Migler.

At higher values of the output rate, the flow ceases to be constant; instead, the pressure oscillates between two limiting values and the extrudate

(a)          (b)          (c)          (d)

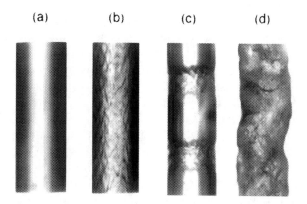

FIGURE 1.2  Typical extrudates showing: (a) smooth surface; (b) shark skin melt fracture; (c) stick–slip melt fracture; and (d) gross melt fracture.

surface alternates between relatively smooth and distorted portions (Fig. 1.2c). These phenomena are known as *oscillating, stick–stick*, or *cyclic melt fracture*, and are discussed in Chapter 6 by Georgiou. The chapter focuses on the critical stresses for the instability as well as the effects of operating conditions and molecular structure. A one-dimensional phenomenological model successfully describes much of the data.

At even higher values of the output rate, a new instability known as *gross melt fracture (GMF)* occurs, which is described in Chapter 7 by Dealy and Kim. Whereas sharkskin and stick–slip instability are associated with the capillary tube, GMF originates in the upstream region where the polymer is accelerated from a wide-diameter barrel to a narrow-diameter capillary or orifice. The extensional stress on the polymer associated with such a flow can rupture it, leading to a chaotic appearance when it finally emerges from the capillary. This chapter presents a number of observations and contains a critical review of our understanding of GMF from the perspective of our limited understanding of what causes the rupture of molten polymers.

To increase the rate of production by eliminating or postponing the melt fracture phenomena to higher shear rates, processing additives/aids must be used. These are mainly fluoropolymers and stearates, which are widely used in the processing of polyolefins such as high-density polyethylene (HDPE) and linear low-density polyethylene (LLDPE). They are added to the base polymer at low concentrations (approximately 0.1 %), and they effectively act as die lubricants, modifying the properties of the polymer–wall interface (increasing slip of the molten polymers). Chapter 8 by Kharchenko, Hatzikiriakos, and Migler discusses processing additives with particular emphasis on fluoropolymer additives. These fluoropolymer additives have been known for a long time, but until recently, one could only speculate as to precisely how they function. Recently, through visualization methods, the precise nature of the coating and how it is created through the flow have become clear. Although these fluoropolymer additives are effective for sharkskin and stick–slip instability, they remain ineffective for the case of GMF.

It has been recently discovered that compositions containing boron nitride can be successfully used as processing aids to not only eliminate surface melt fracture, but also to postpone gross melt fracture and thereby permit the use of significantly higher shear rates. These processing aids can be used for a variety of important extrusion processes, namely, tubing extrusion, film blowing, blow molding, and wire coating. The mechanisms by which boron nitride affects the processability of molten polymers and other important experimental observations related to the effects of boron nitride-based processing aids on the rheological behavior and processability of polymers can be found in Chapter 9 by Hatzikiriakos.

## 1.4  PART C: APPLICATIONS

### 1.4.1  Draw Resonance in Film Casting

In film casting as well as in extrusion coating of polymeric sheet, a polymeric melt curtain is extruded through a narrow die slot, across an air gap or a liquid bath, and then onto a pair of (or just a single) take-up or chill rolls (see Fig. 1.3). Efforts to increase production speed and/or reduce film thickness by going to higher draw ratio (take-up speed/extrudate speed) are hampered by *edge neck-in* and *bead formation*, but mainly by process instabilities (draw resonance and edge weave) (7), which give rise to spontaneous thickness and width oscillations (Fig. 1.4). Due to the high melt viscosity, the liquid curtain is usually pulled downstream by the drum, resulting in a long effective casting span. As a result, the curtain necks-in at the edges giving rise to a nonuniform gauge profile with a characteristic "dog bone" shape (Fig. 1.4). On the other hand, draw resonance is accompanied by spontaneous thickness and width oscillations—effects that are undesirable in film production.

Chapter 10 by Co summarizes experimental observations of such instabilities that might occur in film casting and extrusion coating of polymeric materials. In addition, modeling and stability analysis of draw resonance with the aim of predicting such instabilities are also thoroughly discussed. As will be seen in this chapter, computational fluid mechanics modeling can help predict such instabilities and extend the parameter range of stable and defect-free operation (so-called "process operability window").

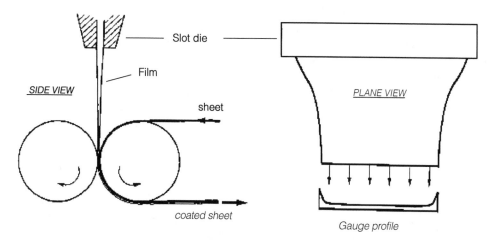

FIGURE 1.3  Schematic of a typical film casting process.

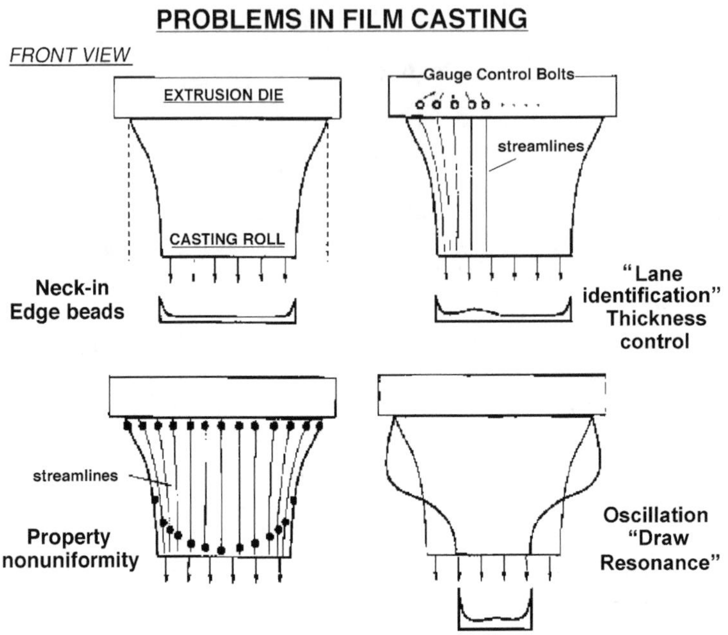

## PROBLEMS IN FILM CASTING

*FRONT VIEW*

**FIGURE 1.4**   Instabilities associated with the film casting process of polymers.

### 1.4.2   Fiber Spinning Instabilities

A simple schematic of the fiber spinning process is depicted in Fig. 1.5. The polymer melt is pumped by an extruder and flows through a plate containing many small holes—the spinneret. Fig. 1.5 shows the flow of the fiber through one such hole. The extruded filament is air-cooled by exposure to ambient air, and it is stretched by a rotating take-up roll at a point lower to its solidification point.

As in all other polymer processing operations discussed before, the rate of production is limited by the onset of instabilities, and these are discussed in Chapter 11 by Jung and Hyun. Three types of instability might occur in fiber spinning (8). The first instability is called *spinnability*, which is defined as the ability of the polymer melt to stretch without breaking (8). *Necking* might occur due to capillary waves or a *brittle* type of fracture due to crystallization induced by stretching. The second type of instability is referred to as *draw resonance*, which manifests itself as periodic fluctuation of the cross-sectional area in the take-up area. This latter instability is similar to draw resonance occurring in the film casting process of polymers (see Fig. 1.4). Finally, *melt fracture* phenomenon, as with other types of instabilities that may limit the

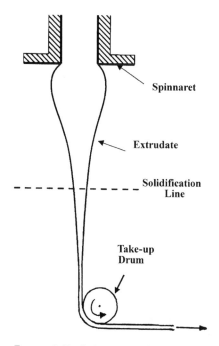

FIGURE 1.5   Schematic of a typical fiber spinning process.

rate of production such as fiber spinning, involves extrusion through dies at high shear rates.

Experimental observations in fiber spinning application for various resins are discussed in conjunction with modeling techniques of the process with a focus on draw resonance. Attempts to predict such instabilities with the aim of predicting operability windows for increasing the rate of production are also discussed. The rheological properties of the melts that play a role in stabilizing/destabilizing the process are also relevant in such discussions.

### 1.4.3   Film Blowing Instabilities

A simple schematic of the film blowing process is depicted in Fig. 1.6. An extruder melts the resin and forces it to flow through an annular die. The melt extruded in the form of a tube is stretched in the machine direction by means of nip roles above the die, as shown in Fig. 1.6. Air flows inside the film bubble to cool down the hot melt. Thus, a frost line is established. The ratio of the film velocity to the average velocity of the melt in the die is referred to as the *draw down ratio* (DDR).

The film blowing instabilities are discussed in great detail in Chapter 11 by Jung and Hyun. As in all polymer processes discussed so far, several

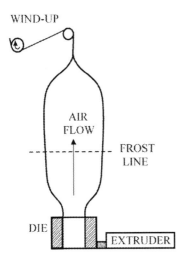

**FIGURE 1.6**   Schematic of a typical film blowing process.

instabilities might result as the extrusion rate (production rate) and DDR are increased. First, as the extrusion rate is increased (shear rate in the die), melt fracture might result depending on the molecular characteristics of the resin. These are manifested as loss of gloss of the film, small-amplitude periodic distortions, or more severe form of gross distortions. These instabilities influence seriously the optical and mechanical properties of the final product. Melt fracture is discussed in detail in Chapters 5–7 (present volume). Ways of postponing these instabilities to higher shear rates by means of using processing aids are discussed in Chapters 8 and 9 (present volume).

Draw resonance also occurs in this process as the DDR ratio increases (9). In the present process, draw resonance appears as periodic fluctuations of the bubble diameter known also as bubble instability shapes (10). Another type of draw resonance that might occur in film blowing is in the form of film gauge (film thickness) nonuniformity. This typically occurs at high production rates when the cooling rate requirements are increased (11). Molecular orientation during stretching of the film and the extensional properties of the resin in the melt state are important factors influencing the quality of the final products (11).

### 1.4.4   Coextrusion and Interfacial Instabilities

Coextrusion instabilities are discussed in depth in Chapter 12 by Dooley. Coextrusion refers to the process when two or more polymer liquid layers are extruded from a die to produce either a multilayer film or a fiber. There are three main problems associated with coextrusion. First, depending on the materials to be processed, *melt fracture* phenomenon might appear. This has already been discussed as these phenomena occur in most polymer processes that produce extrudates at high production rates.

More importantly are the instabilities caused by differences in the viscous and elastic properties of the components (see Fig. 1.7). If there is a significant viscosity difference between two liquids (i.e., two-layer flow), then the fluid having the lower viscosity will tend to encapsulate the fluid having the higher viscosity (9). For this to occur, the length-to-gap ratios should be relatively high, which gives enough time for rearrangement. Finally, even in the absence of a viscosity mismatch, the interface can become wavy. These instabilities are collectively known in coextrusion as *convective interfacial instabilities*, or simply *interfacial instabilities*. These instabilities are demonstrated in Chapter 12 to be caused by viscoelastic mismatches between the fluids coextruded, together with the geometry used for the process (11–14). For example, the point of layer merging or the point on the interface at the exit might cause interfacial waviness. The phenomena are complicated and not completely well understood as there are multiple factors involved.

### 1.4.5  Injection Molding and Tiger Stripes Instability

In injection molding, the polymer melt is softened first in an extruder and pumped forward through a runner to fill in a mold that is in the shape of the product article. The challenge is to produce a product that is free of voids, has a smooth and glossy surface, exhibits no warpage, and has sufficient mechanical strength and stiffness for its end use. The latter is significantly influenced by residual stresses due to the viscoelastic nature of flow, as well as due to

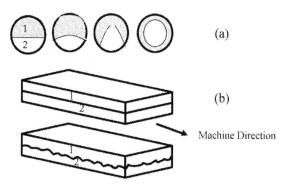

(a)

(b)

Machine Direction

**Figure 1.7**  Instabilities in coextrusion: (a) gradual encapsulation of a viscous fluid (2) by a less viscous fluid (1) as both flow in a circular tube; and (b) interfacial instability in the form of a wavy interface.

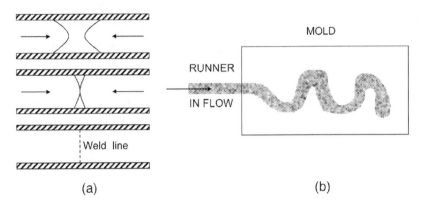

**FIGURE 1.8** Instabilities in injection molding: (a) formation of a weld line; and (b) the phenomenon of jetting.

shrinkage during cooling of the mold. As in most polymer processing operations, several problems (instabilities) might appear.

First, if the cavity to be filled has an insert, then the melt flow splits and flows around the obstacle. Consequently, these moving fronts meet again to result a *weld line* or *knit line*. In other cases, multiple runners are used to fill in the mold and, therefore, multiple *weld lines* result (see Fig. 1.8a). Lack of sufficient reentanglement of molecules across this line would lead to poor mechanical strength of the final product.

Jetting is another problem that might occur in injection molding. When the size (gap or diameter) is much smaller than the mold gap, the melt does not properly wet the entrance to the mold in order to fill it gradually. Instead, it "snakes" its way into the gap. This is shown in Fig. 1.8b.

Flow instabilities during injection molding (mold filling) also result in surface defects on polymer parts. For example, in filled polypropylene systems, the regular dull part of finished parts is broken by periodic shiny bands perpendicular to the flow direction (15). The appearance of these striped surface defects is known as *tiger stripes*. At the leading edge of the flow during mold filling, the polymer undergoes what is known as "fountain flow." Great recent progress in this area is summarized in Chapter 13 by Bogaerds, Peters, and Baaijens, in which they demonstrate that there is a viscoelastic flow instability at this air–polymer–wall juncture that leads to the tiger stripe phenomenon.

## REFERENCES

1. Drazin, P.G.; Reid, W.H. *Hydrodynamic Stability*; Cambridge University Press: New York, 1981.

2.  Bird, R.B.; Armstrong, R.C.; Hassager, O. *Dynamics of Polymeric Liquids: Vol. 1. Fluid Mechanics*; John Wiley and Sons: New York, 1987.
3.  Pearson, J.R.A. *Mechanics of Polymer Processing*; Elsevier: New York, 1985.
4.  de Gennes, P.G. Viscometric flows of tangled polymers. C. R. Acad. Sci. Paris, Ser. B 1979, *288*, 219–222.
5.  Brochard-Wyart, F.; de Gennes, P.G. Shear-dependent slippage at a polymer/solid interface. Langmuir 1992, *8*, 3033–3037.
6.  Migler, K.B.; Hervet, H.; Leger, L. Slip transition of a polymer melt under shear stress. Phys. Rev. Lett. 1993, *70*, 287–290.
7.  Wang, S.-Q.; Drda, P.A. Stick–slip transition in capillary flow of polyethylene. 2. Molecular weight dependence and low-temperature anomaly. Macromolecules 1996, *29*, 4115–4119.
8.  Petrie, C.J.S.; Denn, M.M. Instabilities in Polymer Processing. AIChE 1976, *22*, 209.
9.  Baird, G.B.; Kolias, D.I. *Polymer Processing*; Butterworth-Heinemann: Toronto, 1995.
10. Kanai, T.; White, J.L. Kinematics, dynamics and stability of the tubular film extrusion of various polyethylenes. Polym. Eng. Sci. 1984, *24* (15): 1185–1201.
11. Rincon, A.; Hrymak, A.N.; Vlachopoulos, J.; Dooley, J. Transient simulation of coextrusion flows in coat-hanger dies. Society of Plastics Engineers. Proc. Annu. Tech. Conf. 1997, *55*, 335–350.
12. Tzoganakis, C.; Perdikoulias, J. Interfacial instabilities in coextrusion flows of low-density polyethylenes: experimental dies. Polym. Eng. Sci. 2000, *40*, 1056–1064.
13. Martyn, M.T.; Gough, T.; Spares, R.; Coates, P.D. Visualization of melt interface in a co-extrusion geometry. Proceedings of Polymer Process Engineering Conference; Bradford, 2001, 37–45.
14. Dooley, J. *Viscoelastic Flow Effects in Multilayer Polymer Coextrusion*. Ph.D. Thesis, Technical University of Eindhoven, 2002.
15. Grillet, A.M.; Bogaerds, A.C.B.; Bulters, M.; Peters, G.W.M.; Baaijens, F.P.T. An investigation of flow mark surface defects in injection molding of polymer melts. Proceedings of the XIIIth International Congress on Rheology: Cambridge, 2000; Vol. 3, 122.

# 2

# Elements of Rheology

**John M. Dealy**
McGill University, Montreal, Quebec, Canada

## 2.1 RHEOLOGICAL BEHAVIOR OF MOLTEN POLYMERS

Flow instabilities that occur in melt processing arise from a combination of polymer viscoelasticity and the large stresses that occur in large, rapid deformations. This is in contrast to flows of Newtonian fluids, where inertia and surface tension are usually the driving forces for flow instabilities. Both the elasticity and the high stresses that occur in the flow of molten polymers arise from their high molecular weight (i.e., from the enormous length of their molecules). The high stresses are associated with the high viscosity of molten commercial thermoplastics and elastomers. Typical values range from $10^3$ to $10^6$ Pa s, whereas the viscosity of water is about $10^{-3}$ Pa sec. An easily observed manifestation of melt elasticity is the large swell in cross section that occurs when a melt exits a die.

The way melts behave in very small or very slow deformations is described quite adequately by the theory of *linear viscoelasticity* (LVE), and one material function, the linear relaxation modulus $G(t)$, is sufficient to describe the response of a viscoelastic material to any type of deformation history, as long as the deformation is very small or very slow. This information is useful in polymer characterization and determination of the relaxation spectrum. However, there is no general theoretical framework that describes viscoelastic behavior in the large, rapid deformations that arise in polymer processing (i.e., there exists no general theory of *nonlinear viscoelasticity*). Nonlinear phenomena that arise in large, rapid flows include the strong dependence of viscosity on shear rate and the very large tensile stresses that arise when long-chain branched polymers are subjected to extensional deformations.

Flow instabilities nearly always occur in response to large, rapid deformations. This means that a flow simulation able to predict when a flow will become unstable must be based on a reliable model of nonlinear viscoelastic behavior. The lack of such a model at the present time greatly limits our ability to model instabilities. Also lacking are generally accepted, quantitative criteria for the occurrence of instabilities, for example, in terms of a dimensionless group.

Although there remain major difficulties in the modeling of melt flow instabilities, many experimental observations have been reported, and empirical correlations with various rheological properties have been proposed. In order to understand these, some knowledge of the rheological behavior of molten polymers is needed. It is the purpose of the present chapter to provide a brief overview to prepare the uninitiated reader to understand the later chapters of this book and the current stability literature.

## 2.2  VISCOELASTICITY—BASIC CONCEPTS

Molten polymers are elastic liquids, and it is their elasticity that is the root cause of most flow instabilities. Thus, these are usually *hydroelastic* rather than *hydrodynamic* instabilities. Melts of high polymers can store elastic energy and, as a result, they retract when a stress that has been applied to them is suddenly released. However, they do not recover all the strains undergone as a result of this stress, as they are liquids, and their deformation always entails some viscous dissipation. These materials are thus said to have a *fading memory*.

The elasticity of polymers is intimately associated with the tendency of their molecules to become oriented when subjected to large, rapid deformations. This molecular orientation gives rise to anisotropy in the bulk polymer, and manifestations of this anisotropy include flow birefringence and normal stress differences.

A variable of central importance in the rheological behavior of polymers is time. Any system whose response to a change in its boundary conditions involves both energy storage and energy dissipation must have at least one material property that has units of time. Examples of such systems are resistance–capacitance electrical circuits and the suspension systems of automobiles, which include springs (to store energy) and shock absorbers (to dissipate energy). A system often used to demonstrate this fact consists of a linear (Hookean) spring and a linear dashpot, connected in series as shown in Fig. 2.1. This system is called a *Maxwell element*, and we will see that its behavior has proven very useful in describing the viscoelastic behavior of polymers.

The force in the spring is proportional to the distance through which it is stretched ($F_s = K_e \Delta X_s$), and the force in the dashpot is proportional to the velocity with which its ends are separated [$F_d = K_v(dX_d/dt)$]. In the absence of inertia, the forces in the two elements are equal, and the displacement of one end relative to the other is $\Delta X = \Delta X_s + \Delta X_d$. From this, it is easy to demonstrate that if the assembly is subjected to sudden stretching in the amount of $\Delta X_0$, the force will rise instantaneously to $K_e \Delta X_0$ and then decay exponentially, as shown by Eq. (2.1):

$$F(t) = K_e \Delta X_0 \exp\left[-t(K_e/K_v)\right] \tag{2.1}$$

We see that the ratio ($K_v/K_e$) is a parameter of the system and has units of time. It is thus the *relaxation time* of the mechanical assembly.

## 2.3  LINEAR VISCOELASTICITY

Linear viscoelasticity is a type of rheological behavior exhibited by polymeric materials in the limit of very small or very slow deformations. Although not directly applicable to the deformations that give rise to gross melt fracture,

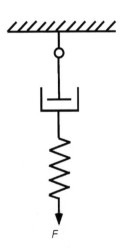

$F$

**FIGURE 2.1**  The simplest mechanical analog of a viscoelastic liquid; the single Maxwell element consisting of a linear spring in series with a linear dashpot.

LVE behavior is important for two reasons. First, information about linear viscoelastic behavior is very useful in the characterization of a polymer (i.e., in determining its molecular structure). Second, it provides information about the linear relaxation spectrum, which is an essential element of a model of nonlinear viscoelastic behavior. A comprehensive discussion of the linear viscoelastic properties of polymers can be found in the book by Ferry (1).

The most basic experiment in viscoelasticity is the measurement of the transient stress following a sudden "step" deformation. In the case of molten polymers, this is nearly always a shearing deformation, and the measured stress is reported in terms of the relaxation modulus $G(t)$. This function of time is defined as the stress divided by the amount of shear strain $\gamma$ imposed on the sample at $t = 0$:

$$G(t) \equiv \sigma(t)/\gamma \tag{2.2}$$

A function such as this, which is a characteristic of a particular material, is called a *material function*.

The basic axiom of linear viscoelasticity, the Boltzmann superposition principle, tells us that this function is independent of $\gamma$ and contains all the information needed to predict how a viscoelastic material will respond to any type of deformation, as long as this deformation is very small or very slow. This principle can be stated in terms of an integral equation as follows for simple shear deformation:

$$\sigma(t) = \int_{t'=-\infty}^{t} G(t-t')\,d\gamma(t') \tag{2.3}$$

where $\sigma$ is the shear stress at time $t$, and $d\gamma(t')$ is the shear strain that occurs during the time interval $dt'$. This can easily be generalized for any kinematics by use of the *extra stress tensor* and the *infinitesimal strain tensor*, whose components are represented by $\sigma_{ij}$ and $\gamma_{ij}$, respectively:

$$\sigma_{ij}(t) = \int_{t'=-\infty}^{t} G(t-t')\,d\gamma_{ij}(t') \tag{2.4}$$

and, in terms of the rate-of-strain tensor, this is:

$$\sigma_{ij}(t) = \int_{t'=-\infty}^{t} G(t-t')\dot{\gamma}_{ij}(t')\,dt' \tag{2.5}$$

The extra stress is that portion of the total stress that is related to deformation. We recall that for in an incompressible fluid, an isotropic stress (i.e., one that has no shear components and whose normal components are the same in all directions) does not generate any deformation. For example, the

Earth's atmosphere at sea level, when completely still, is under a compressive stress of one atmosphere. Thus, there is an isotropic component of the total stress tensor that is not related to deformation in an incompressible fluid. And the infinitesimal strain tensor is a very simple measure of deformation that is valid only for very small deformations. The rate-of-deformation tensor is related to the velocity components by Eq. (2.6):

$$\dot{\gamma}_{ij} \equiv \left( \frac{\partial v_i}{\partial x_j} + \frac{\partial v_j}{\partial x_i} \right) \tag{2.6}$$

Eqs. (2.4) and (2.5) are alternative, concise statements of the Boltzmann superposition principle. A very simple model of linear viscoelastic behavior can be obtained by inserting into Eq. (2.5) the relaxation modulus that is analogous to a single Maxwell element (i.e., a single exponential):

$$G(t) = Ge^{-t/\tau_r} \tag{2.7}$$

where $G$ is the instantaneous modulus and $\tau_r$ is a *relaxation time*. This is called the *Maxwell model* for the relaxation modulus. Inserting this into Eq. (2.4), we obtain the simplest model for linear viscoelasticity:

$$\sigma_{ij}(t) = G \int_{t'=-\infty}^{t} e^{-(t-t')/\tau_r} d\dot{\gamma}_{ij}(t') \tag{2.8}$$

The stress relaxation in an actual polymer can only very rarely be approximated by a single exponential, and $G(t)$ is usually represented either by a discrete or a continuous *relaxation spectrum*. In the case of a discrete spectrum, the relaxation modulus is represented as a sum of weighted exponentials, as shown by Eq. (2.9):

$$G(t) = \sum_{i=1}^{N} G_i e^{-t/\tau_i} \tag{2.9}$$

Comparing Eqs. (2.1) and (2.9), we see that this is analogous to the response of a mechanical assembly consisting of a series of Maxwell elements connected in parallel, so that the displacements of all the elements are the same, and the total force is the sum of the forces in all the elements. Thus, Eq. (2.9) is called the *generalized Maxwell model* for the relaxation modulus.

The continuous spectrum $F(\tau)$ is defined in terms of a continuous series of exponential decays, as shown by Eq. (2.10):

$$G(t) = \int_0^\infty F(\tau)e^{-t/\tau}d\tau \tag{2.10}$$

It turns out to be more convenient to work with the time-weighted spectrum function $H(\tau)$, which is defined as $\tau F(\tau)$, as shown by Eq. (2.11):

$$G(t) = \int_{-\infty}^{\infty} H(\tau)e^{-t/\tau} \mathrm{d} \ln \tau \tag{2.11}$$

It has been found that frequency (oscillatory shear) domain experiments are much easier to perform than time domain (step strain) experiments, and these provide an alternative means of establishing the linear behavior of a polymer. In oscillatory shear, the input strain is sinusoidal in time:

$$\gamma(t) = \gamma_0 \sin(\omega t) \tag{2.12}$$

As long as the response is linear (i.e., if $\gamma_0$ is sufficiently small), the shear stress is also sinusoidal, but with a phase shift $\delta$, as shown by Eq. (2.13):

$$\sigma(t) = \sigma_0 \sin(\omega t + \delta) \tag{2.13}$$

The response could be characterized by the amplitude ratio $(\sigma_0/\gamma_0)$ and the phase shift $\delta$ as functions of frequency, but it is more informative to represent the response in terms of frequency-dependent in-phase ($G'$) and out-of-phase ($G$) components, as shown by Eq. (2.14):

$$\sigma(t) = \sigma_0(G' \sin \omega t + G'' \cos \omega t) \tag{2.14}$$

The two material functions of this relationship are $G'(\omega)$, the *storage modulus*, and $G''(\omega)$, the *loss modulus*. The most common way of reporting linear viscoelastic behavior is to show plots of these two functions. Ferry (1) provides formulas for converting one LVE material function into another.

An alternative approach to viscoelastic characterization is to impose a prescribed stress on the sample and monitor the strain. If a stress $\sigma_0$ is imposed instantaneously at time zero, and the resulting shear strain is divided by this stress, the resulting ratio is the creep compliance $J(t)$:

$$J(t) \equiv \gamma(t)/\sigma_0 \tag{2.15}$$

If the imposed stress is sufficiently small, the result will be governed by the Boltzmann superposition principle, and the creep compliance can, in principle, be calculated if the relaxation modulus $G(t)$ is known.

One can also impose a sinusoidal stress and monitor the time-dependent deformation as an alternative technique to determine the storage and loss moduli.

## 2.4  VISCOSITY

Viscosity is defined as the steady-state shear stress divided by the shear rate in a steady simple shear experiment. For a Newtonian fluid, the viscosity is independent of shear rate, and this is also the prediction of the theory of linear viscoelasticity. This means that a molten polymer will exhibit a shear rate-independent viscosity at a sufficiently small shear rate, although such a low-shear rate may be difficult to access experimentally, especially in the case of polymers with broad molecular weight distribution or long-chain branching. Outside this range, the viscosity is a strong function of shear rate, and the dependency of viscosity on shear rate $\eta(\dot{\gamma})$ is an example of a nonlinear material function.

### 2.4.1 Dependence of Viscosity on Shear Rate, Temperature, and Pressure

At sufficiently high-shear rates, the viscosity often approaches a power–law relationship with the shear rate. Figure. 2.2 is a plot of viscosity vs. shear rate for a molten polymer, and it shows both a low-shear rate Newtonian region and a high-shear rate power–law region. These data were reported by Meissner (2) some years ago and represent the ultimate in rheometrical technique. The variation of $\eta$ with $\dot{\gamma}$ implies the existence of at least one material property with units of time. For example, the reciprocal of the shear rate at which the extrapolation of the power–law line reaches the value of $\eta_0$ is a characteristic time that is related to the departure of the viscosity from its zero-shear value. Let us call such a nonlinearity parameter $\tau_n$.

The viscosity falls sharply as the temperature increases, as shown in Fig. 2.2. Two models are widely used to describe this dependency: the Arrhenius

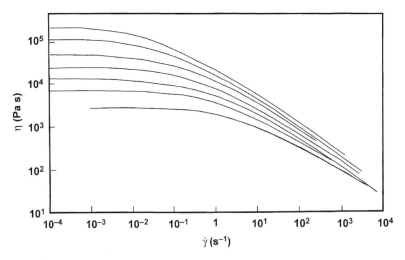

**FIGURE 2.2**  Viscosity vs. shear rate at several temperatures for a low-density polyethylene. (From Ref. 2.)

equation and the Williams-Landel-Ferry (WLF) equation. These are given in standard rheology references (1,3,4).

## 2.4.2 Dependence on Molecular Structure

The limiting low-shear-rate viscosity $\eta_0$, the *zero-shear viscosity*, increases linearly with weight-average molecular weight when this is below *an entanglement molecular weight* $M_C$, whereas above this value, the viscosity increases in proportion to $(M_w)^\alpha$, where $\alpha$ is usually around 3.5. This is one of several dramatic rheological manifestations of the phenomenon called *entanglement coupling*, which has a very strong effect on the flow of high-molecular-weight polymers. Although it was once thought that these effects were caused by actual physical entanglements between molecules, it is now recognized that they are not the result of localized restraints at specific points along the chain. Instead, it is understood that they result from the severe impediment to lateral motion that is imposed on a long molecule by all the neighboring molecules, although the term *entanglement* continues to be used to describe this effect. At higher shear rates, the effect of molecular weight on viscosity decreases, so it is the zero-shear value $\eta_0$ that is most sensitive to molecular weight.

## 2.5 NORMAL STRESS DIFFERENCES

The shear rate-dependent viscosity is one manifestation of nonlinear visco-elasticity, and there are two additional steady-state material functions associated with steady simple shear when the shear rate is not close to zero. These are the *first and second normal stress differences* $N_1(\dot{\gamma})$ and $N_2(\dot{\gamma})$. The three material functions of steady simple shear $\eta(\dot{\gamma})$, $N_1(\dot{\gamma})$, and $N_2(\dot{\gamma})$ are known collectively as the *viscometric functions*.

These are defined using the standard frame of reference for simple shear shown in Fig. 2.3. The shear stress $\sigma$ is $\sigma_{21}$ (equal to $\sigma_{12}$), and the three normal stresses are: $\sigma_{11}$, in the direction of flow ($x_1$); $\sigma_{22}$, in the direction of the gradient ($x_2$); and $\sigma_{33}$, in the *neutral* ($x_3$) direction. As this is, by definition, a

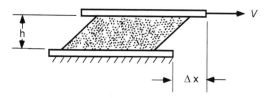

**FIGURE 2.3** Standard coordinate definitions for simple shear.

two-dimensional flow, there is no velocity and no velocity gradient in the $x_3$ direction.

In an incompressible material, normal stresses are themselves of no rheological significance because as long as they are equal in all directions, they cause no deformation. However, differences between normal stress components are significant, because they do cause deformation. For steady simple shear, the two rheologically significant differences are defined by Eq. (2.16a,b):

$$N_1(\dot{\gamma}) \equiv \sigma_{11} - \sigma_{22} \tag{2.16a}$$

$$N_2(\dot{\gamma}) \equiv \sigma_{22} - \sigma_{33} \tag{2.16b}$$

In the limit of zero-shear rate (i.e., for linear viscoelastic behavior), these two material functions approach zero, and as the shear rate increases, they are at first proportional to the square of the shear rate. Thus, although the shear stress becomes linear with shear rate as the shear rate approaches zero, $N_1$ and $N_2$ are second order in $\dot{\gamma}$ in this limit. This dependency inspired the definition of the two alternative material functions defined by Eq. (2.17a,b):

$$\Psi_1(\dot{\gamma}) \equiv N_1/\dot{\gamma}^2 \tag{2.17a}$$

$$\Psi_2(\dot{\gamma}) \equiv N_2/\dot{\gamma}^2 \tag{2.17b}$$

In other chapters of this book, it will be shown that these functions are related to several types of flow instability that occur in flows of viscoelastic melts.

## 2.6 TRANSIENT SHEAR FLOWS USED TO STUDY NONLINEAR VISCOELASTICITY

The response of a molten polymer to any transient shear flow that involves a large or rapid deformation is a manifestation of nonlinear viscoelasticity. Some examples of flows used to characterize nonlinear melt behavior are described in this section.

For large-step stress relaxation, the relaxation modulus is a function of the imposed strain as well as time: $G(t,\gamma)$. Except at very short times, the nonlinear relaxation modulus is often found to exhibit *time–strain separability*, which means that it can be represented as the product of the linear relaxation modulus and a function of strain $h(\gamma)$ called the *damping function*, as shown by Eq. (2.18):

$$G(t, \gamma) = G(t)h(\gamma) \tag{2.18}$$

The damping function is thus a material function that is wholly related to nonlinearity in a step strain test. As the strain approaches zero, $h(\gamma)$ obviously approaches one, and as the strain increases, it decreases.

The damping effect can be understood in terms of the *tube model* of the dynamics of polymeric molecules. In this model, the constraints imposed on a given polymer chain by the surrounding molecules and that give rise to entanglement effects are modeled as a tube. In response to a sudden deformation, the tube is deformed, and the relaxation of the molecule of interest is constrained by its containment in its tube. When the imposed deformation is very small, two mechanisms of relaxation occur: *equilibration* and *reptation*. Equilibration involves the redistribution of stress along the chain within the tube. Further relaxation can only occur as a result of the molecule escaping the constraints of the tube, and this requires it to slither along or *reptate* out the tube. This is a much delayed mechanism, and this is the cause of the plateau in the relaxation modulus for polymers with narrow molecular weight distributions. If the molecular weight is not narrow, the shorter molecules making up the tube will be able to relax fast enough to cause a blurring of the tube. This phenomenon is called *constraint release* and speeds up the relaxation of a long molecule in its tube.

The relaxation processes described above apply to linear viscoelastic behavior. If the deformation is not small, there is an additional relaxation mechanism—*retraction within the tube*. This is a fast relaxation, and once it is completed, the remainder of the relaxation process occurs as in the case of a linear response. It thus results in a relaxation modulus curve that has an early, rapid decrease due to retraction, followed by a curve that has the same shape as that for linear behavior. Thus, except for the very short-term relaxation, the relaxation modulus can be described as the linear modulus multiplied by a factor that accounts for the relaxation by retraction. This factor is the damping function.

In *start-up of steady simple shear*, the measured stress is divided by the imposed constant shear rate to obtain the *shear stress growth coefficient*, which is defined as follows:

$$\eta^+(t) \equiv \sigma(t)/\dot{\gamma} \tag{2.19}$$

And the similarly defined *shear stress decay coefficient* $\eta^-(t)$ describes stress relaxation following the cessation of steady simple shear.

## 2.7 EXTENSIONAL FLOWS

Most experimental studies of melt behavior involve shearing flows, but we know that no matter how many material functions we determine in shear, outside the regime of linear viscoelasticity, they cannot be used to predict the behavior of a melt in any other type of flow (i.e., for any other flow kinematics). A type of flow that is of particular interest in commercial processing and the instabilities that arise therein is extensional flow. In this type of flow, material points are stretched very rapidly along streamlines. An important example is the flow of a melt from a large tube into a much smaller one. In order to satisfy mass continuity, the velocity of a fluid element must increase markedly as it flows into the smaller tube, and this implies rapid stretching along streamlines. This is shown schematically in Fig. 2.4.

Although entrance flow subjects some fluid elements to large rates of elongation, the rate of elongation is not uniform in space, so this flow field is not useful in determining a well-defined material function that describes the response of a material to extensional flow. It turns out to be quite difficult to subject a melt to a uniform stretching deformation, and this is why reports of such measurements are much rarer than those of shear flow studies.

The experiment usually carried out to study the response of a melt to uniaxial extension is start-up of steady simple extension at a constant *Hencky strain rate ε̇*. The Hencky strain rate is defined as follows, in terms of the length $L$ of a sample:

$$\dot{\varepsilon} = \mathrm{d}\ln L/\mathrm{d}t \qquad (2.20)$$

Note that the length of a sample subjected to a constant Hencky strain rate increases exponentially with time. This strain rate is a measure of the speed with which material particles are separated from each other. The nonlinear material function most often reported is the *tensile stress growth coefficient*, which is defined as the ratio of the tensile stress to the strain rate, as shown by Eq. (2.21):

$$\eta_E^+(t,\dot{\varepsilon}) \equiv \sigma_E(t,\dot{\varepsilon})/\dot{\varepsilon} \qquad (2.21)$$

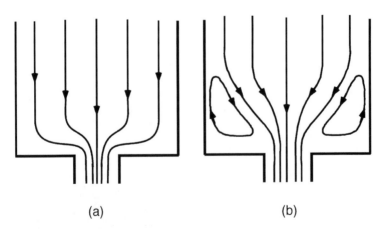

(a)                                              (b)

FIGURE 2.4  Sketch of entrance flow showing stretching along streamlines: (a) without corner vortex; (b) with corner vortex.

## 2.8 DIMENSIONLESS GROUPS GOVERNING THE BEHAVIOR OF VISCOELASTIC FLUIDS

When making a general statement about the flow behavior of polymers, it is useful to represent the results of theoretical treatments and experiments in terms of dimensionless variables. Two dimensionless groups are often used to describe the rate or duration of an experiment. One of these, the Weissenberg number ($Wi$), is a measure of the degree of nonlinearity or anisotropy exhibited by the fluid in a particular deformation, and the second, the Deborah number ($De$), is a measure of the degree to which a material exhibits elastic behavior. We will see in later chapters that these groups are useful for describing flow instabilities.

The *Deborah number* ($De$) is a measure of the degree to which the fluid will exhibit elasticity in a given type of deformation. More specifically, it reflects the degree to which stored elastic energy either increases or decreases during a flow. In steady simple shear, at steady state, when all stresses are constant with time, the amount of stored elastic energy is constant with time, so the Deborah number is zero. Thus, it is only in transient flows that the Deborah number has a nonzero value. Here "transient" means that the state of stress in a fluid element changes with time. This can arise as a result of a time-varying boundary condition, in a rheometer, or of the flow from one channel into a smaller one so that acceleration is involved. This dimensionless group is the ratio of a time arising from the fluid's viscoelasticity (i.e., its relaxation time) to a time that is a measure of the duration of the deformation.

For example, consider the response of a fluid to the start-up of steady simple shear in which a shear rate of $\dot{\gamma}$ is applied to a fluid initially in its rest state. In the initial stages of the deformation, the shear stress will increase with time and will eventually reach a steady value. At times $t$ after the shearing is begun, the stress will continue to change with time until $t$ is long compared to the relaxation time of the fluid. Thus, the Deborah number for this deformation is $\tau_r/t$; when this ratio is large (short times), the amount of stored elastic energy is increasing, but when it is small (long times), it becomes steady, and $De$ approaches zero. In this example, the Deborah number varies with time.

The Deborah number is only zero in deformations with constant stretch history (steady from the point of view of a material element). It is difficult to write a concise definition of a time constant that governs the rate at which stored elastic energy changes in a given deformation without reference to a specific rheological model, and we therefore give a definition of the Deborah number in general terms:

$$De \equiv \frac{\tau_r}{\text{characteristic time of transient deformation}} \tag{2.22}$$

Turning to the Weissenberg number, its use can be demonstrated by reference to steady simple shear flow. We have seen that any fluid that has a shear rate-dependent viscosity or is viscoelastic must have at least one material constant that has units of time and is characteristic of rheological nonlinearity, and we have called this time $\tau_n$. Furthermore, the type of behavior exhibited by a particular material depends on how this time constant compares with the reciprocal of the rate of the deformation. For example, let us say that a non-Newtonian fluid has a nonlinearity time constant $\tau_n$ of 1 sec. This implies that if the shear rate $\dot{\gamma}$ in steady simple shear is much smaller than the reciprocal of this time, its viscosity will be independent of shear rate. This suggests the definition of the *Weissenberg number* (*Wi*) as follows:

$$Wi \equiv \dot{\gamma}\tau_n \tag{2.23}$$

(The symbols *We* and *Ws* are also used for the Weissenberg number.) Thus, this dimensionless group is a measure of the degree to which the behavior of a fluid deviates from Newtonian behavior. For single-phase, low-molecular-weight fluids, the time constant of the material is extremely short, so that the Weissenberg number is always very small for the flows that occur under normal circumstances. But for molten, high-molecular-weight polymers, $\tau_n$ can be quite large. For polymeric liquids, the Weissenberg number also indicates the degree of anisotropy generated by the deformation. And in the case of steady simple shear, the normal stress differences are manifestations of anisotropy and thus of nonlinear viscoelasticity. Therefore, the Weissenberg number also governs the degree to which the normal stress differences will differ from zero.

To summarize, when the Weissenberg number is very small in a simple shear flow, the viscosity will be independent of shear rate, and the normal stress differences will be negligible. These phenomena reflect the fact that in the limit of vanishing shear rate, the shear stress is first order in the shear rate, whereas the normal stress differences increase with the square of the shear rate. The Weissenberg number is easily defined for any flow with constant stretch history. For example, for steady uniaxial extension, we need only replace the shear rate by the Hencky strain rate $\dot{\varepsilon}$.

In general, the degree to which behavior is nonlinear depends on the Weissenberg number. At very small deformation rates, *Wi* is much less than one, and the stress is governed by the Boltzmann superposition principle. At higher deformation rates, *Wi* increases and nonlinearity appears, as reflected, for example, in the dependence of viscosity on shear rate and the appearance of normal stress differences. The two characteristic times defined above ($\tau_n$ and $\tau_r$) are closely related, and in recognition of this, the subscripts will be dropped in the rest of this discussion.

There are several pitfalls in the use of the two dimensionless groups defined above to characterize the response of a material to a given type of deformation. First, there are hardly any materials whose viscoelastic behavior can be described using a single relaxation time. More typically, a spectrum of times is required, and this causes difficulty in the choice of a time for use in defining the Deborah number. The "longest relaxation time" is often identified as the appropriate one for defining these groups, but for highly polydispersed or branched polymers, it may be impossible to identify a "longest time."

Another problem in the use of dimensionless groups to characterize deformations is that for a melt consisting of a single polymer, in several flows of practical interest, $Wi$ and $De$ are directly related to each other. This causes confusion, as authors of books and research papers tend to use the two groups interchangeably in all circumstances. Some examples will help to illustrate the correct use of the Weissenberg and Deborah numbers.

A flow that has been of great interest to both experimental and theoretical polymer scientists is the entrance flow from a circular reservoir into a much smaller tube or capillary. A Weissenberg number can readily be defined for this flow as the product of the characteristic time of the fluid and the shear rate at the wall of the capillary. However, entrance flow is clearly not a flow with constant stretch history, and the Deborah number is thus nonzero as well. Furthermore, as demonstrated by Rothstein and McKinley (5), in such flows, the two groups are directly related. This can lead to confusion when experimental data and flow simulation results are compared (6). Note that for fully developed capillary flow, $De = 0$.

On the other hand, a flow in which the two numbers can be varied independently is oscillatory shear. As used for the determination of the linear viscoelastic behavior of polymers, this deformation is carried out at very small strain amplitudes, so that the Weissenberg number will be much less than one. As the frequency is increased from zero, the Deborah number, defined here as $\omega\tau$, is at first very small, and the response is purely viscous. Because $Wi$ is also very small, the melt behaves like a Newtonian fluid. If now the frequency is increased, the Deborah number increases and the importance of elasticity grows, and at very high $De$, the behavior becomes almost purely elastic.

If, however, the strain amplitude $\gamma_0$ is increased, then the strain rate amplitude $\dot{\gamma}_0 \equiv \omega\gamma_0$, and the Weissenberg number, which is equal to $\dot{\gamma}_0\tau$, also increases. By changing the frequency or the amplitude, $Wi$ and $De$ can be varied independently. A convenient way of representing the parameter space of large-amplitude oscillatory shear is a Pipkin diagram, which is a graph of Weissenberg number vs. Deborah number (4,7). In such a diagram, the behavior in the lower left-hand corner ($Wi \ll 1$; $De \ll 1$) is that of a Newtonian fluid. Near the vertical axis ($De \ll 1$), the behavior is nonlinear but inelastic and is governed by the viscometric functions. Near the horizontal axis ($Wi \ll 1$), linear viscoelastic behavior is expected, and far to the right ($De \gg 1$), the behavior becomes indistinguishable from that of an elastic solid.

## 2.9  EXPERIMENTAL METHODS IN RHEOLOGY—RHEOMETRY

The objective of rheometry is to make measurements whose data can be interpreted in terms of well-defined material functions, without making any assumptions about the rheological behavior of the material. Examples of material functions are the viscosity as a function of shear rate and the relaxation modulus as a function of time. These functions are physical properties of a material, and the detailed method by which they are determined need not be reported in order for them to be properly interpreted. In contrast to this type of measurement are empirical industry tests that yield numbers that can be used to compare materials but are not directly related to any one physical property. Such test data are widely used as specifications for commercial products and for quality control. An important example that involves the flow of molten plastics is the *melt flow rate*, often called the *melt index* (8).

In an experiment designed to determine a material function, it is necessary to generate a deformation in which the streamlines are known a priori (i.e., that are independent of the rheological properties of the material). Such a deformation is called a *controllable* flow, and the number of such deformations is very limited. In fact, the only practically realizable controllable flows are simple shear, simple (uniaxial) extension, biaxial extension, and planar extension. And the last two of these are sufficiently difficult to generate that they are rarely used. Additional limitations on our ability to determine rheological material functions in the laboratory are imposed by various instabilities that occur even in these very simple flows.

There are two ways of generating a shear deformation in a rheometer: drag flow and pressure flow. In drag flow, one surface in contact with the sample moves relative to another to generate shearing. In pressure flow, pressure is used to force the fluid to flow through a straight channel, which may be a capillary or a slit. Drag flow can be used to determine a variety of material functions, including the storage and loss moduli as functions of frequency, the creep compliance as a function of time, the viscosity and normal stress differences as functions of shear rate, and various nonlinear transient material functions. Pressure-driven rheometers are useful primarily for the measurement of viscosity at high-shear rates.

Extensional rheometers are most often designed to generate uniaxial (tensile) extension in which either the tensile stress or the strain rate is maintained constant.

Below are presented brief overviews of the way melt rheometers are used. More detailed information about experimental rheology can be found in various rheology books (3,4).

## 2.9.1  Rotational Rheometers

Rotational rheometers can be classified according to the type of fixture used and by the variable that is controlled (i.e., the independent variable). Two types of fixture are commonly used with molten polymers: cone-plate and plate-plate. The flow between a cone and a plate, one of which is rotating with respect to the other at an angular velocity $\Omega$, closely approximates uniform simple shear, as the shear rate at a radius $r$ is the local rotational speed $\Omega r$ of the rotating fixture at $r$, divided by the gap between the fixtures at this value of $r$, which we call $h$. In the cone and plate geometry, this distance is linear with $r$, with the result that $r\Omega/h$ (i.e., the local shear rate) is uniform. This feature of cone-plate flow makes it useful for studies of nonlinear viscoelasticity. Cone-plate fixtures are used to determine the viscosity and first normal stress difference as functions of shear rate at low-shear rates, as well as the response of the shear and normal stresses to various transient shearing deformations.

The equations for calculating the strain rate and stresses of interest are as follows for the case of a small cone angle $\Theta_o$:

$$\dot{\gamma} = \Omega/\Theta_o \tag{2.24}$$

$$\sigma = 3M/2\pi R^3 \tag{2.25a}$$

$$N_1 = 2F/\pi R^2 \tag{2.26}$$

where $\Omega$ is the angular velocity of the rotating fixture; $\sigma$ is the shear stress; $M$ is the torque measured on either the rotating or the stationary shaft; $F$ is the total normal thrust on the fixtures; and $R$ is the radius of the fixtures.

Uniform simple shear is very well approximated by the flow between a cone and a plate; if the cone angle is small, and the shear rate is not very high. Starting at moderate shear rates, however, various types of instability render the flow unsuitable for rheological measurements. Thus, cone-plate fixtures are useful for determining moderate departures from linear viscoelasticity. The major concerns in using this technique are calibrating the sensors, avoiding degradation of the sample, and recognizing when an instability has occurred.

Although cone-plate fixtures can be used to determine the material functions of linear viscoelasticity, it is more convenient to use plate-plate fixtures for this application because the preparation and loading of samples, as well as the setting of the gap between the two fixtures, are much simplified. Although the local shear strain between the parallel plates varies linearly with radius, as long as the viscoelastic behavior is linear, the local stress is proportional to the strain, and the variation poses no problem in the interpretation of the data. The equations for calculating the storage and loss moduli are as follows:

$$G' = \frac{2M_o h \cos \delta}{\pi R^4 \phi_o} \tag{2.25b}$$

$$G'' = \frac{2M_o h \sin \delta}{\pi R^4 \phi_o} \tag{2.25c}$$

where $M_o$ is the amplitude of the torque signal; $\phi_o$ is the amplitude of the angular displacement of the oscillating shaft; and $\delta$ is the phase angle between the angular displacement and the measured torque.

The primary concerns in making reliable measurements are calibrating the sensors, avoiding thermo-oxidative degradation of the sample, and ensuring that the amplitude selected for the deformation or torque does not take the sample out of its linear range of response.

Rotational rheometers can also be classified according to which variable is controlled. Thus, there are controlled strain (actually controlled angular motion) and controlled stress (actually controlled torque) instruments. Both can be used to determine linear viscoelastic properties, although controlled stress instruments give better results at low frequencies, whereas controlled strain instruments are preferred at high frequencies. Thus, the two types of rheometer provide information that is, to some degree, complementary.

### 2.9.2 Sliding Plate Rheometers

Sliding plate melt rheometers were developed to make measurements of nonlinear viscoelastic behavior under conditions under which cone-plate flow is unstable (i.e., in large, rapid deformations) (9). The sample is placed between two rectangular plates, one of which translates relative to the other, generating, in principle, an ideal rectilinear simple shear deformation. There are significant edge effects at large strains, and to minimize the effect of these on the measurement, the stress should not be inferred from the total force on one of the plates but measured directly in the center of the sample by a shear stress transducer. Such instruments have been used to determine the shear stress response to large, transient deformations (9,10). In addition, they have been used to determine the effect of pressure on the viscosity and nonlinear behavior of melts (11,12). Their advantage over capillary instruments for high-pressure measurements is that the pressure and shear rate in the sample are uniform.

Phenomena that limit their use are slip, cavitation, and rupture, which interrupt experiments at sufficiently high strains and strain rates. At the same time, however, sliding plate rheometers have been found to be useful tools for the study of melt slip (13).

### 2.9.3  Capillary and Slit Rheometers

Pressure-driven rheometers, particularly capillary instruments, are the work horses of plastics rheology, as they are relatively simple and easy to use. In most capillary rheometers, the flow is generated by a piston moving in a round reservoir to drive the melt through a small capillary, which often has a diameter of about 1 mm. After a short entrance length, the flow becomes fully developed (i.e., the velocity profile and shear stress become independent of distance from the entrance). For a Newtonian fluid, the fully developed velocity distribution is parabolic, and the wall shear rate can be calculated by knowing only the volumetric flow rate $Q$ and the radius of the capillary $R$. However, in the case of a non-Newtonian fluid, the velocity distribution depends on the viscosity function, which is initially unknown. It has proven useful in polymer characterization, however, to use the formula for Newtonian fluids to define an *apparent wall shear rate* $\dot{\gamma}_A$. The wall shear rate $(\partial v/\partial r)_{r\,=\,R}$ is actually negative because the velocity is zero at the wall and positive at the center, but it is the magnitude of this quantity that is universally used in discussing capillary and slip flow. Thus, the apparent wall shear rate is the positive quantity defined as follows:

$$\dot{\gamma}_A \equiv \frac{4Q}{\pi R^3} \tag{2.27a}$$

Techniques for determining the true wall shear rate $\dot{\gamma}_w$ can be found in various rheology books (3,4).

   If the pressure gradient in the region of fully developed flow is known, the wall shear stress can be readily calculated. It is possible to make such a measurement in a slit rheometer by mounting two or more pressure transducers in the wall of the slit. But in a capillary, this is not practical, and it is only the driving pressure $P_d$ that is known, either from the force required to drive the piston, or from a pressure transducer mounted in the reservoir. The pressure at the exit of the capillary (1 atm) is practically negligible for high-viscosity melts, so the driving pressure is essentially equal to the overall pressure drop. However, in addition to the pressure drop resulting from flow in the region of fully developed flow, $P_d$ also includes a substantial *entrance pressure drop*. Two methods have been developed to determine the portion of the pressure drop due to entrance flow so that the magnitude of the true wall shear stress $\sigma_w$ can be determined. These are the Bagley plot method and the orifice method (4). Once the wall shear stress has been calculated, it can be divided by the wall shear rate to yield the viscosity. Because the shear rate is not uniform along or across a capillary or slit, these flows are not useful for the determination of viscoelastic material functions. However, they complement steady-shear, cone-plate instruments by extending the viscosity curve to much higher shear rates.

The major concerns in using a capillary or slit to measure melt viscosity include: viscous heating, effect of pressure on viscosity, slip, and entrance flow instabilities. Hay et al. (14) have explained in detail some of the problems associated with the use of pressure flow rheometers, whereas slip and entrance flow instabilities are discussed in Chapters 4–7 of the present volume.

### 2.9.4  Extensional Rheometers

Melt behavior has been studied using uniaxial (also called simple or tensile), biaxial, and planar extensional flows (3,4). However, only the first of these is in general use and will be discussed here. A uniaxial extensional rheometer is designed to generate a deformation in which either the tensile stress $\sigma_E$ or the Hencky strain rate $\dot{\varepsilon}$ [defined by Eq. (2.21)] is maintained constant. The material functions that can, in principle, be determined are the tensile stress growth coefficient $\eta_E^+(t,\dot{\varepsilon})$, the tensile creep compliance $D(t,\sigma_E)$, and the tensile (uniaxial) extensional viscosity $\eta_E(\dot{\varepsilon})$, which is the long-time, steady-state value of $\eta_E^+(t,\dot{\varepsilon})$.

The major challenges in this type of measurement are: supporting the sample, applying a traction to stretch it, and maintaining a uniform deformation and temperature throughout the experiment. The sample is supported by immersing it in—or floating it on—an oil bath, or by means of gas flowing through a porous metal plate. Over the past 40 years, much effort by several outstanding rheologists has been put into the development of reliable extensional rheometers for melts. Traction is applied by either a rotary clamp (15,16), adhesive clamp (17), or a moving-belt clamp (18). Because they are liquids, melts pose major challenges for the designer and user of an extensional rheometer. In the adhesive clamp (or "end separation")-type instrument, the sample length increases exponentially with time to maintain the constant strain rate, whereas in the rotary and belt clamp instruments, the "clamps" maintain a constant velocity at a fixed points in space, and the *gauge length* (i.e., the length of the portion of the sample that is actually stretched) remains constant. An obvious difficulty in measuring extensional flow properties is that the cross section of the sample becomes very small at significant strains. For example, to reach a Hencky strain of seven, the diameter of a cylindrical sample having an initial diameter of 1 cm is reduced to 0.3 mm. Recent advances in the development of constant gauge length techniques have been reported by Sentmanat (19) and Gendron et al. (20).

Obviously, it is much more difficult to measure the response of a melt to a stretching deformation than to a shearing deformation. However, without a reliable rheological constitutive equation, there is no way to predict how a melt will behave in a large, rapid extensional flow, based only on knowledge of its response to shearing deformations. In other words, extensional tests reveal aspects of the nonlinear viscoelasticity of melts that cannot be predicted from shear data. Therefore, the essential criterion for the success of an extensional flow technique is the degree of departure from linear viscoelastic behavior that can be observed.

We know that for very slow or very small deformations, the behavior will be linear. We also know that linear behavior in flows of any kinematics can be predicted by the Boltzmann superposition principle. For example, Eq. (2.4) can be used to show that the tensile stress start-up coefficient is given by:

$$\eta_E^+(t) = 3 \int_0^t G(s)\mathrm{d}s \tag{2.27b}$$

It can readily be seen by setting $t = \infty$ that the low strain rate limiting value of $\eta_E(t)$ is $3\eta_0$. Thus, if we have determined $G(s)$ for a melt using a plate-plate rotational rheometer, we can calculate the response of that melt in a very slow or very small uniaxial extension. It is almost universal practice when reporting the results of an extensional flow experiment to compare the nonlinear material function $\eta_E^+(t,\dot{\varepsilon})$ with the linear response given by Eq. (2.27a,b). If both data sets are accurate, the nonlinear response should agree with the linear one at very short times or very low strain rates. Furthermore, the way in which the nonlinear response departs from the linear one is used to classify the extensional flow behavior. If the nonlinear data rise above the linear curve at some point, the melt is said to be strain-hardening, and if they fall below, it is said to be strain-softening. Figure. 2.5 shows the data of Münstedt and Laun (21) for a low-density polyethylene, and we see that this material is strain-hardening. Linear polymers containing no very-high-molecular-weight components exhibit strain softening.

Strain-softening polymers are very prone to ductile failure in extension, and this poses a major challenge for the experimentalist. If there is a small variation in diameter along the sample, the resistance to further deformation will be reduced at this point, leading to instability and failure. For this reason, it is very rarely that one is able to carry an experiment to steady state in such a material. This instability has been treated theoretically by McKinley and Hassager (22).

In spite of the best efforts of some exceptionally capable polymer scientists, the many pitfalls involved in determining extensional flow properties limit such measurements to quite small strain rates. Thus, even though a candidate constitutive equation is able to fit extensional flow data for a particular polymer, there is no assurance that it can predict the response of

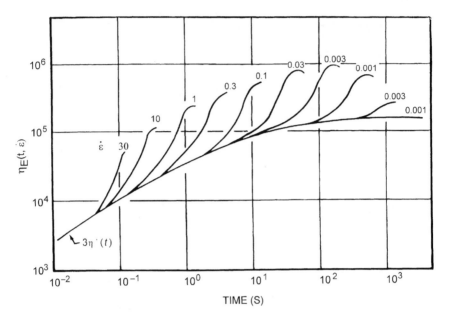

**FIGURE 2.5** Extensional stress growth function for a low-density polyethylene at 150°C. (From Ref. 21.)

that material to the rapid stretching that is often associated with flow instabilities.

## 2.10  MODELING NONLINEAR VISCOELASTIC BEHAVIOR

### 2.10.1  Introduction

If a general model to describe nonlinear viscoelastic behavior were available, it could be used to calculate the response of a melt to all of the simple deformations discussed above as well as the more complicated ones that arise in polymer processing. Although a universally applicable theory of nonlinear viscoelasticity does not exist at present, some proposed models have been used to simulate the instabilities that occur in the flow of viscoelastic materials.

A very fundamental approach to calculating the stresses arising from a given deformation is to use a molecular dynamics model based on the principles governing the behavior of individual polymer molecules (23). As can be readily imagined, this is an extremely intensive operation from a computational point of view, and it is likely to be many years before this technique can be used to model nonhomogeneous flows. Thus, molecular dynamics does not, at present, provide a useful basis for the simulation of the complex flows involved in flow instabilities.

Much of the enormous complexity of molecular dynamics models can be eliminated if we make use of averaging to produce a *mean field model*. Here, instead of starting from a detailed picture of the interactions between individual molecules, we focus attention on a single molecule and represent the effect of all the surrounding molecules by an average field of constraints surrounding it. Such models yield equations for calculating, for example, the linear relaxation modulus $G(t)$ and the damping function. For flow simulation, however, what is needed is an equation that can be used in a complex flow field to predict the velocity and stress fields with given boundary conditions.

Such a model is called a *constitutive equation* or a *rheological equation of state* (REOS). In the following discussion, such a model will be referred to as an REOS. Nearly all of the many REOS models that have been proposed over the past 50 years are basically empirical equations, and only very recently have such models been developed on the basis of mean field molecular theories. The earlier REOS models are called continuum models or continuum mechanics models, and these are described in Section 2.10.2.

## 2.10.2 Continuum Models

Continuum models can usually be written in the form of closed-form, integral, or differential equations. These are empirical equations whose tensorial form is inspired by the general principles of continuum mechanics (24). Like all empirical equations, their validity must be established by the comparison of model predictions with experimental data. However, the experimental techniques now available only allow us to study the response of melts to a limited number of types of deformation (e.g., simple shear and uniaxial extension at very low deformation rates). Nevertheless, most modeling of flow instabilities up to the present time is based on constitutive equations of this type, and it is important to describe some of their features.

### 2.10.2.1. Integral Continuum Models

A constitutive equation for a viscoelastic material is a relationship between the stress in a particular element of a material at a time $t$ and the deformation that element experienced during previous times $t'$, up to the time $t$. Both the stress and the strain are tensorial quantities having nine components, which are functions of time and position in space. However, both are symmetrical tensors, so only six of these components are independent. Unlike stress, strain is a *relative* tensor that compares the shape and size of a material element at one time $t_2$ with that at some time $t_1$, when the element is in its reference state. For example, if you walk into a room and see someone holding a stretched strip of rubber with a spring scale attached to it at one end, you can see at a glance the force in the strip by looking at the spring scale. But you cannot say anything about the strain unless you know how its present length compares with that before it was stretched. In this simple example, only one component of force is involved, but a general model must give all the components of the stress tensor.

We can condense the above discussion into the following statement:

$\sigma_{ij}(t)$ depends on : $S_{ij}$ (relative to reference state at times $t'$ from $-\infty$ to $t$)

In this statement, $\sigma_{ij}$ is the $i$–$j$ component of the stress tensor, and $S_{ij}$ is the $i$–$j$ component of a relative strain tensor. To move toward a specific model, we must first select an appropriate reference state and then define the strain tensor.

For a crosslinked rubber, the obvious reference state is that in which the material has been unstretched for a long-enough time that all anisotropic stresses have completely relaxed. However, for a viscoelastic fluid, such as a molten polymer or an uncured elastomer, there is no unique unstretched state because irreversible flow occurs along with elastic deformation. In fact, the only time when a fluid element is in a uniquely defined state is the time $t$ at which we wish to evaluate the stress. Therefore, the reference state for a material element in a constitutive equation for a melt is the state of the element at time $t$, and the relative strain at time $t'$ is measured relative to the reference configuration at time $t$. We can then write our general strain tensor as $S_{ij}(t,t')$.

To arrive at a specific equation for calculating stress, a number of choices must be made, and these are guided mainly by observations of the general behavior of viscoelastic materials and by the need to produce a model that is convenient for calculations. First, we must select a strain tensor for large-strain deformations from among the many that can be defined for this purpose. Finally, we have to decide on a particular kernal function to give a weighting to the strain that has occurred in a fluid element according to how long ago it occurred (i.e., according to the value of $t-t'$ ). These selections are guided by some of the following considerations:

1. The need to satisfy the principle of material indifference
2. The requirement that the nonlinear model converge to the Boltzmann superposition principle for very small or very slow deformations
3. The principle of fading memory

4. Our general understanding of how polymers behave
5. Agreement with experimental data for simple shear and uniaxial extension at low strain rates
6. The need to determine all empirical constants or functions using only measurable rheological properties
7. An energy principle governing the storage of elastic energy
8. The need for an equation that is simple enough to be used to calculate stresses, at least in simple flow fields.

As a result, the final model is very much empirical in nature, and we have no way of knowing how valid it is for flows other than those used to evaluate the empirical functions or constants that it contains.

A class of integral REOS that has been popular involves the following choices:

1. The Finger tensor is chosen as the nonlinear strain measure.
2. A single integral is chosen as the form of the dependency.
3. The time weighting is described by a product of a *memory function* derived from the relaxation modulus of linear viscoelastic behavior and a *damping function* that takes into account the nonlinearity of the material's behavior.

The final model has the following form:

$$\sigma_{ij}(t) = \int_{t'=-\infty}^{t} m(t-t')h(I_1,I_2)B_{ij}(t,t')dt' \qquad (2.28)$$

The damping function $h$ depends on the strain tensor through its two scalar invariants $I_1$ and $I_2$, and $m(t-t')$ is related to the linear relaxation modulus $G(t-t')$ as follows:

$$m(t-t') = \frac{dG(t-t')}{dt'} \qquad (2.29)$$

The popularity of this choice is explained mainly by the fact that anything more complicated renders the resulting model too complex for practical use.

Although we have already made a number of major assumptions about the form of the constitutive equation, we still need to establish a form for the damping function. Lodge (25) introduced his *rubberlike liquid* model to demonstrate viscoelastic phenomena. It is equivalent to Eq. (2.28) with $h$ equal to one. If a single Maxwell element is used for the memory function in the rubberlike liquid, we obtain the *Lodge equation* (Eq. (2.30)):

$$\sigma_{ij}(t) = G\int_{t'=-\infty}^{t} \frac{1}{\tau}\exp\left[-(t-t')/\tau\right]B_{ij}(t,t')dt' \qquad (2.30)$$

Lodge also derived a transient network model that can be represented as a rubberlike liquid with its memory function equal to that of a generalized Maxwell fluid:

$$\sigma_{ij}(t) = \int_{t'=-\infty}^{t} \sum_{i=1}^{N} \frac{G_i}{\tau_i} \exp\left[ -(t - t')/\tau_i \right] B_{ij}(t, t') dt' \qquad (2.31a)$$

This model is able to describe, qualitatively, a number of viscoelastic phenomena, but it is not useful for the quantitative description of polymer behavior. For example, it predicts that the viscosity is constant with shear rate, and the second normal stress difference is zero.

Wagner proposed letting the damping function be an empirical quantity determined by fitting experimental data. The problem here is that reliable experimental data are available only for uniform flows that can be readily generated in the laboratory. These are primarily simple shear and uniaxial extension, and the strain rates at which stable flows can be generated are limited to rather low values. Thus, it is not possible to carry out experiments that involve more complex flow fields and high strain rates. This seriously limits the conditions under which data can be used to establish empirical the material functions, such as $h(I_1, I_2)$, that arise in a continuum mechanics rheological model. And even if data at higher strains and strain rates were available, they would not be sufficient to establish a general damping function valid for every possible type of deformation.

Lodge's Eq. (2.30) and Wagner's Eq. (2.28) are special cases of a more general integral model called the K-BKZ model. In this model, the strain is described by a linear combination of the Finger and Cauchy tensors and, as a result, the predicted second normal stress difference is nonzero.

If we are objective, we must conclude that it would be truly miraculous if nature had conspired to reduce a very complex system whose rheological behavior is, in principle, describable by an infinite sum of multiple integrals, each with its own kernal function, to a single integral involving the linear relaxation modulus and a single, scalar descriptor of nonlinearity. McLeish (26) has discussed how unlikely this is.

## 2.10.2.2. Differential Continuum Models

Equation. (2.8) can be written in the form of a differential equation as shown below:

$$\tau \frac{d\sigma}{dt} + \sigma = G\tau\dot{\gamma} \qquad (2.31b)$$

Note that for a step shear strain of $\gamma_0$ at $t = 0$, the resulting shear stress is:

$$\sigma(t) = G\gamma_0 e^{-t/\tau} \qquad (2.32)$$

And for steady simple shear, the long-time limiting stress is:

$$\sigma = G\tau\dot{\gamma} \tag{2.33}$$

But, from the definition of the viscosity, we know that:

$$\sigma = \eta\dot{\gamma} \tag{2.34}$$

Thus, the viscosity $\eta_0$ in this simple LVE model is $G\tau$.

To generalize Eq. (2.31b) to describe flows having any kinematics, we replace the shear stress and shear rate by the corresponding tensorial quantities to obtain the Maxwell model:

$$\tau\frac{d\sigma_{ij}}{dt} + \sigma_{ij} = G\tau\dot{\gamma}_{ij} \tag{2.35}$$

And this model can be generalized to accommodate a discrete spectrum of relaxation times by writing Eq. (2.35) for each relaxation mode and summing the stresses resulting from solving each equation.

Just as there are various possible finite strain tensors, there are various time derivatives that can be used in place of the ordinary derivative of stress in Eq. (2.31b) to satisfy the continuum mechanics requirements for a model that is to describe large, rapid deformations. The derivative that [used in place of the time derivative in Eq. (2.35)] yields a differential model equivalent to Eq. (2.30) is the *upper convected time derivative*.

Other possibilities include the lower-convected Maxwell derivative and the corotational derivative. Furthermore, a weighted sum of two of these derivatives can be used to formulate a differential constitutive equation for polymeric liquids. In particular, the Gordon–Schowalter convected derivative is defined in this manner. For example, the Johnson–Segalman model is obtained by replacing the ordinary time derivative in Eq. (2.35) by the Gordon–Schowalter derivative.

Differential models obtained by replacing the ordinary time derivative in Eq. (2.35) by one that can describe large, rapid deformations are able to describe some viscoelastic phenomena, but only qualitatively. To improve on such models, it is necessary to introduce nonlinearity into the equation itself. In the widely used Phan–Thien–Tanner model, the Gordon–Schowalter convected derivative is used, and nonlinearity is introduced by multiplying the stress term by a function of the trace of the stress tensor. The Giesekus and Leonov models are other examples of nonlinear differential models. All of the models mentioned above are described in the monograph by Larson (24).

### 2.10.3 Constitutive Equations from Tube Models

Tube models are based on a picture in which the constraints imposed on a highly entangled polymer molecule (*test chain*) by the surrounding ones are modeled as a *tube* having a characteristic length and diameter (27). This is an example of a mean field theory, in which the effects of surrounding molecules on the test chain are averaged, which drastically reduces the computational effort that would be required by a full molecular dynamics model. Tube models have shown promise in the prediction of linear viscoelastic behavior and some types of nonlinear behavior for certain types of molecular structure. Although the behavior of materials containing certain types of branching structure has been successfully modeled, polydispersity poses a major problem that has not yet been completely solved (28). Thus, the ability of tube models to predict quantitatively the material functions of typical commercial polymers is, as yet, somewhat limited.

For use in flow simulation, what is needed is a constitutive equation (REOS), not just a procedure for calculating the relaxation modulus, and considerable effort has been made to develop general constitutive equations from tube models. Because these models arise from a picture of polymer behavior at the molecular level, they are less empirical than continuum models, and one might hope that they would thus have more universal validity. Ideally, one would like to arrive at a model in which all the parameters can be predicted from polymer molecular physics, without fitting rheological data.

The effort to develop such a model started with the original Doi–Edwards tube model, from which a constitutive equation can be derived if several additional simplifying assumptions are made (27). However, only rather recently has a rheological EOS developed from a tube model found some success in the simulation of a complex flow. Some of the difficulties involved in developing such models have been exposed by Wapperom et al. (29) and Marrucci (30). Likhtman et al. (31) proposed a procedure for deriving simple differential constitutive equations using tube model ideas. Curiously, the most successful model to date was developed for *pom-pom* molecules, which have a central backbone with branch points at both ends and two or more free arms attached at each branch point. This model was originally proposed by Larson and McLeish (32), and they also proposed a constitutive equation based on this picture. Verbeeten et al. (33) modified the original pom-pom constitutive equation, and several promising flow simulations based on this modification have been reported (34).

### 2.10.4  Mathematical vs. Hydroelastic Instability

A flow simulation cannot be carried out without a REOS, and if fluid elasticity plays a significant role in the instability, this model must describe with some accuracy the nonlinear viscoelastic behavior of the fluid when subjected to the deformation that gives rise to the instability. However, as we have seen, all the REOS models described above are more or less empirical and have never been verified in flows of significant complexity. Even in the case of models inspired by the tube concept, sufficient simplifying assumptions must be made to arrive at a model convenient for flow simulations that are, at best, semiempirical.

Another major challenge is that the use of a nonlinear REOS in a flow simulation leads to serious computational difficulties, with the result that a convergent solution is only possible under rather mild flow conditions. This is often referred to as the "high Weissenberg number problem." It is now understood that this type of problem arises because of the nonlinear nature of the REOS, and there have been a number of studies of the mathematical stability of these equations. For example, Grillet et al. (35) studied the stability of the exponential Phan–Thien–Tanner and the Giesekus models when used to describe plane Couette and Poiseuille flows. They had previously examined several other models, and their studies revealed "a surprising weakness in computational rheology—the lack of models which can predict linearly stable pressure driven flow of polymer melts. Even in this relatively simple flow, many of the most common rheological models break down in an array of . . . instabilities which can be expected to wreak havoc on numerical simulations." They go on to observe that these instabilities could be "related to experimentally observed instabilities." However, instabilities corresponding to those that arise in the simulations have never been observed. Some instabilities that arise in the use of the pom-pom model for flow simulation have been described by Lee et al. (36).

We conclude that our efforts to carry out stability analyses able to predict the instabilities that are observed in the flow of molten polymers are, at present, hampered by the unavailability of a model that can describe accurately highly nonlinear rheological behavior and is suitable for use in numerical computations.

## 2.11  SUMMARY

The instabilities that occur in the flow of entangled polymeric liquids are nearly always manifestations of their viscoelasticity, and because they occur in flows at fairly high deformation rates, a full understanding and modeling of them requires a nonlinear theory of viscoelasticity. Although there exists, at present, no generally valid theory governing all nonlinear behaviors, there has been some success in either establishing empirical relationships between specific rheological properties and instabilities, or interpreting a particular observation using an empirical model of viscoelastic behavior.

However, the limited information available from practically realizable rheological measurements is not adequate to evaluate constitutive equations for use in modeling the large, rapid transient flows that are involved in melt flow instabilities. This means that when several models give equally good fittings of measurable rheological properties but different predictions of flow instabilities, there is little, if any, basis for choosing among them. It is not even possible to know whether a predicted instability really predicts a flow situation or a mathematical instability caused by the structure of the model.

## REFERENCES

1. Ferry, J.D. *Viscoelastic Properties of Polymers*, 3rd Ed. New York: John Wiley and Sons, 1980. Chapters 3 and 4.
2. Meissner, J. Deformationsverhalten der Kunststoffe im flüssigen und im festen Zustand. Kunststoffe 1971, *61*, 576–582.
3. Macosko, C.W. *Rheology: Principles, Measurements, and Applications*; VCH: New York, 1994.
4. Dealy, J.M.; Wissbrun, K.F. Melt Rheology and Its Role in Plastics Processing. corrected edition. Amsterdam: Kluwer Academic Publishers, 1999.
5. Rothstein, J.P.; McKinley, G.H. Extensional flow of a polystyrene Boger fluid through a 4:1:4 axisymmetric contraction/expansion. J. Non-Newton. Fluid Mech. 1999, *86*, 61–88.
6. Boger, D.V.; Crochet, M.J.; Keiller, R.A. On viscoelastic flows through abrupt contractions. J. Non-Newton. Fluid Mech. 1992, *44*, 267–279.
7. Pipkin, A.C. *Lectures in Viscoelastic Theory*; New York: Springer-Verlag, 1972.
8. Dealy, J.M.; Saucier, P. *Rheology for Plastics Quality Control*; Hanser Publications: New York, 2000.
9. Dealy, J.M.; Giacomin, A.J. Sliding plate and sliding cylinder rheometers. *Rheological Measurement*, 2nd Ed.; Collyer, A.A., Clegg, D.W., Eds.; Chapman and Hall: London, 1998. Chapter 8.
10. Giacomin, A.J.; Dealy, J.M. Using large-amplitude oscillatory shear. *Rheological Measurement*, 2nd Ed.; Collyer, A.A., Clegg, D.W., Eds.; Chapman and Hall: London, 1998. Chapter 11.
11. Koran, F.; Dealy, J.M. A high-pressure sliding plate rheometer for polymer melts. J. Rheol. 1999, *43*, 1279–1290.
12. Koran, F.; Dealy, J.M. Wall slip of polyisobutylene: interfacial and pressure effects. J. Rheol. 1999, *43*, 1291–1306.
13. Hatzikiriakos, S.G.; Dealy, J.M. Wall slip of molten high density polyethylene: I. Sliding plate rheometer studies. J. Rheol. 1991, *35*, 497–523.
14. Hay, G.; Mackay, M.E.; Awati, K.M.; Park, Y. Pressure and temperature effects in slit rheometry. J. Rheol. 1999, *43*, 1099–1116.
15. Meissner, J. Dehnungsverhalten von polyäthylen-Schmelzen. Rheol. Acta 1971, *10*, 230–242.
16. Laun, H.M.; Münstedt, H. Comparison of elongational behavior of a polyethylene melt at constant stress and constant strain rate. Rheol. Acta 1976, *15*, 517–524.

17. Münstedt, H. New universal extensional rheometer for polymer melts. Measurements on a polystyrene sample. J. Rheol. 1979, *223*, 421–436.

18. Meissner, J.; Hostettler, J. A new elongational rheometer for polymer melts and other highly viscoelastic liquids. Rheol. Acta 1994, *33*, 1–21.

19. Sentmanat, M.L. A novel device for characterizing polymer flows in uniaxial extension. Soc. Plast. Eng., Annu. Tech. Conf. (ANTEC Tech. Pap.) 2003, *49*, 992–996.

20. Gendron, R.; Sammut, P.; Dufour, M.; Gauthier, B. Low-coherence interferometry applied to uniaxial rheometry. Soc. Plast. Eng., Annu. Tech. Conf. (ANTEC Tech. Pap.) 2003, *49*, 1008–1012.

21. Münstedt, H.; Laun, H.M. Elongational behavior of a low density polyethylene melt: II. Transient behavior in constant stretching rate and tensile creep experiments. Rheol. Acta 1979, *18*, 492–504.

22. McKinley, G.H.; Hassager, O. The Considère condition and rapid stretching of linear and branched polymer melts. J. Rheol. 1999, *43*, 1195–1212.

23. Suen, J.K.C.; Loo, Y.L.; Armstrong, R.C. Molecular orientation effects in viscoelasticity. Annu. Rev. Fluid Mech. 2002, *34*, 417–444.

24. Larson, R.G. *Constitutive Equations for Polymer Melts and Solutions*; Butterworth: Boston, 1988.

25. Lodge, A.S. *Elastic Liquids*; Academic Press: New York, 1964. Chapter 6.

26. McLeish, T. The hitchhikers guide to polymer rheology. Bull. Br. Soc. Rheol. 2000, *43* (1), 8–15.

27. Doi, M.; Edwards, S.F. *The Theory of Polymer Dynamics*; Oxford University Press: Oxford, 1986.

28. Dealy, J.M.; Larson, R.G. *Rheology and Structure of Molten Polymers*; Hanser Publishers: New York, 2004.

29. Wapperom, P.; Keunings, R.; Ianniruberto, G. Prediction of rheometrical and complex flows of linear polymers using the double-convected reptation model with chain stretch. J. Rheol. 2003, *47*, 247–265.

30. Marrucci, G.; Ianniruberto, G. A note added to "Prediction of rheometrical and complex flows". J. Rheol. 2003, *47*, 267–268.

31. Likhtman, A.E.; Graham, R.S.; McLeish, T.C.B. How to get simple constitutive equations for polymer melts from molecular theory. Proceedings of the 6th European Conference on Rheology, 2002; 237–238.

32. Larson, R.; McLeish, T.C.B. Molecular constitutive equations for a class of branched polymers: the pom-pom polymer. J. Rheol. 1998, *42*, 81–112.

33. Verbeeten, W.M.H.; Peters, G.W.M.; Baaijens, F.P.T. Differential constitutive equations for polymer melts: the extended pom-pom model. J. Rheol. 2001, *45*, 823–843.

34. Verbeeten, W.M.H. Computational polymer melt rheology. Doctoral dissertation, Eindhoven University of Technology, 2002.

35. Grillet, A.; Bogaerds, A.; Peters, G.; Baaijens, F. Stability analysis of constitutive equations for polymer melts in viscometric flows. J. Non-Newton. Fluid Mech. 2002, *103*, 250–321.

36. Lee, J.W.; Kim, D.; Kwon, Y. Mathematical characteristics of the pom-pom model. Rheol. Acta 2002, *41*, 223–231.

# 3

# Secondary Flow Instabilities

**Evan Mitsoulis**
National Technical University of Athens, Athens, Greece

## 3.1 INTRODUCTION

*Secondary flows* are defined as those flows which have components in a plane orthogonal to the main direction of flow (1). In a two-dimensional representation, these secondary flows appear as *vortices* or *eddies*, circulating in a direction opposite to the main ($x$) flow direction (see Fig. 3.1a from simulations in entry flow in a contraction, where the flow as seen by velocity vectors is coming from left) (2). In a three-dimensional representation, it is possible that these secondary flows take a helical path traveling also in the third ($z$) direction (Fig. 3.1b from simulations in entry flow in a contraction) (2).

*Instabilities* are defined as space-dependent and time-dependent disturbances around the main flow components, which usually have a detrimental effect on the flow. In the present chapter, we are dealing exclusively with non-Newtonian polymeric fluids and their secondary flow instabilities. Therefore, any discussion on Newtonian fluids or inertial instabilities is excluded. Also the discussion will be directed toward polymer processing flows, which are of interest in the plastics industry.

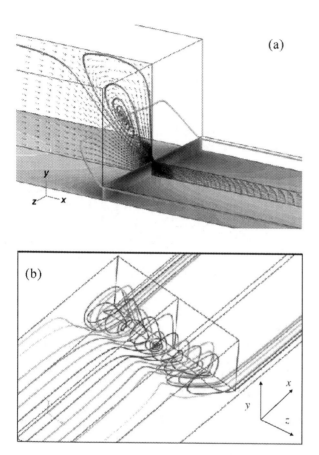

**FIGURE 3.1** Typical secondary flows in a contraction: (a) vortex formation in a midplane x–y for entry flow in a contraction; (b) helical flow in a y–z plane for entry flow in a contraction. The flow, as described by velocity vectors, is coming from the left. (From Ref. 2.)

Non-Newtonian fluids (polymer solutions and melts) are rheologically complex materials, which exhibit both viscous and elastic effects, and are called, therefore, *viscoelastic* (3). Their viscoelastic character is usually assessed by a dimensionless number. There are several dimensionless numbers used within the context of viscoelasticity, but these can be seen as being equivalent (4). For example, the Deborah number (*De*) is defined as:

$$De = \frac{\lambda}{\theta} = \lambda \dot{\gamma} \qquad (3.1)$$

where $\lambda$ is a material relaxation time, $\theta$ is a process relaxation time usually taken to be equal to $1/\dot{\gamma}$, and $\dot{\gamma}$ is a shear rate usually evaluated at the channel wall. The Weissenberg number ($Wi$) is defined as:

$$Wi = \lambda \frac{U}{H} \tag{3.2}$$

where $U$ is a characteristic velocity (usually taken as the average speed) and $H$ is a characteristic length (usually taken as the channel gap). The recoverable shear or stress ratio ($S_R$) is defined as:

$$S_R = \frac{N_{1,w}}{2\tau_w} \tag{3.3}$$

where $N_{1,w} = \tau_{11} - \tau_{22}$ is the first normal stress difference and $\tau_w$ is the shear stress, both evaluated at the channel wall. The equivalence is evident when we take:

$$\dot{\gamma} = U/H, \qquad N_{1,w} = \Psi_1 \dot{\gamma}^2, \qquad \tau_w = \eta\dot{\gamma}, \qquad \lambda = \Psi_1/2\eta \tag{3.4}$$

where $\Psi_1$ is the first normal stress difference coefficient and $\eta$ is the shear viscosity. The case of $De = Wi = S_R = 0$ corresponds to inelastic fluids ($\lambda = \Psi_1 = 0$), whereas it is understood that $De = Wi = S_R = 1$ corresponds to the elastic effects being as important as the viscous effects, and for $De = Wi = S_R > 1$, the elastic effects dominate the flow over the viscous effects.

In a manner very similar to that of Newtonian fluids and the transition from laminar flow to turbulent flow around a critical Reynolds number ($Re$), the creeping flow ($Re \approx 0$) of polymer solutions and melts shows a transition from a laminar stable flow to a still laminar flow that is unstable. This occurs at some critical value of the corresponding viscoelastic number ($De$, $Wi$, $S_R$) depending on the polymer at hand. This behavior, which is reminiscent of the inertial turbulence for inelastic fluids and is primarily due to the viscoelastic character of the polymeric fluids, has been given the term *elastic turbulence* in some recent works (5,6).

The present chapter reviews some secondary flows that appear in polymer processing and discusses several issues (usually still unresolved) and their influence in polymer processing. Some secondary flows arising in elementary viscometric flows are also discussed to some extent.

## 3.2  GOVERNING EQUATIONS

### 3.2.1  Conservation and Constitutive Equations

In order to study secondary flow instabilities, it is essential to consider first the governing flow equations. The flow of incompressible fluids (such as polymer solutions and melts, at least in situations where they are considered as incompressible) is governed by the usual conservation equations of mass, momentum, and energy (3,4), that is,

$$\nabla \cdot \mathbf{v} = 0 \tag{3.5}$$

$$\rho \mathbf{v} \cdot \nabla \mathbf{v} = -\nabla p + \nabla \cdot \tau \tag{3.6}$$

$$\rho c_\mathrm{p} \mathbf{v} \cdot \nabla T = k\nabla^2 T + \tau : \nabla \mathbf{v} \tag{3.7}$$

where $\mathbf{v}$ is the velocity vector, $p$ is the scalar pressure, $\tau$ is the extra stress tensor, $\rho$ is the density, $c_\mathrm{p}$ is the heat capacity, $k$ is the thermal conductivity, and $T$ is the temperature.

The above system of conservation equations (usually called the Navier–Stokes equations in fluid mechanics) is not closed for non-Newtonian fluids due to the presence of the stress tensor $\tau$. The required relationship between the stress tensor $\tau$ and the kinematics (velocities and velocity gradients) is given by appropriate *rheological constitutive equations*, and this is an eminent subject in theoretical rheology (3,7,8).

For purely viscous fluids, the rheological constitutive equation that relates the stresses $\tau$ to the velocity gradients is the generalized Newtonian model (3,4,7) and is written as:

$$\tau = \eta(\dot{\gamma})\dot{\gamma} \tag{3.8}$$

where $\dot{\gamma} = \nabla \mathbf{v} + \nabla \mathbf{v}^T$ is the rate-of-strain tensor and $\eta(\dot{\gamma})$ is the apparent viscosity given—among others—by the Carreau model (7):

$$\eta\left(\dot{\gamma}\right) = \eta_\infty + \eta_0 \left[1 + \left(\lambda \dot{\gamma}\right)^2\right]^{\frac{n-1}{2}} \tag{3.9}$$

In the above, $\eta_0$ is the zero shear rate viscosity, $\eta_\infty$ is the infinite shear rate viscosity, $\lambda$ is a characteristic time, and $n$ is the power–law index. The magnitude $\dot{\gamma}$ of the rate-of-strain tensor is given by:

$$\dot{\gamma} = \sqrt{\frac{1}{2} II_{\dot{\gamma}}} = \sqrt{\frac{1}{2}\left(\dot{\gamma} : \dot{\gamma}\right)} \tag{3.10}$$

where $II_{\dot{\gamma}}$ is the second invariant of the rate-of-strain tensor. The Carreau model describes well the shear-thinning behavior of polymer solutions and melts.

Regarding viscoelasticity, a plethora of constitutive equations with varying degrees of success and popularity exists. Standard textbooks on the subject (3,7–11) list categories of these equations and their predictions in several types of flow and deformation. There are constitutive equations of differential type, integral type, molecular models, etc. This subject matter is more fully explored in Chapter 2 by Dealy (present volume).

As an example of a popular viscoelastic constitutive equation used in the last 20 years, which possesses enough degree of complexity so as to capture as accurately as possibly the complex nature of polymeric liquids, we present here the K-BKZ integral constitutive equation with multiple relaxation times proposed by Papanastasiou et al. (12) and further modified by Luo and Tanner (13). This is often referred to in the literature as the K-BKZ/PSM model and is written as:

$$
\tau = \frac{1}{1-\theta} \int_{-\infty}^{t} \sum_{k=1}^{N} \frac{G_k}{\lambda_k} \exp\left(-\frac{t-t'}{\lambda_k}\right) H\left(I_{C^{-1}}, II_{C^{-1}}\right)
$$
$$
\times \left[ C_t^{-1}(t') + \theta C_t(t') \right] dt'
\tag{3.11}
$$

where $\tau$ is the stress tensor for the polymer; $\lambda_k$ and $G_k$ are the relaxation times and relaxation moduli, respectively; $N$ is the number of relaxation modes; $\theta$ is a material constant, $C_t$ is the Cauchy–Green tensor, $C_t^{-1}$ is the Finger strain tensor, and $I_{C^{-1}}, II_{C^{-1}}$ are its first and second invariants, respectively. $H$ is a strain–memory (or damping) function, and the following formula was proposed by Papanastasiou et al. (12):

$$
H\left(I_{C^{-1}}, II_{C^{-1}}\right) = \frac{\alpha}{(\alpha-3) + \beta I_{C^{-1}} + (1-\beta) II_{C^{-1}}}
\tag{3.12}
$$

where $\alpha$ and $\beta$ are nonlinear model constants to be determined from shear and elongational flow data, respectively. The $\theta$ parameter (a negative number) relates the second normal stress difference $N_2 = \tau_{22} - \tau_{33}$ to the first $N_1$ according to:

$$
\frac{N_2}{N_1} = \frac{\theta}{1-\theta}
\tag{3.13}
$$

The linear viscoelastic storage and loss moduli, $G'$ and $G''$, can be expressed as a function of frequency $\omega$ as follows:

$$
G'(\omega) = \sum_{k=1}^{N} G_k \frac{(\omega\lambda_k)^2}{1 + (\omega\lambda_k)^2}
\tag{3.14a}
$$

$$
G''(\omega) = \sum_{k=1}^{N} G_k \frac{(\omega\lambda_k)}{1 + (\omega\lambda_k)^2}
\tag{3.14b}
$$

These functions are independent of the strain–memory function (i.e., the type of constitutive model) and only $\lambda_k$ and $G_k$ can be determined from these data.

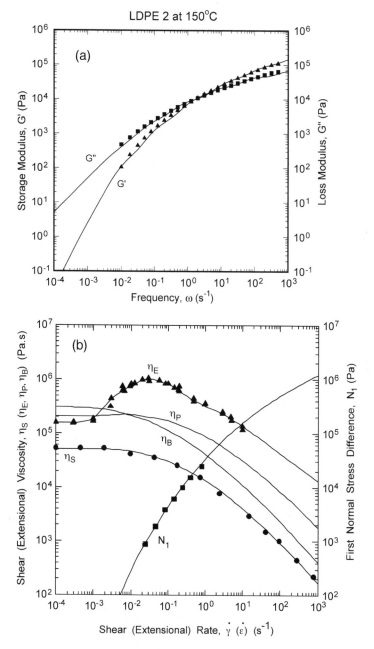

**FIGURE 3.2** Experimental data (symbols) and model predictions (lines) for the IUPAC-LDPE melt A using the K-BKZ/PSM model (Eq. (3.11)) with the parameters of Table 3.1: (a) dynamic data; (b) steady data. (From Ref. 17.)

The strain–memory function is derived from the first and second invariants of the Finger strain tensor. For simple shear flow, the strain–memory function is given as:

$$H(I_{C^{-1}}, II_{C^{-1}}) = \frac{\alpha}{\alpha + \gamma^2} \tag{3.15}$$

where $\gamma$ is the shear strain. The strain–memory function in simple shear flow is dependent on $\alpha$ but not on $\beta$. This is expected because $\alpha$ is viewed as a *shear* parameter, whereas $\beta$ is viewed as an *elongational* parameter.

The above constitutive equation has been used successfully for fitting data for many polymer solutions and melts (12,14). A good example is the IUPAC-LDPE (low-density polyethylene) melt A, which has been widely studied experimentally by several groups around the world (15). The fitting of the rheological data with the K-BKZ/PSM model is shown in Fig. 3.2, with the values of the model given in Table 3.1. The fitting involves the determination of the relaxation spectrum (parameters $N$, $\lambda_k$, and $G_k$) from experimental data on the storage and loss moduli $G'$ and $G''$. Then the nonlinear parameters $\alpha$ and $\beta$ are determined from shear and elongational data, in this case from shear viscosity $\eta_S$, first normal stress difference $N_1$, and uniaxial elongational viscosity $\eta_E$. The value of $\theta$ is usually set to be a small negative number (around $-0.1$) according to experimental evidence (13). Then the other extensional viscosities in planar extension $\eta_P$ and biaxial extension $\eta_B$ are predicted by the model. These predictions can also be extended to transient effects for all rheological functions at different times (12).

This integral model has been used in numerical flow simulations for a number of flow problems with more or less good success (e.g., see Refs. 13, 16–18). A recent review (19) on the subject has a list of problems solved with

**TABLE 3.1** Material Parameter Values Used in Eq. (3.11) for Fitting Data of the IUPAC-LDPE (sample A) Melt at 150°C ($\alpha = 14.38$, $\theta = -1/9$)

| $k$ | $\lambda_k$ (sec) | $G_k$ (Pa) | $\beta_k$ (—) |
|---|---|---|---|
| 1 | $10^{-4}$ | $1.29 \times 10^{+5}$ | 0.018 |
| 2 | $10^{-3}$ | $9.48 \times 10^{+4}$ | 0.018 |
| 3 | $10^{-2}$ | $5.86 \times 10^{+4}$ | 0.08 |
| 4 | $10^{-1}$ | $2.67 \times 10^{+4}$ | 0.12 |
| 5 | $10^{0}$ | $9.80 \times 10^{+3}$ | 0.12 |
| 6 | $10^{+1}$ | $1.89 \times 10^{+3}$ | 0.16 |
| 7 | $10^{+2}$ | $1.80 \times 10^{+2}$ | 0.03 |
| 8 | $10^{+3}$ | $1.00 \times 10^{0}$ | 0.002 |

*Source*: Ref. 13.

this model through numerical simulation, including many flows from polymer processing operations, with or without secondary flows. Other flows solved with a number of different constitutive equations can be found in a recent book on computational rheology (20).

## 3.3 STABILITY ANALYSIS

### 3.3.1 Linear Stability Analysis

For the traditional linear stability analysis, we consider the stability of the system with respect to small perturbations, which are ever present in real systems (21,22). The steady-state flow is considered as the *base flow* and a starting point for the analysis. A given solution to the governing equations and boundary conditions is subjected to a basic set of disturbance nodes of infinitesimal amplitude from which any such disturbance could be composed. If all the disturbances decay in time, the steady state is said to be *asymptotically stable*. In the opposite case, the steady state is *unstable*. In the intermediate condition where the disturbances neither grow nor decay in time, the system is *marginally* or *neutrally stable*. In mathematical terms, this is written as:

$$f_j(x, y, t) = f_{sj}(y) + \varepsilon \hat{f_j}(x, y, t) \tag{3.16}$$

where $f$ represents variables such as velocities, pressure, and the location of the interface in multilayer flows. The subscript "s" denotes the steady state, the hat indicates the perturbation to the base flow, and $\varepsilon$ is a small parameter proportional to the amplitude of the perturbation. The $x$-direction is the direction along the flow, and the $y$-direction is the direction along the thickness, whereas $t$ is the time. The time dependence of the disturbances can always be represented in the form of an exponential function for these linear equations. Thus, the disturbance can be represented as:

$$\hat{f_j}(x, y, t) = Re\left[\bar{f}(y)\exp(i\alpha x)\exp(ct)\right] \tag{3.17}$$

where $Re[\ ]$ is the real part of the term in the brackets, the complex function $\bar{f}(y)$ represents the spatial variation in the amplitude of the disturbance, the wave number $\alpha$ is a real quantity, and the wave propagation speed $c$ is a complex quantity given by:

$$c = c_r + ic_i \tag{3.18}$$

The real part of $c$ ($c_r$) provides the growth or decay of the disturbance, whereas the imaginary part of $c$ ($c_i$) provides the oscillating frequency. An eigenvalue problem is formed with the complex wave speed (disturbance velocity) as the eigenvalue, and the amplitude of the disturbances as the eigenfunction. Depending on the sign of the imaginary part of $c$, the flow may or may not be stable. When $c_i \gg 0$, the flow is temporarily unstable, and when $c_i \ll 0$, the flow is stable. If $c_i = 0$, the flow is neutrally stable (21,22).

### 3.3.2 Nonlinear Stability Analysis

In order to find out how nonlinearities influence characteristics of the secondary flow, a nonlinear analysis is necessary. The formulation of the nonlinear stability analysis is straightforward for the governing equations and boundary conditions discussed earlier. For a given spatial discretization technique in the numerical analysis, the resulting equations give rise to a set of ordinary nonlinear differential equations with respect to time, which can be integrated by any standard integration method for any initially imposed disturbance. The response of the flow to the initial disturbance provides a clue not only about the stability of the flow, but also about the complete characteristics of the response in terms of amplitudes and frequencies over a long period of time.

The advantages of the nonlinear stability analysis are as follows (21):

1. The response of the process can be investigated for realistic finite disturbances.
2. The frequencies and amplitudes of the ongoing disturbances can be determined.
3. The formulation of the nonlinear stability analysis is much easier compared to the formulation of the linear stability analysis.

However, computations of solving nonlinear equations are very expensive compared to solving the linear equations. When the problems are stiff—in other words, when the growth rates are much smaller than the frequencies of the disturbances—the computation time increases rapidly for the nonlinear analysis. Besides, there are virtually infinite types of initially imposed disturbances for which the transient response of the process can be studied. Examples of such nonlinear stability analyses can be found in Ref. 21.

## 3.4 POLYMER PROCESSING FLOWS

Various examples of polymer processing flows where secondary flows occur can be classified as *flows with fixed boundaries* (e.g., flows in contractions—expansions) and *flows with moving boundaries* (e.g., coating flows, roll coating or wire coating, calendering, cavity flows in extruder channels, etc.). The combination of drag and pressure-driven flow, as it exists inside processing equipment, easily generates vortices even for inelastic fluids obeying the generalized Newtonian model. Viscoelasticity can have either a stabilizing or a destabilizing effect on these flows depending on the material at hand, the process, and the operating conditions (22).

### 3.4.1 Polymer Processing with Fixed Boundaries

#### 3.4.1.1. Flow in a Contraction

Few flows have received so much interest both experimentally and computationally than flow in a contraction. It is related to the flow in the final stage of the extruder, that of the forming die, which in the simplest of cases is just a tubular orifice through which the polymer exits to the atmosphere from a larger reservoir. The simplified version of the problem is depicted in Fig. 3.3. The polymeric fluid passes from a larger reservoir of diameter $D_{res}$ to a smaller tube of diameter $D_0$. In the analogous sudden *planar* contraction, there is a step change in slit width. The presence of the abrupt contraction gives rise to many interesting flow patterns, including secondary flows in the reservoir, whereas the singularities at the salient corners render the problem extremely difficult to solve with any of the known numerical techniques.

For any fluid, Newtonian or non-Newtonian, there is in these geometries a region of extensional flow near the centerline and the contraction plane, whereas shear dominates near the walls (see Fig. 3.4). The well-known and different entry patterns of LDPE and high-density polyethylene (HDPE) melts (Fig. 3.5) in a 20:1 circular contraction with their distinct behavior have beguiled researchers in the scientific community for more than 40 years, since they appeared in 1960 in the work of Bagley and Birks (23). Although, for Newtonian fluids, fluid inertia can change the flow and the shape of the salient corner vortex (24), much more interesting transitions occur for viscoelastic fluids as the flow rate is increased and the flow becomes more elastic. The early

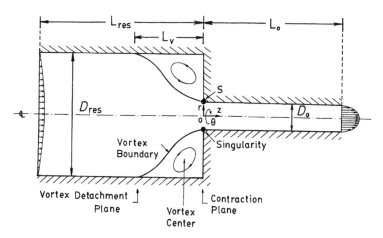

**FIGURE 3.3** Schematic representation of an abrupt contraction. (From Ref. 16.)

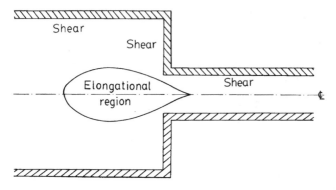

**FIGURE 3.4** Schematic representation of an abrupt contraction with highlighted regions of shear and elongation.

studies by Giesekus (7), have been followed by studies by Boger and Walters (1), Boger et al. (24), Cable and Boger (25–27), Nguyen and Boger (28), and Boger and Binnington (29), with some very interesting transitions shown in Figs. 3.6 and 3.7. In these and later studies (30–32), a complicated series of transitions has been found, the details of which depend on the specific fluid and the geometric parameters of the contraction, especially the contraction ratio $\beta = D_{res}/D_0$.

For example, for the test fluid M1 (a 0.244% polyisobutylene in a mixed solvent consisting of 7% kerosene in polybutene), the experimental flow patterns in a 4:1 circular contraction (29) are shown in Fig. 3.8 together with steady-state simulations using the K-BKZ/PSM model (16). For the lowest

LDPE          HDPE

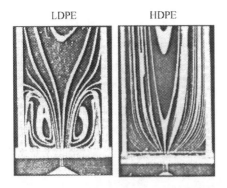

**FIGURE 3.5** Flow patterns of LDPE and HDPE melts flowing through a 20:1 axisymmetric contraction before unstable conditions set in. Note the big vortex for LDPE and the absence of any vortex activity for HDPE. (From Ref. 23.)

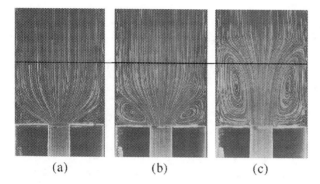

(a)            (b)            (c)

**FIGURE 3.6** Vortex growth for creeping flow in a 4:1 circular contraction for a Boger fluid (0.04% polyacrylamide in water and corn syrup solution). (a) $\dot{\gamma} = 3.4\ \text{sec}^{-1}$, $Re = 1.76 \times 10^{-3}$, $Wi = 0.120$; (b) $\dot{\gamma} = 9.3\ \text{sec}^{-1}$, $Re = 4.8 \times 10^{-3}$, $Wi = 0.179$; (c) $\dot{\gamma} = 24.2\ \text{sec}^{-1}$, $Re = 1.25 \times 10^{-2}$, $Wi = 0.204$. (From Ref. 24; reproduced in Ref. 1.)

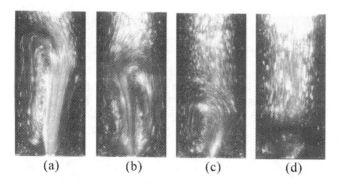

(a)        (b)        (c)        (d)

**FIGURE 3.7** Periodic helical flow (sequence a–d) in a 7.67:1 circular contraction for a Boger fluid (0.05% polyacrylamide in a glucose solution) at $\dot{\gamma} = 300\ \text{sec}^{-1}$, $Re = 2.9 \times 10^{-2}$. The flow lines descend into the downstream tube like a tornado (i.e., when viewed in two dimensions, the points of contact between the flow lines at the wall, at fixed flow rate, gradually move downstream, with the secondary flow vortex diminishing in size until it disappears into the downstream tube, with the fluid in the larger upstream tube surging into the smaller tube). A large new vortex is then formed and the process is repeated. (From Ref. 28; reproduced in Ref. 1.)

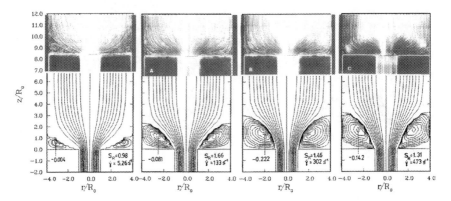

**FIGURE 3.8** Vortex behavior for creeping flow in a 4:1 circular contraction for fluid M1 (0.244% polyisobutylene in a mixed solvent consisting of 7% kerosene in polybutene). Upper part: experiments. (From Ref. 29.) Lower part: steady-state simulations. (From Ref. 16.) The experiments show a steady stable flow for low shear rates ($\dot{\gamma}$ = 5.26 sec$^{-1}$, $S_R$ = 0.98). For high shear rates, there is a transition into unstable flow, with a pulsating lip vortex (A,B,C), which the steady-state simulations cannot capture.

shear rate, there is agreement between experiments and simulations. For the higher shear rates, there is a distinct disagreement between experiments and simulations. The experiments show a pulsating flow with a lip vortex that oscillates near the entry to the die (A,B,C), whereas the simulations show a stable flow with strong vortex growth. On the other hand, the same fluid in a 22:1 circular contraction shows stable experimental patterns (29) in agreement with the steady-state simulations (16) as evidenced in Fig. 3.9. The differences in behavior are summarized in Fig. 3.10, which show that the contraction ratio has a strong influence on the stability of the flow for the same fluid. Obviously, any steady-state, two-dimensional, axisymmetric simulations are incapable of reproducing three-dimensional, time-dependent effects, such as the pulsating flow patterns exhibited by fluid M1 in a 4:1 circular contraction.

The works of Lawler et al. (30) and McKinley et al. (31) have shown that at $De \approx 0(1)$, there is an enlargement of either the corner vortex or the second vortex that often appears near the lip of the contraction. At higher $De$ numbers, the greatly enlarged vortex either pulsates regularly or becomes asymmetric and spirals around the upstream tube. Finally, at $De \approx 15$, the oscillations become aperiodic (31).

In the Laser–Doppler velocimetry (LDV) studies of Lawler et al. (30) and McKinley et al. (31), the first transition with increasing $De$ produced a

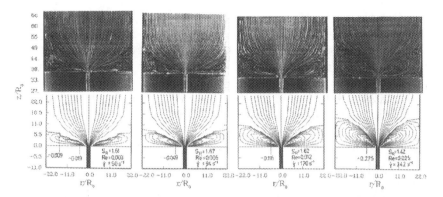

**FIGURE 3.9** Vortex growth for creeping flow in a 22:1 circular contraction for fluid M1 (0.244% polyisobutylene in a mixed solvent consisting of 7% kerosene in polybutene). Upper part: experiments. (From Ref. 29.) Lower part: simulations. (From Ref. 16.) Both the experiments and the simulations show a steady stable flow with vortex enhancement.

time-dependent flow. It occurred before the onset of the visible pulsing of the enlarged vortex. These LDV studies show that, at least in some cases, three-dimensional, time-periodic oscillations in velocity occur in a confined region near the lip at a critical number $De_{crit}$ below the regime of vortex enhancement. As $De$ is increased above $De_{crit}$, higher harmonics and even a sub-harmonic can appear in the frequency spectrum. At still higher $De$, in the

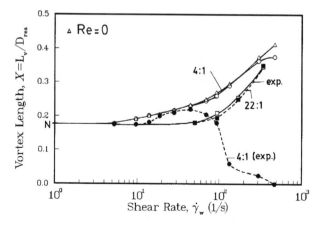

**FIGURE 3.10** Simulated vortex length $X$ vs. $\dot{\gamma}_w$ for fluid M1 (From Ref. 16.) compared with the experimental observations (solid symbols). (From Ref. 29.)

work of McKinley et al. (31), a lip vortex appeared. The velocity field associated with this lip vortex was quasi-periodic for small $\beta$ ($\leq$5) and was steady for larger $\beta$. At still higher $De$, in the vortex growth regime, the frequency spectrum showed that the flow was aperiodic for $\beta \leq 5$. Figure 3.11 reproduces a diagram showing the flow transitions determined in the LDV and visualizations studies of McKinley et al. (31). For the Boger fluid studied in the earlier work by Lawler et al. (30), time-periodic flow occurred at a $De_{crit}$, but then mysteriously disappeared when $De$ was raised to another critical Deborah number above $De_{crit}$. The sequence of transitions that occur in axisymmetric contraction flow is sensitive to both the contraction ratio $\beta$ and the fluid rheology (31,32).

In a planar contraction geometry, corner vortex growth does not seem to occur for Boger fluids, but does occur for shear-thinning polyacrylamide solutions (33,34). As in axisymmetric contractions, these vortices show instabilities to three-dimensional, time-dependent motions at high flow rates and are sensitive to the contraction geometry. There is a possible connection between these entry flow instabilities and extrudate distortion (see other chapters of this book, e.g., Chapter 6 by Migler, Chapter 7 by Georgiou, and Chapter 8 by Dealy, present volume).

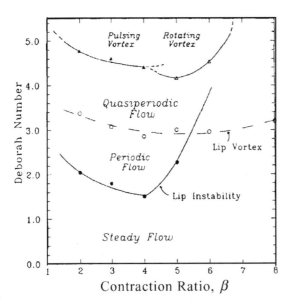

FIGURE 3.11 Flow transition diagram for sudden contraction flow of a Boger fluid (solution of high-molecular weight polyisobutylene in low-molecular weight polybutene with a little added tetradecane). (From Ref. 31.)

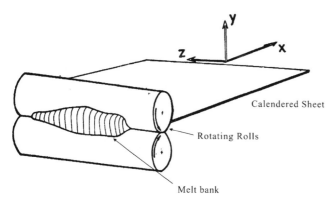

**FIGURE 3.12** Schematic representation of calendering a plastic sheet. (From Ref. 36.)

## 3.4.2 Polymer Processing with Moving Boundaries

### 3.4.2.1. Calendering

In the process of calendering, a molten polymer enters usually as a sheet on one of two rotating rolls and leaves on the other with a reduced thickness. The process is schematically shown in Fig. 3.12. It is seen that due to the reduced area, a melt bank is created before the nip region. In this melt bank, a very interesting flow pattern develops with multiple recirculation regions, as shown in Fig. 3.13 from experiments on rigid polyvinyl chloride (PVC) by Agassant and Espy (35). A huge vortex appears in the melt bank, whereas smaller ones develop near the entering sheet and near the nip region.

A purely viscous nonisothermal model (power–law with slip at the wall) for PVC is able to capture this vortex behavior as evidenced in Fig. 3.14a (36). The simulations also provide the temperature field (Fig. 3.14b) (36,37). When

**FIGURE 3.13** Experimental stable flow pattern in the melt bank of calendered rigid PVC. (From Ref. 35.)

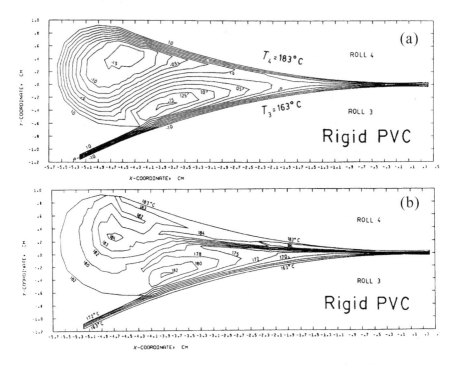

**FIGURE 3.14** Simulations of stable flow in calendering rigid PVC: (a) flow pattern seen through streamlines; (b) isotherms. (From Ref. 36.)

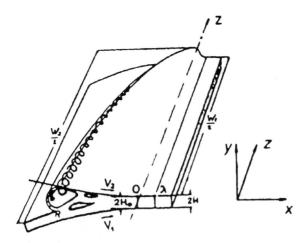

**FIGURE 3.15** Schematic helical flow pattern in the melt bank of calendered rigid PVC. (From Ref. 38; reproduced in Ref. 35.)

the calendering process is stable (i.e., when the melt bank is stable), the experimental streamlines are very similar to the computed ones (presence of two stable recirculating regions). When the calendering process is unstable (high output rate or a melt bank that is too big), the bank and the recirculating regions vary in the third direction, and a spiral flow ensues. The spiral flow in the third direction shown in Fig. 3.15 was observed by Unkrüer (38) and gives rise to several instabilities as pointed out also by Agassant and Espy (35). The input rate is locally greater than the output rate. Thus, the first vortex region becomes bigger and gradually encloses the second region (Fig. 3.16). When

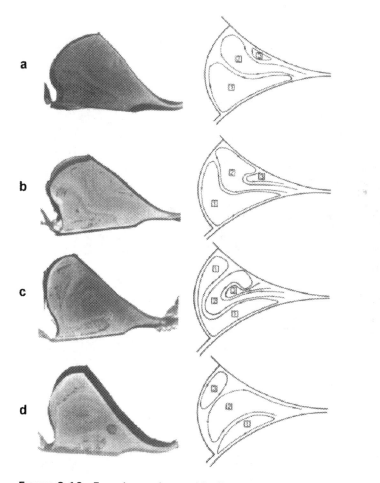

**FIGURE 3.16** Experimental unstable flow patterns in the melt bank of calendered rigid PVC. (From Ref. 35.)

the first vortex comes into contact with the upper roll, the flow kinematics is incompatible with the speed of the roll, and a part of the first vortex (no. 3) is abruptly separated and flows in the gap along the upper roll (Fig. 3.16). These instabilities take the shape of Vs, as shown in Fig. 3.17 (from Ref. 39). These types of instabilities are detrimental to the sheet produced and must be repressed. This is usually done by reducing the roll speed or increasing the minimum gap so as to allow for less severe conditions to take place in the nip region (35,37,39). At the present time, no attempt has been made to predict such an unstable development of the melt bank. This subject is still open in the literature for further research and definite conclusions.

### 3.4.2.2. Roll Coating

In the process of roll coating, a sheet is produced much as in calendering but usually for polymeric solutions. In such flows, surface tension becomes im-

**FIGURE 3.17** V-shape sheet defects in calendering rigid PVC. (From Ref. 39.)

portant. Both forward and reverse roll-coating operations are used in practice. The important work by Coyle et al. (40–42) has done much to increase our understanding of the fluid dynamics in both forward and reverse roll coating. The reverse process is depicted schematically in Fig. 3.18 (from Ref. 42). Experiments coupled with theory show interesting flow patterns with vortices and various types of instabilities, such as cascade, ribbing, seashore, etc. For example, in reverse roll coating of Newtonian and shear-thinning inelastic fluids, vortices appear as evidenced in Fig. 3.19 (42). Typical roll-coating instabilities observed experimentally are shown schematically in Fig. 3.20 (4). The stability analysis was based on Newtonian fluids, with the Capillary number as the nonlinear effect causing the instabilities. Experimental work based on shear-thinning inelastic fluids has shown that shear thinning also plays an important role, as evidenced in Fig. 3.21 (42). For viscoelastic polymer solutions, the experimental work of Coyle et al. (42) showed that the ribbing phenomenon becomes irregular and time-dependent. There is no sharp transition to cascade instability, accompanied by the steep upturn in coating thickness as speed ratio is increased. Rather, the coated film becomes mottled in appearance and the average coating thickness stays relatively constant (42). These instabilities are due to secondary helical flows in the third dimension and must be resolved before a better coating is obtained.

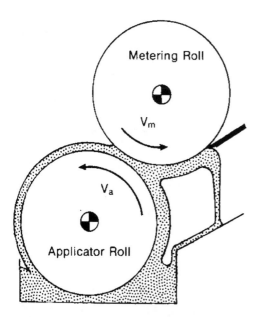

**FIGURE 3.18**  Schematic representation of reverse roll coating to study the flow in the metering gap. (From Ref. 42.)

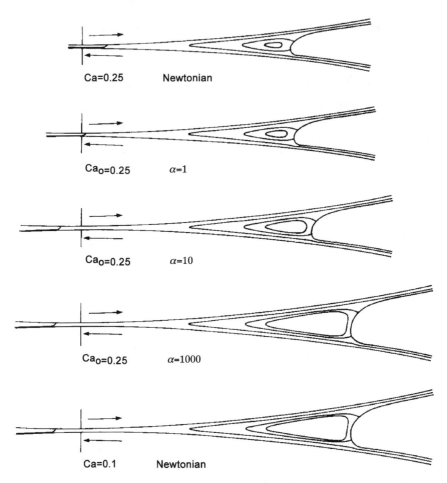

$Ca=0.25$     Newtonian

$Ca_o=0.25$     $\alpha=1$

$Ca_o=0.25$     $\alpha=10$

$Ca_o=0.25$     $\alpha=1000$

$Ca=0.1$     Newtonian

**FIGURE 3.19** Predicted streamlines in stable flow for Newtonian and shear-thinning inelastic fluids in reverse roll coating. (From Ref. 42.)

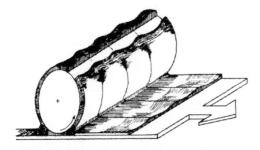

**FIGURE 3.20** Schematic instabilities in roll coating. (From Ref. 4.)

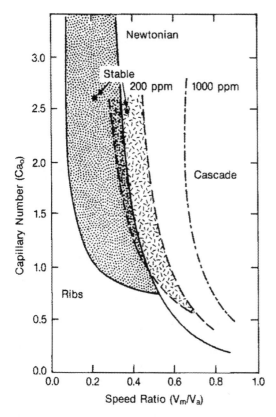

**FIGURE 3.21** Effect of shear thinning on the stability in reverse roll coating. (From Ref. 42.)

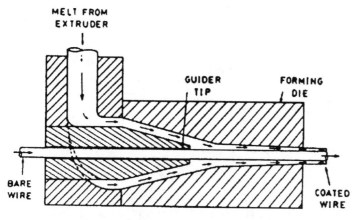

**FIGURE 3.22** Schematic representation of flow in a wire-coating die. (From Ref. 43.)

### 3.4.2.3. Wire Coating

The process of wire coating is depicted in Fig. 3.22. A molten polymer, usually under the influence of a pressure-driven flow, is extruded through a wire-coating die having inside a "torpedo" (guider) to guide the wire. The position of the torpedo and its guider tip is crucial for a good design, which avoids recirculation and limits the level of stresses generated in the melt as it enters from the annular channel into the die region under the dragging action of the moving wire (43).

A design that generates a big vortex for a Newtonian fluid may not generate one for a polymer melt. This is evidenced in Fig. 3.23, where the full solution of the steady-state flow shows just that (44). On the other hand, for a

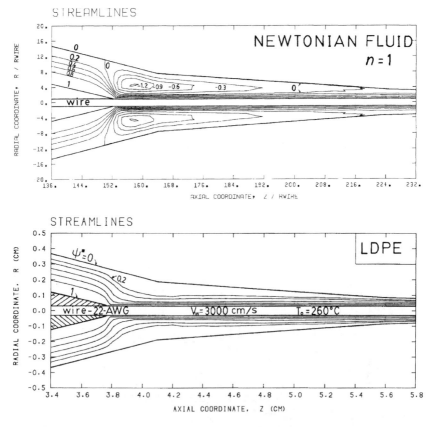

**Figure 3.23** Stable flow patterns of Newtonian and shear-thinning polymer melts (LDPE) in Fenner's die. The Newtonian fluid exhibits a big vortex, whereas in the same geometry, the LDPE melts flow without recirculation. (From Ref. 44.)

viscoelastic polymer melt, a bad design that leaves enough "gum space" between the torpedo and die wall may generate a vortex that is detrimental to the exiting melt (43). Good designs based on the appropriate "gum space" have been put forward, and these eliminate both the unwanted vortices and stress "jumps" along the die walls (43). Vortices may also be produced in wire-coating coextrusion, depending on the viscosities of the two fluids, as evidenced in Fig. 3.24 (45). Accordingly, they can be eliminated by a judicious choice of the polymer viscosities and/or temperatures (45). The analysis of Tadmor and Bird (46) has shown that a nonzero second normal stress difference $N_2$ exhibited by polymer melts has a stabilizing effect on the wire-coating process, by reducing the eccentricity that may appear in high-speed operations. A full stability analysis of the process is still missing.

### 3.4.2.4. Coextrusion

In the coextrusion process, two or more fluids come together in a die and flow with a common interface. This type of flow can quite easily generate all sorts of interface instabilities, and many papers have appeared over the years to

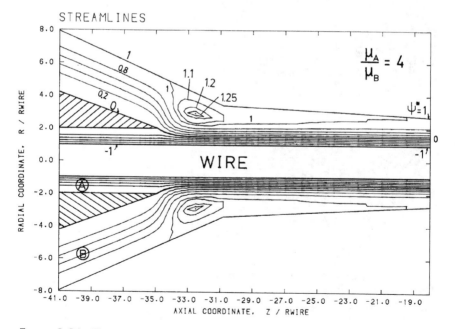

**FIGURE 3.24** Flow pattern of two Newtonian fluids in wire-coating coextrusion. Depending on the viscosity ratio, a big vortex can be produced in this die design. (From Ref. 43.)

deal with interface instabilities (e.g., see review by Larson (22) and Chapter 12 by Dooley in the present volume).

In the present work, because we are dealing with secondary flow instabilities, we are concentrating in the generation of vortices in the steady-state flow. This is evidenced in Fig. 3.25, where, depending on the flow rate, two layers may flow either in a stable or an unstable configuration with the presence of a vortex, as seen in the experiments by Strauch (47). The simulations of the base flow also show this vortex activity, as seen in Fig. 3.26 (from Ref. 48). The experiments by Strauch (47) show a window of stable/unstable operation based on vortex activity. The stability analysis of such types of flows can be carried out along the lines put forward by Anturkar et al. (21).

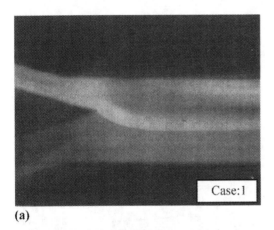

(a)

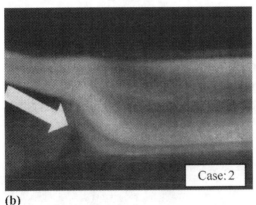

(b)

FIGURE 3.25 Flow patterns in coextrusion dies: (a) stable configuration; (b) unstable configuration with recirculation. (From Ref. 48.)

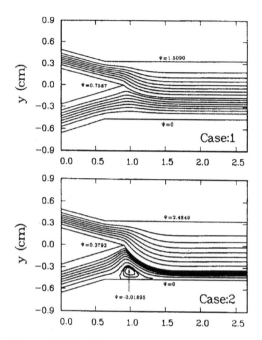

**FIGURE 3.26**  Simulated flow patterns in coextrusion dies. (From Ref. 48.)

For more details on the subject, the reader is also referred to Chapter 12 (by Dooley, present volume), on coextrusion instabilities, where a more in-depth analysis is given.

## 3.5  CONCLUSIONS

In the present chapter, some secondary flow instabilities have been reviewed, which arise in polymer processing operations with viscoelastic polymer melts. The topic is vast and this chapter has dealt exclusively with secondary flows resulting in vortices and helical flows. Although experimental evidence has accumulated over the years from various researchers and for various processes, the theory and predictions have lagged behind, due mainly to the difficulty of the subject. Many steady-state flows show vortices in stable operation, which may spiral out of control in an unstable operation. On top of viscoelasticity, which is still very much an active topic of research, one must consider linear and nonlinear stability analyses, which can give operating windows, which are ever so crucial for good operations. This is still a much sought-after subject, especially for such complicated flows as calendering and coating flows, flows through extruder channels, and injection molds. It is guaranteed to keep the scientific community busy for many years to come.

## ACKNOWLEDGMENTS

Financial assistance from the National Technical University of Athens (NTUA) is gratefully acknowledged.

## REFERENCES

1. Boger, D.V.; Walters, K. *Rheological Phenomena in Focus*; Elsevier: New York, 1993; 35–72.
2. Sirakov, I. Étude par Éléments Finis des Écoulements Viscoélastiques des Polymères Fondus dans des Géométries Complexes: Résultats Numériques et Expérimentaux. Ph.D. Thesis, Université Jean Monnet, Saint-Etienne, France, 2000.
3. Barnes, H.A.; Hutton, J.F.; Walters, K. *An Introduction to Rheology. Rheology Series*; Elsevier: Amsterdam, 1989, 11–35.
4. Middleman, S. *Fundamentals of Polymer Processing*; McGraw-Hill: New York, 1977; 8–68.
5. Groisman, A.; Steinberg, V. Elastic turbulence in a polymer solution flow. Nature 2000, *405*, 53–55.
6. Balkovsky, E.; Fouxon, A.; Lebedev, V. Turbulence of polymer solutions. Phys. Rev. E 2001, *64*, 056301-1-14.
7. Bird, R.B.; Armstrong, R.C.; Hassager, O. Dynamics of Polymeric Liquids. *Fluid Mechanics*, 2nd Ed.; Wiley: New York, 1987; Vol. 1, 1–52.
8. Larson, R.G. *Constitutive Equations for Polymer Melts and Solutions*; Butterworth: Boston, 1988; 59–89.
9. Dealy, J.M.; Wissbrun, K.F. *Melt Rheology and Its Role in Polymer Processing. Theory and Applications*; Van Nostrand-Reinhold: New York, 1990, 1–41.
10. Larson, R.G. *The Structure and Rheology of Complex Fluids*; Oxford University Press: Oxford, 1999; 1–56.
11. Tanner, R.I. *Engineering Rheology*, 2nd Ed.; Oxford University Press: Oxford, 2000; 33–75.
12. Papanastasiou, A.C.; Scriven, L.E.; Macosko, C.W. An integral constitutive equation for mixed flows: rheological characterization. J. Rheol. 1983, *27*, 387–410.
13. Luo, X.L.; Tanner, R.I. Finite element simulation of long and short circular die extrusion experiments using integral models. Int. J. Numer. Methods Eng. 1988, *25*, 9–22.
14. Kajiwara, T.; Barakos, G.; Mitsoulis, E. Rheological characterization of polymer solutions and melts with an integral constitutive equation. Int. J. Polym. Anal. Charact. 1995, *1*, 201–215.

15.  Meissner, J. Basic parameters, melt rheology, processing and end-use properties of three similar low density polyethylene samples. Pure Appl. Chem. 1975, *42*, 551–612.

16.  Park, H.J.; Mitsoulis, E. Numerical simulation of circular entry flows of fluid M1 using an integral constitutive equation. J. Non-Newton. Fluid Mech. 1992, *42*, 301–314.

17.  Barakos, G.; Mitsoulis, E. Numerical simulation of extrusion through orifice dies and prediction of Bagley correction for an IUPAC-LDPE melt. J. Rheol. 1995, *39*, 193–209.

18.  Sun, J.; Phan-Thien, N.; Tanner, R.I. Extrudate swell through an orifice die. Rheol. Acta 1996, *35*, 1–12.

19.  Keunings, R. Finite element methods for integral viscoelastic fluids. In *Rheology Reviews 2003*; Binding, D.M., Walters, K., Eds.; British Society of Rheology: London, 2003; 167–195.

20.  Owens, R.G.; Phillips, T.N. Computational Rheology; London: Imperial College Press, 2002, 19–47.

21.  Anturkar, N.R.; Papanastasiou, T.C.; Wilkes, J.O. Stability of multilayer extrusion of viscoelastic liquids. AIChE J. 1990, *36*, 710–724.

22.  Larson, R.G. Instabilities in viscoelastic flows. Rheol. Acta 1992, *31*, 213–263.

23.  Bagley, E.B.; Birks, A.M. Flow of polyethylene into a capillary. J. Appl. Phys. 1960, *31*, 556–561.

24.  Boger, D.V.; Hur, D.U.; Binnington, R.J. Further observations of elastic effects in tubular entry flows. J. Non-Newton. Fluid Mech. 1986, *20*, 31–49.

25.  Cable, P.J.; Boger, D.V. A comprehensive experimental investigation of tubular entry flow of viscoelastic fluids: Part 1. Vortex characteristics in stable flow. AIChE J. 1978a, *24*, 869–879.

26.  Cable, P.J.; Boger, D.V. A comprehensive experimental investigation of tubular entry flow of viscoelastic fluids: Part 2. The velocity field in stable flow. AIChE J. 1978b, *24*, 882–899.

27.  Cable, P.J.; Boger, D.V. A comprehensive experimental investigation of tubular entry flow of viscoelastic fluids: Part 3. Unstable flow. AIChE J. 1979, *25*, 152–159.

28.  Nguyen, H.; Boger, D.V. The kinematics and stability of die entry flows. J. Non-Newton. Fluid Mech. 1979, *5*, 353–368.

29.  Boger, D.V.; Binnington, R.J. Circular entry flows of M1. J. Non-Newton. Fluid Mech. 1990, *35*, 339–360.

30.  Lawler, J.V.; Muller, S.J.; Brown, R.A.; Armstrong, R.C. Laser Doppler velocimetry measurements of velocity fields and transitions in viscoelastic fluids. J. Non-Newton. Fluid Mech. 1986, *20*, 51–92.

31.  McKinley, G.H.; Raiford, W.P.; Brown, R.A.; Armstrong, R.C. Nonlinear dynamics of viscoelastic flow in axisymmetric abrupt contraction. J. Fluid Mech. 1991, *223*, 411–456.

32.  Pakdel, P.; McKinley, G.H. Elastic instability and curved streamlines. Phys. Rev. Lett. 1996, *77*, 2459–2462.

33.  Evans, R.E.; Walters, K. Flow characteristics associated with abrupt changes in

geometry in the case of highly elastic liquids. J. Non-Newton. Fluid Mech. 1986, *20*, 11–29.

34. Evans, R.E.; Walters, K. Further remarks on the lip-vortex mechanism of vortex enhancement in planar-contraction flows. J. Non-Newton. Fluid Mech. 1989, *32*, 95–105.

35. Agassant, J.F.; Espy, M. Theoretical and experimental study of the molten polymer flow in the calender bank. Polym. Eng. Sci. 1985, *25*, 118–121.

36. Mitsoulis, E.; Vlachopoulos, J.; Mirza, F.A. Calendering analysis without the lubrication approximation. Polym. Eng. Sci. 1985, *25*, 6–18.

37. Vlachopoulos, J.; Mitsoulis, E. Fluid Flow and Heat Transfer in Calendering: A Review. Transport Phenomena in Polymeric Systems 2: Vol. VI. *Advances in Transport Processes*; Kamal, M.R., Mashelkar, R.A., Mujumdar, A.S., Eds.; Wiley Eastern: New Delhi, 1988; 79–104.

38. Unkrüer, W. Beitrag zur Ermittlung des Drucksverlaufes und der Fliessvorgänge im Walzspalt bei der Kalanderverarbeitung von PVC Hart zu Folien. Ph.D. Thesis, IKV, TU Aachen, 1970.

39. Agassant, J.F. Le Calandrage des Matières Thermoplastiques. Doctoral Thesis, Paris 6, France, 1980.

40. Coyle, D.J.; Macosko, C.W.; Scriven, L.E. Film-splitting flows of shear-thinning liquids in forward roll coating. AIChE J. 1987, *33*, 741–746.

41. Coyle, D.J.; Macosko, C.W.; Scriven, L.E. The fluid dynamics of reverse roll coating. AIChE J. 1990, *36*, 161–174.

42. Coyle, D.J.; Macosko, C.W.; Scriven, L.E. Reverse roll coating of non-Newtonian liquids. J. Rheol. 1990, *34*, 615–636.

43. Mitsoulis, E. Fluid flow and heat transfer in wire coating: a review. Adv. Polym. Tech. 1986a, *6*, 467–487.

44. Mitsoulis, E. Finite element analysis of wire coating. Polym. Eng. Sci. 1986b, *26*, 171–186.

45. Heng, F.L.; Mitsoulis, E. Numerical simulation of wire-coating coextrusion. Int. Polym. Proc. 1989, *4*, 44–56.

46. Tadmor, Z.; Bird, R.B. Rheological analysis of stabilizing forces in wire-coating dies. Polym. Eng. Sci. 1974, *14*, 124–136.

47. Strauch, T. Ein Betrag zur Rheologischen Auslegung von Coextrusionswerkzeugen. Doctoral Thesis, IKV, TH Aachen, Germany, 1986.

48. Hannachi, A.; Mitsoulis, E. Sheet coextrusion of polymer solutions and melts: comparison between simulation and experiments. Adv. Polym. Technol. 1993, *12*, 217–231.

# 4

## Wall Slip: Measurement and Modeling Issues

**Lynden A. Archer**
Cornell University, Ithaca, New York, U.S.A.

### 4.1 INTRODUCTION

The no-slip hydrodynamic condition states, without proof, that the tangential velocity component of a liquid in contact with a solid substrate is the velocity of the substrate. This hypothesis is important because it provides a vital link between continuum equations that govern flow of bulk liquids and the complex molecular processes responsible for momentum transport across fluid–solid interfaces. A good summary of the early history surrounding the no-slip condition can be found in Ref. (1). Pioneering experiments by Coulomb, for example, are believed to have provided critical early support for the no-slip condition. Coulomb measured the drag force exerted by liquid water on oscillating metal disks smeared with grease and with grease covered by powdered sandstone and found no difference. This finding is significant because it indicates that the friction coefficient of the solid substrate has no influence on the fluid velocity at the fluid–solid interface, a requirement for the no-slip condition to be universally valid. Early analyses by Stokes and others of pressure-driven flow of simple liquids through tubes with no-slip at the walls also yielded results consistent with experimental observations for flow of water through glass tubes (1).

After nearly two centuries, it is now accepted that, but for motion at developing liquid–solid interfaces, the no-slip condition accurately describes the flow of simple liquids near solid substrates (2,3). If the substrates are rendered nonwetting by chemical modification (4,5) or are smooth on length scales comparable to the size of fundamental fluid elements (e.g., molecules in simple liquids or solid particles in a granular material) (6–8), flow predictions based on the no-slip condition are not always accurate, even for simple liquids. Apparent violations of the no-slip condition have also been reported in simple liquids when the dimensions of flow channels become comparable to the molecular size (9). In the simplest situations, a relationship between shear stress and slip velocity $V_s$ of the form $\sigma = kV_s$ follows from linear response theory (1). Here $k$ is the dynamic friction coefficient per unit area between liquid and solid. Thus if a liquid is subject to a simple laminar shear flow, $v = v_x(y)e_x$ between parallel planes with surface normal $n_y$ (Fig. 4.1),

$$V_s = \frac{\eta}{k} \frac{\partial}{\partial y} v_x(y) \tag{4.1}$$

where $\eta$ is the apparent viscosity, and the quantity $\eta/k = b$ has dimensions of length. The slip length $b$ is physically the extrapolated distance into the stationary substrate where the tangential component of the liquid's velocity vanishes. The severity of slip violations at a particular level of stress may therefore be specified either in terms of deviations of $b$ or $V_s$ from zero.

Claims of slip violations are more commonplace for polymeric fluids (10–15). Velocimetry measurements using large and small particle tracers (16–18), hot film methods (19–21), and fringe pattern fluorescence recovery after photobleaching FPFRAP (22–24) reveal non-zero tangential velocities in a wide variety of polymer liquids during laminar flow near stationary solid boundaries. Polymer viscosities measured in pressure-driven and drag flows have been reported to systematically decrease as the dimensions of the flow channel is reduced (25–27). Other flow phenomena, such as a sudden increase in volumetric flow rate observed when the pressure on a polymer liquid in

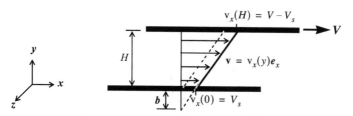

FIGURE 4.1  Planar Couette shear flow with slip at the wall.

Poiseuille flow is increased beyond a critical value and aperiodic oscillations of shear stress, optical birefringence, and flow rate reported in polymer liquids undergoing laminar flow between parallel planes, have also been credited to interfacial slip and stick–slip flow (11–15,26,28,29).

Unlike simple liquids, where slip violations are generally limited to unusual wetting scenarios or uncommonly low values of surface roughness, polymeric fluids appear to slip by such a variety of mechanisms that locating the plane of slip is difficult (Fig. 4.2). Work by several groups, particularly during the last decade, has established that both $V_s$ and $b$ depend on surface conditions (12–14,30–35). This clearly indicates that fundamental under-standing of polymer adsorption and interactions between polymer chains near the polymer–solid interface are important for resolving the molecular-scale processes responsible for slip violations. In some cases, slip violations appear to coincide with adhesive failure at the interface. This type of slip has been termed *true slip* because relative tangential motion between an adsorbed liquid polymer and a solid substrate occurs at the solid–liquid interface (Fig. 4.2), in direct violation of the no-slip condition. In other situations, slip arises from failure at a polymer–polymer interface displaced one or more molecular diameters away from the polymer–solid interface. This latter type of slip is broadly termed *apparent slip* because the tangential velocity of fluid in molecular contact with the substrate is that of the substrate, as required by the no-slip condition.

In this chapter, we will review current understanding of interfacial slip violations in polymer liquids. Various experimental methods for quantifying

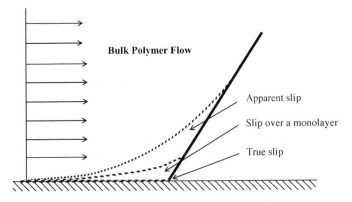

FIGURE 4.2 Locating the plane of slip for interfacial slip by adhesive failure (true slip), by disentanglement of surface-attached and bulk molecules (slip over a monolayer), and by polymer depletion or constitutive instabilities (apparent slip).

slip, key findings, and theoretical approaches for understanding polymer flow and dynamics near interfaces will be discussed. It will quickly emerge that polymer systems provide a rich setting for studying the general validity of the no-slip condition. The chapter is organized as follows: In the first section, we consider measurement techniques for quantifying slip violations in polymers and discuss the most salient types of information obtained from these measurements. This will be followed by a mostly qualitative discussion about failure mechanisms in polymer liquids to highlight the different theoretical approaches that can be taken in studying interfacial slip. In the third section, we will introduce several theories for interfacial slip in polymers, primarily focusing on molecular theories that allow slip parameters to be related to fundamental material and interfacial properties. When possible, we will provide quantitative predictions for features of slip that are accessible by many measurement techniques.

## 4.2 EXPERIMENTAL METHODS FOR QUANTIFYING SLIP VIOLATIONS: INDIRECT METHODS

### 4.2.1 Gap-Dependent Measurements in Planar Couette Shear Flow

The most common methods for quantifying slip violations utilize gap-dependent measurements of stresses and/or flow rates in simple flow geometries. Couette shear flow between parallel planes provides a good setting to illustrate the basic measurement principle. This geometry is also advantageous because it can be used to create an ideal, rectilinear shear flow in which the shear stress is independent of position. Consider a fluid subject to uniform laminar shear flow in the planar Couette flow apparatus depicted in Fig. 4.1. Shear flow is created by translating one plane with velocity $Ve_x$ while the other is held stationary. Assuming the plane dimensions are large enough that edge effects are unimportant and that the no-slip condition holds at the shear cell walls, the velocity field in the fluid is

$$v = v_x(y)e_x = \dot{\gamma}ye_x \tag{4.2}$$

$\dot{\gamma} \equiv V/H$ is the shear rate, $H$ is the plane separation or gap, and $Ve_x$ is the fluid velocity at the upper plane. There is then only one non-zero component of the velocity gradient tensor, $\nabla v = \dot{\gamma}e_ye_x$, which is independent of location in the gap.

The flux of momentum or shear stress $\sigma_{yx} = \sigma$ between the top and bottom planes is therefore the same across all fluid planes. Further, if the materials of construction and surface roughness of the two planes are the same, slip violations are equally likely at $y = 0$ and $y = H$ in this geometry.

Taking the relative velocity between fluid and solid at either interface to be $V_s$, the true shear rate $\dot{\gamma}_T$ experienced by the fluid is then given by

$$\dot{\gamma}_T = \dot{\gamma}_n - 2\frac{V_s}{H} \tag{4.3}$$

where $\dot{\gamma}_n = V/H$ is the *nominal shear rate*. Introducing the definition of the slip length, Eq. (4.3) becomes

$$\frac{\dot{\gamma}_n}{\dot{\gamma}_T} = 1 + \frac{2b}{H} \tag{4.4}$$

Slip violations should therefore have a negligible effect on the shear field when $b \ll H$. It is also apparent from Eqs. (4.3) and (4.4) that the shear rate required to maintain a constant shear stress in a fluid subject to planar Couette shear flow increases inversely with the gap. Thus a plot of $\dot{\gamma}_n(H)$ vs. $1/H$ at a fixed shear stress yields a straight line with slope $2V_s$ and intercept $\dot{\gamma}_T$. By Eq. (4.4), a plot of $\dot{\gamma}_n(H)/\dot{\gamma}_n(H \to \infty)$ vs. $1/H$ also yields a straight line at a fixed shear stress, except with a slope $2b$. Provided that the plane separation is small enough that $\dot{\gamma}_n(H) \neq \dot{\gamma}_T$, $V_s$ and $b$ can also be determined at a fixed shear stress $\sigma$ from nominal shear rate measurements at any two gaps

$$b = \frac{(\alpha - 1)H_1 H_2}{2(H_1 - \alpha H_2)} \tag{4.5}$$

$$V_s = \frac{\dot{\gamma}_n(H_1)H_1 H_2}{2(H_2 - H_1)}(1 - \alpha) \tag{4.6}$$

Here $\alpha = \dot{\gamma}_n(H_2)/\dot{\gamma}_n(H_1)$ and $H_i$ are the plane separations.

The study by Hatzikiriakos and Dealy (26) appears to be the first to successfully quantify slip violations in polymers using gap-dependent planar Couette shear flow measurements. These authors used a sliding plate rheometer equipped with a centrally mounted shear force transducer to measure slip violations in commercial high-density polyethylene (HDPE) resins. Plots of $\dot{\gamma}_a$ vs. $1/H$ were found to indeed be linear with slopes that increase with shear stress $\sigma$. Slip velocities obtained from straight-line fits to the data suggest two flow regimes. At shear stresses below a value $\sigma_c \cong 0.09$ MPa, the no-slip condition holds. On the other hand, for $\sigma > \sigma_c$ very large levels of slip $V_s \sim O$ $(10^4 \ \mu m/sec)$ are observed. At high shear stresses, the slip velocity is found to be a strong function of stress and temperature, $V_s = \xi\sigma^m$, where $m$ is of order 3.3 and essentially independent of temperature, and $\xi$ varies with temperature and substrate surface chemistry. Hatzikiriakos and Dealy found that the Williams–Landel–Ferry (WLF) equation, commonly used to correlate temperature dependences of bulk polymer rheological properties, adequately describe the temperature dependence of $\xi$, pointing to a relationship between $V_s$ and fluid rheology.

Recently, Dao and Archer used planar Couette flow measurements to quantify slip violations in entangled melts of narrow molecular weight distribution 1,4-polybutadienes (35). As in the earlier study by Hatzikiriakos and Dealy (26), the flow device was equipped with a centrally mounted shear force transducer and slip violations were determined from gap-dependent stress measurements. Figure 4.3 summarizes steady state shear stress vs. shear rate data for a polymer with $M_w = 3.15 \times 10^5$ g/mol ($M_w/M_n = 1.04$). It is immediately apparent from the figure that, but for the lowest shear stresses studied, progressively larger shear rates are required to generate the same shear stress at a lower gap. Using Eq. (4.5), the slip length for the material is found to be $b = (85 \pm 9)\,\mu m$, independent of the shear stress. The slip velocity calculated using Eq. (4.6) increases linearly with shear stress. At shear stresses above a critical value of around 0.2 MPa, Dao and Archer (35) observed large

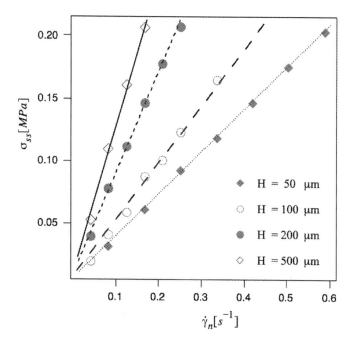

**FIGURE 4.3** Plot of steady state shear stress $\sigma_{ss}$ vs. nominal shear rate $\gamma_n$ for a high-molecular weight 1,4-polybutadiene melt ($\overline{M}_w = 3.15 \times 10^5$ g/mol) that manifests massive levels of wall slip. Measurements were performed using a planar Couette shear cell with a centrally mounted shear force transducer (35). $H$ is the plane separation or gap.

random differences between shear stress–shear rate data measured at variable gaps. This behavior coincides with previous observations of a transition to a regime of massive slip $[V_s \sim O\,(5 \times 10^2\,\mu\text{m/sec})]$ in similar 1,4-polybutadiene melts using velocimetry measurements (34).

Stress measurements in planar Couette shear flow have also been used to study various dynamic features of slip violations in polymers. Hatzikiriakos and Dealy (26), for example, reported complex (possibly chaotic) stress transients during large amplitude oscillatory shear (LAOS) flow of commercial high-density polyethylene melts sheared between stainless-steel substrates. An analysis by Black and Graham (36) suggests that these observations can result from loss of stability of planar Couette shear flow when the slip velocity is a nonlinear function of shear stress. Time-dependent stress decay and complex mixing of high- and low-frequency harmonics have also been reported by other authors during large-amplitude oscillatory shear measurements using sliding plate rheometry (37–38). Dhori et al. (38), for example, recorded optical micrographs of their polyethylene samples at different times following inception of shear. Their work indicates that the shape of the material dramatically changes during LAOS and that motion and fracture at the three-phase contact line (common line) could play a role in the unusual time-dependent stresses recorded during LAOS of polymers.

Dao and Archer (35) recently reported well-defined asymmetric stress oscillations in 1,4-polybutadiene melts at shear stresses in the vicinity of $\sigma_c$ during start-up of steady shear flow between aluminum and stainless-steel substrates in a planar Couette flow geometry. Figure 4.4 illustrates the basic time-dependent structure of the stress waveforms reported by these authors. It is apparent from the figure that both the amplitude and frequency of the oscillations are functions of the imposed shear rate. Significantly, at high shear rates, the stress oscillations are observed to disappear entirely. The time-dependent stress profiles are also seen to be stable over long periods of time and can be separated into stages of gradual growth followed by rapid collapse, reminiscent of stick–slip behavior in solid–solid sliding contacts. Dao and Archer also found that the stress oscillations are induced at shear rates close to the reciprocal terminal relaxation time of the polymer and discussed their findings in terms of a transition from a state of no-slip, to stick–slip, and ultimately to continuous slip at the polymer–solid interface.

Similar transient stress oscillations have also been reported by Leger et al. (39) near the slip transition in poly(styrene-butadiene) copolymer melts and by Mhetar and Archer (40) in PBD melts. Work by Mhetar and Archer (34,40) show that tracer particles near the polymer–solid interface execute periods of fast and slow motion parallel to the driving velocity that coincide with the stress oscillations. Both the regularity and amplitude of the stress oscillations were also observed to increase when the surface of the shearing

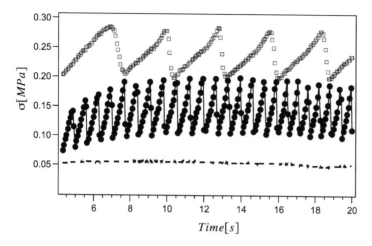

**FIGURE 4.4** An example of dynamic stick–slip flow in a high-molecular weight 1,4-polybutadiene melt during planar Couette shear between aluminum substrates. At the two low shear rates, $\dot{\gamma}_n = 0.63 \text{ sec}^{-1}$ (square symbols) and $\dot{\gamma}_n = 1.5 \text{ sec}^{-1}$ (circles) the stick–slip period decreases nearly in proportion to the nominal shear rate (35). At the highest shear rate $\dot{\gamma}_n = 4.5 \text{ sec}^{-1}$ (broken line), stick–slip flow gives way to continuous slip.

planes were modified by end-tethering a single layer of a high-molecular weight 1,4-polybutadiene. Finally, Dao and Archer reported that an approximately 10-fold reduction in steady state shear stress and complete elimination of stress oscillations at all shear rates can be achieved if freshly polished $\alpha$-Brass is used for the shearing surfaces (35). Mhetar and Archer (34) reported similar absence of shear stress oscillations in entangled PBD melts sheared between silica glass planes modified by low-energy perfluorinated silane coatings. These authors also found an approximately 10- to 20-fold increase in $V_s$ at shear stresses below $\sigma_c$ when unmodified, clean silica glass planes were replaced by fluorinated ones.

Despite the simplicity of planar Couette shear flow for studying slip violations in polymers, quantitative measurements of slip are only possible in this geometry if several complicating factors are addressed. These include secondary flow in elastic liquids, lack of plane parallelism, and viscous heating in continuous shear flow of high-viscosity polymer melts. The limited travel distances possible in the geometry also restricts the shear rates and shear strains that can be accessed at normal gaps $H \sim O\,(5 \times 10^2\ \mu m)$. Secondary flow effects in planar Couette shear can be minimized by any of three procedures: (1) Stress measurements should be performed well away from the fluid meniscus (41). This can be most conveniently accomplished by a shear force transducer of the type developed by Dealy et al. (26,37,42) or by using an optical polarimetry technique with a laser source (43,44). (2) Using a very low sample aspect ratio (gap/width < 0.04) for planar Couette shear measurements involving highly elastic materials (41). (3) The most important secondary flow in this geometry has been shown to be driven by imbalance of shear-induced normal stresses at the free edges. Lubricated side walls with surface normals along the neutral direction ($z$) can therefore be used to minimize measurement errors caused by secondary flow. A simple design based on PTFE (Teflon) has been used for this purpose in birefringence experiments in planar Couette shear flow of concentrated polystyrene solutions (44).

Measurement errors caused by absence of parallelism between the planes used to produce shear flow in polymers can be minimized using either of two approaches. One method proposed by Larson and Mead (45) uses a pivot mount with micrometer positioners to directly adjust the parallelism between planes. This method can be used to align glass planes with parallelisms down to the wavelength of light. A simpler, perhaps more accessible approach is to use a high-strength adhesive with a long cure time to permanently affix each plane to the rheometer assembly. If a moderate normal force is applied during attachment, such that the opposed faces (shear surfaces) of the planes are forced to make complete contact with each other before the adhesive cures, the two planes will self-align by automatically adjusting the thickness of the adhesive layer between the back-side of each plane and the rheometer assembly. This method can obviously only yield levels of plane parallelism comparable to the mean surface roughness of shear surfaces. If the planes are constructed using high-grade optical quality silica glass, polished metals, or laser-cut silicon wafers, slip lengths down to approximately 1 μm can be reliably measured using a planar Couette shear device.

If the rate of heat generation per unit volume $\dot{q}_v = \sigma\dot{\gamma}$ in a viscous fluid subjected to steady shear flow in a planar Couette geometry exceeds the rate of heat removal $\dot{q}_c = k\Delta T/H^2$ per unit volume of fluid, by conduction to the shear surfaces, the fluid's temperature will rise. Here $k$ is the thermal conductivity of the fluid, $H$ is the plane separation, and $\Delta T$ is the temperature change. If the shear surfaces are maintained at a constant temperature and viscosity changes with temperature as $\eta(T) = \eta(T_{\mathrm{ref}})\exp[\beta(T_{\mathrm{ref}} - T)]$, the fluid's viscosity will change by a relative amount

$$\frac{\Delta\eta}{\eta} = \beta\Delta T = \frac{\sigma\dot{\gamma}H^2\beta}{k} \tag{4.7}$$

The last term on the R.H.S. is the Nahme number $Na$ and can be used to estimate the severity of viscous heating for a material of interest under typical flow conditions (46). For most liquids, $\beta$ is a positive number; viscous heating will therefore cause the fluid viscosity to decrease in a time-dependent fashion. It is evident from Eq. (4.7) that viscous heating is most important in fast flows of viscous liquids with large viscosity temperature coefficients. Equation (4.7) also indicates that viscous heating can be substantially reduced by performing measurements in narrow gaps; underscoring the need for good control of gap parallelism in planar Couette shear devices. It is also possible in some cases to minimize measurement errors caused by viscous heating by terminating shear immediately after a steady state fluid response is observed.

### 4.2.2 Gap-Dependent Measurements in Torsional Shear Flow

Gap-dependent torque measurements in fluids subjected to torsional shear flow may also be used to quantify slip violations in polymers (47). This approach is particularly attractive because it can be implemented using most commercial mechanical rheometers. In the case where shear is generated between parallel disks, the slip measurement principle is similar to that developed in the last section. An important difference is that for small gaps $H$ the shear rate in a parallel disk geometry (Fig. 4.5) is a function of radial position

$$\dot{\gamma} = Ar \tag{4.8}$$

where $A = \Omega/H$. The total torque generated in a fluid subject to laminar shear flow in this geometry is therefore given by, $M = 2\pi \int_0^{\dot{\gamma}_R} (R/\dot{\gamma}_R)^3 \dot{\gamma}^2 \sigma_{\theta z} d\dot{\gamma}$, where $\dot{\gamma}_R$ is the shear rate at the rim ($r = R$) and $\sigma_{\theta z}$ is the local shear stress. Differentiating this expression using Leibnitz rule yields the well-known relationship between the measured torque and shear stress at $r = R$

$$\sigma_{\theta z}(R) = \frac{M}{2\pi R^3} \left[ 3 + \frac{d}{d\ln\dot{\gamma}_R} \ln M \right] \tag{4.9}$$

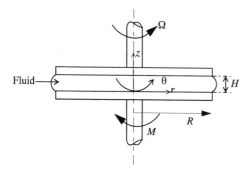

**FIGURE 4.5** Schematic of apparatus for performing torsional shear flow between parallel plates.

Measurements of $M$ vs. $\dot{\gamma}_R$ at variable gap $H$ can be used to determine how the shear rate $\dot{\gamma}_R$, required to produce a given stress, changes with disk separation (46).

It is reasonable to expect the slip velocity at the upper and lower substrate to be the same. Equations (4.8) and (4.9) are therefore expected to hold even in the presence of slip. For a given shear stress at the rim, $\sigma_{\theta z}(R)$, the slip velocity and slip length at $r = R$ can be obtained using expressions analogous to Eqs. (4.3) and (4.4), $\dot{\gamma}_{RT} = \dot{\gamma}_{Rn} - 2V_s/H$ and $\dot{\gamma}_n/\dot{\gamma}_{RT} = 1 + 2b/H$. A plot of $\dot{\gamma}_R(H)/\dot{\gamma}_R(H \to \infty)$ vs. $1/H$ at a fixed shear stress therefore yields a straight line with slope $2b$. Also, provided that $\dot{\gamma}_R(H) \neq \dot{\gamma}_{RT}$, $V_s$ and $b$ at a particular shear stress can be determined from shear rate data at any two gaps, using expressions analogous to Eqs. (4.5) and (4.6). $b = (\alpha - 1)H_1 H_2/(2(H_1 - \alpha H_2))$ and $V_s = (\dot{\gamma}_R(H_1)H_1 H_2/2(H_2 - H_1))(1 - \alpha)$, where $\alpha = \dot{\gamma}_R(H_2)/\dot{\gamma}_R(H_1)$ and $H_i$ are the disk separations.

A similar approach can be used in principle for quantifying slip violations from shear flow measurements using the cone-and-plate geometry. For small cone-angles $\beta$ and no-slip at the walls, the shear rate and shear stress developed in this geometry are both independent of radial position $r$. However, the gap $H$ is a function of $r$ and $\beta$, $H = r \tan \beta$. Changing the cone angle has the same effect as changing $H$. Thus in the absence of slip, plots of shear stress vs. shear rate made using cone-and-plate fixtures with variable $\beta$ will overlap. Conversely, when slip violations are important, the shear rate required to achieve a given stress level in the fluid will increase as the cone angle is reduced. Unfortunately, a recent analysis by Brunn and Ryssel shows that if the no-slip condition is violated, the shear rate and shear stress in cone-and-plate shear flow may no longer be taken to be independent of radial position (48,49); evidently forfeiting most benefits for interfacial slip measurements in this geometry. Add to this the narrow range of cone-angles ($\beta <$ 8°) where secondary flows can be avoided and well-known difficulties in machining and characterizing cones with small differences in $\beta$, and it becomes apparent why torsional shear between cone-and-plate fixtures is not a popular method for quantifying slip violations in polymers.

Gap-dependent torsional shear measurements between parallel disks have been used by a variety of researchers to quantify slip violations in fluids. Yoshimura and Prud'homme (47) used the technique to measure and correct for slip violations in clay suspensions and emulsions. Henson and Mackay (27,50) and Hay et al. (51) used this method to quantify wall slip in various narrow molecular weight distribution polystyrene (PS) and commercial linear low-density polyethylene (LLDPE) melts. These authors found that weak slip violations ($V_s \sim O$ μm/sec to mm/sec) occur in both materials irrespective of the shear stresses studied. For shear stresses in the Newtonian flow regime, a linear relationship between the slip velocity and shear stress was observed for both polymers. At higher shear stresses, but still lower than the critical stress

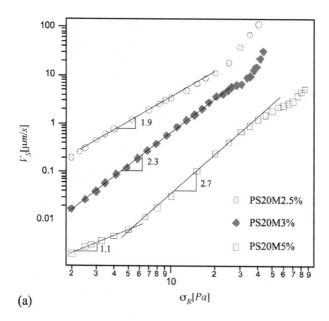

(a)

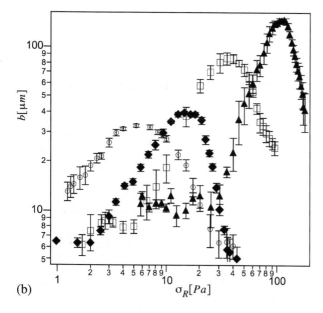

(b)

for macroscopic slip, a nonlinear relationship is observed. For LLDPE, an approximate quadratic relationship holds over most of the nonlinear slip regime, $V_s \sim \sigma^2$, but for PS the slip velocity–shear stress relationship is more complicated.

Sanchez-Reyes and Archer (52) quantified interfacial slip in entangled solutions of high-molecular weight polystyrene in diethylphthalate (DEP) using gap-dependent torsional shear flow between parallel disks. Root mean-squared surface roughness $R_q$ of the disks was systematically varied in the range 9 nm $\leq R_q \leq 1.4 \times 10^4$ nm using narrow size distribution silica beads. Sanchez-Reyes and Archer reported weak slip violations $V_s \sim O(0.1 \, \mu\text{m/sec})$, $b < 1 \, \mu\text{m}$, at low shear stresses and that the slip velocity is nearly proportional to the shear stress in this weak slip regime. At higher shear stresses, both the slip velocity and slip length manifest complex dependences on stress. The slip velocity, for example, first manifests a regime of power-law stress dependence, $V_s \sim \sigma^\vartheta$, then above a concentration-dependent critical stress value, transitions to a regime of stronger stress dependence. For PS/DEP solutions with $3 \times 10^5 \leq \phi M_w \leq 1.6 \times 10^6$ g/mol, where $\phi$ is the volume fraction of polymer, the exponent $\vartheta$ ranges from 1.8 to 2.7, but the power-law slip regime extends for roughly one decade of shear stress irrespective of the solution concentration (Fig. 4.6(a)).

The slip length on the other hand initially increases with stress, but then falls off rapidly at high shear stress (Fig. 4.6(b)). This last behavior was previously reported by Mhetar and Archer (18) using a tracer particle method for measuring slip and follows from Eq. (4.1) if the ratio of the fluid viscosity to the interface friction coefficient decrease with shear rate. Thus if the interface friction coefficient is unaffected by shear but the bulk fluid is shear thinning, it would be expected to manifest the type of stress dependence seen in Fig. 4.6(b). That the slip length can decrease with stress at the same time

---

**FIGURE 4.6** (a) Slip velocity $V_s$ vs. steady state shear stress $\sigma_R = \sigma (r = R)$ for solutions of a high-molecular weight polystyrene $\overline{M}_w = 2.06 \times 10^7$ g/mol, $\overline{M}_w/\overline{M}_n$ = 1.2, in diethyl phthalate (DEP). The volume fraction $\phi$ of polymer in the solutions range from 2.5% (unfilled circles) to 5% (unfilled squares). Slip velocities were obtained from gap-dependent steady shear measurements between stainless steel substrates (root mean square surface roughness 0.1 μm) (52). (b) Slip length $b$ vs. steady state shear stress $\sigma_R = \sigma (r = R)$ for solutions of a high-molecular weight polystyrene $\overline{M}_w = 2.06 \times 10^7$ g/mol, in diethyl phthalate (DEP). The volume fraction $\phi$ of polymer in the solutions are 8% (filled triangles), 5% (unfilled squares), 3.5% (filled diamonds), and 2.5% (unfilled circles). The decrease in slip length observed at high shear stresses should be contrasted with the increase in slip velocities seen in (a) for the similar PS/DEP solutions at comparable shear stresses. (From Ref. 52.)

that slip violations become worse underscores the danger of relying on the slip length alone for quantifying the severity of interfacial slip in non-Newtonian liquids. Sanchez-Reyes and Archer also reported that microscale surface roughness has a profound effect on slip violations in polystyrene solutions. Specifically, the authors found that slip violations can be virtually eliminated when PS/DEP solutions are sheared between substrates with root mean squared roughness greater than one half the molecular radius of gyration in solution (52).

The most significant difficulties for wall slip measurements using gap-dependent torsional shear flow arises from: (1) complications to the flow field caused by lack of parallelism and concentricity. (2) Edge fracture, particularly at large gaps and high shear rates. (3) Sample expulsion from the gap at high shear rates. (4) Spurious time-dependent behavior produced by viscous heating. Measurement artifacts caused by (1) and (2) can be reduced by painstakingly assembling the disks to ensure parallelism, and by careful mechanical design of the parallel plate mounts to ensure that their centers of rotation are coincident and matches that of the drive mechanism. Alignment problems are particularly important for slip measurements because they can produce *apparent gap* errors (46), wherein the gap, back-calculated from viscosity measurements using standard fluids, appears smaller than the actual plate separation used for the measurements. This error is amplified at small gaps and can therefore be easily confused with slip violations.

Edge fracture arises from an imbalance of stresses at the meniscus and is driven primarily by $N_2$ in the fluid (46,53). Edge fracture reduces the cross-sectional area of the fluid in contact with the plates, which causes the measured torque to decrease (usually in a time-dependent manner) during continuous shearing. A common remedy is to reduce $H$ so that the onset shear rate for edge fracture is displaced outside the range of experimental interest. Sample expulsion from the gap is caused by fluid inertia at high rotation rates. It produces irreversible transient torque reduction during continuous shearing and can be avoided by performing experiments at shear rates well below where fluid inertia becomes significant. Viscous heating is always present during shear of high-viscosity polymer liquids. It becomes a problem when the rate of heat generation is so large that temperature gradients are created in the fluid. Experimental conditions that reduce the heat generation rate, e.g., low shear rates, and enhance the heat dissipation rate, e.g., small $H$ and highly conducting metallic disks, are required to minimize measurement errors.

### 4.2.3 Interfacial Slip Measurements in Pressure-Driven Shear Flow

Slip violations can also be quantified using Poiseuille flow measurements in capillaries. Consider the pressure-driven, laminar flow of a general non-

Newtonian liquid in a capillary tube with diameter $2R$, length $L$, and flow axis along $z$. If the liquid violates the no-slip condition, the volumetric flow rate at an arbitrary shear stress $\sigma = \sigma_{rz}$ can be easily shown to be

$$Q = \pi R^2 V_s + \frac{\pi R^3}{\sigma_R^3} \left( \int_0^{\sigma_R} \dot{\gamma} \sigma_{rz}^2 d\sigma_{rz} \right) \tag{4.10}$$

where $\sigma_R = \sigma_{rz}(r = R) = 0.5R(\partial p/\partial z) = 0.5\Delta p/(L/R + e)$ is the shear stress at the wall and $e$ is the empirical Bagley correction (54), which is generally a complicated function of $\Delta p$ and fluid properties. For capillary tubes with large $L/D$ ratios ($L/D > 50$), where $D = 2R$ is the diameter, the Bagley correction is small and Eq. (4.10) provides a simple method for quantifying slip violations. The second term on the R.H.S. is the volumetric flow rate $Q_\infty$ in absence of slip. Introducing the apparent wall shear rate $\dot{\gamma}_{aR} = 4Q/\pi R^3$ into Eq. (4.10) finally yields

$$\dot{\gamma}_{aR} = \dot{\gamma}_{a\infty} + \frac{4V_s}{R} \tag{4.11}$$

which is the Mooney equation (25). Eq. (4.11) can be rearranged further to give an expression for the slip length $b = V_s/\dot{\gamma}_{a\infty}$

$$\frac{\dot{\gamma}_{aR}}{\dot{\gamma}_{a\infty}} = \left( 1 + \frac{4b}{R} \right) \tag{4.12}$$

A plot of $\dot{\gamma}_{aR}$ vs. $1/R$ at a constant $\sigma_R$, using data obtained with capillaries having the same $L/R$ ratio, yields a straight line with slope $4V_s$ and intercept $\dot{\gamma}_{a\infty}$, the apparent shear rate in the absence of slip. On the other hand, a plot of $\dot{\gamma}_{aR}/\dot{\gamma}_{a\infty}$ vs. $1/R$ at fixed $\sigma_R$ yields a straight line with slope $4b$. Thus provided care is taken to remove the effects of entrance and exit losses, capillary flow measurements using multiple tube diameters can be used to quantify slip violations in polymer liquids.

For a power-law fluid, $\sigma_{rz} = K\dot{\gamma}^n$, and Eq. (4.10) can be integrated to yield

$$\frac{\overline{V}}{R} = \frac{V_s}{R} + \frac{n}{3n + 1} \left[ \frac{\sigma_R}{K} \right]^{1/n} \tag{4.13}$$

where $\overline{V} = Q/\pi R^2$. Thus if the power-law variables $n$ and $K$ are known from separate rheometry experiments, $V_s$ can be directly computed from $Q$ vs. $V_s$ data. The extra convenience of this last procedure comes at a price, however. First, it is essential that entrance and exit losses are properly taken into account in determining $\sigma_R$. These losses can be ignored if capillaries with very large length to diameter ratios are used. Second, to produce flow of a viscous polymer in capillaries with large $L/D$ ratios requires large driving pressures.

The effect of pressure on fluid viscosity must then be taken into account in computing $\sigma_R$ from the measured $\Delta p$. Finally, the power-law model best describes polymer flow behavior at high flow rates. Slip velocities determined using Eq. (4.13) are therefore generally untrustworthy at low and intermediate flow rates.

Several studies have used the dependence of apparent wall shear rate $\dot\gamma_a$ on capillary diameter or slit height to quantify slip violations in pressure-driven flows. A smaller number have used Eq. (4.13) for quantifying wall slip in capillary flows. Ramamurthy, for example, observed that the apparent shear rate $\dot\gamma_{aR}$ in HDPE and LLDPE melts subject to pressure-driven flow in capillary tubes noticeably increase as the capillary diameter $D$ decreases at wall shear stresses above 0.1 and 0.14 MPa, respectively (30). Slip velocities deduced from his data using Eq. (4.11) point to a transition to massive levels of interfacial slip, slip velocities $V_s \sim O$ (cm/sec). Ramamurthy observed that the transition from no-slip to strong slip at the wall coincided with a transition from smooth glossy extrudate surface finishes to extrudates with rough, fractured surfaces. Ramamurthy concluded that slip by loss of polymer adhesion to the die wall was responsible for surface fracture of LLDPE and that die construction materials that suppress slip would be advantageous for preventing melt fracture in polyethylene film blowing processes (30).

Work by Kalika and Denn provide even more convincing connections between extrudate surface appearance and wall slip. These authors studied the flow characteristics and extrudate appearance of a well-characterized LLDPE driven at a controlled flow rate through a series of capillaries with fixed $D$, but variable $L/D$. They calculated slip velocities from $Q$ vs. $\sigma_R$ data using the approach outlined in Eq. (4.13) (31). At wall shear stresses above a first critical value $\sigma_c'$ (Fig. 4.7(a)) of ca. (0.26–0.3) MPa, Kalika and Denn observed significant slip violations $V_s \sim O$ (cm/sec) and that the slip violations also coincide with the appearance of sharkskin melt fracture. At wall shear stresses above a second critical value $\sigma_c$ of approximately 0.4 MPa, the authors observed dramatic enhancement in the level of wall slip, with slip velocities exceeding 80% of the average fluid velocity in the capillary (31). At intermediate stresses, a slip–stick regime was observed, which was evidenced by periodic variations in extrudate surface finish [from glossy (slip) to fractured (stick)] and time-dependent pressure oscillations. Kalika and Denn (31) showed that the relative volume of polymer in the reservoir to the volume in the capillary influenced the output rates of the glossy and fractured sections in a manner consistent with expectations for flow of a compressible fluid under an oscillating pressure driving force.

Hatzikiriakos and Dealy (55) used two commercial high-density polyethylene melts and their binary blends to provide a very detailed characterization of wall slip in polymers during capillary flow. These authors observed

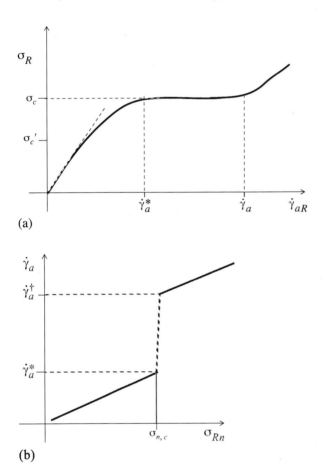

(a)

(b)

**FIGURE 4.7** (a) Schematic of wall shear stress vs. apparent shear rate during capillary flow of a polymer melt at controlled flow rate. The critical stress $\sigma_c'$ is the shear stress where the shear stress–shear rate relationship first becomes nonlinear. The second critical stress $\sigma_c$ is the shear stress where large violations of the no-slip condition are first observed. (b) Schematic of nominal wall shear stress $\sigma_{Rn} = \Delta p/(2L/R)$ vs. apparent shear rate for a polymer melt during capillary flow at controlled pressure. The critical stress $\sigma_{n,c}$ is the shear stress at which the volumetric flow rate of fluid in the capillary increases abruptly. This sudden increase in flow rate is believed to directly reflect a breakdown of the no-slip condition. (From Ref. 13.)

a transition from $R$-independent to $R$-dependent flow curves at wall shear stresses above a molecular weight and concentration-dependent value $\sigma = (0.09-0.18)$ MPa. For capillaries with fixed $L/D$ ratios, slip velocities computed using the Mooney approach and from $Q$ vs. $\sigma_R$ data, using Eq. (4.13), were found to be consistent with each other. However, slip velocities determined from capillary flow data using either method were significantly lower than those measured in planar Couette shear of the same polymer over the same range of wall shear stresses. Recently, Hay et al. (51) also compared slip velocity data for LLDPE melts obtained using parallel plate and pressure-driven flow measurements in slits and capillaries. At low shear stresses, $\sigma < \sigma_c \approx 0.3$ MPa, capillary flow measurements reveal no-slip in LLDPE while gap-dependent parallel plate measurements reveal measurable levels of slip, $V_s \sim O$ ($\mu$m/sec to mm/sec). The authors explain this difference in terms of the lower sensitivity of capillary flow measurement techniques for quantifying interfacial slip of polymers.

At high shear stresses, $\sigma > \sigma_c$, Hatzikiriakos and Dealy reported that while $V_s$ measured using planar Couette shear depends only on shear stress, material properties, and temperature slip velocities obtained from capillary rheometry experiments also varied with capillary geometry. Specifically, $V_s$ for HDPE appeared to decrease with increasing $L/D$, up to a value of about 60 (31,55). Based on these observations and on the differences in slip velocities measured at comparable stress levels using planar Couette shear and capillary rheometry, Hatzikiriakos and Dealy argued that $V_s$ must be a function of the total pressure of the fluid in the capillary (55). Work by Mackay et al. (56,57) revealed large differences in rheological properties of LLDPE obtained using parallel plate and capillary flow measurements. These authors point to the possibility of additional flow complications, e.g., pressure effects on fluid viscosity and viscous heating at high pressure, as the likely source of this behavior.

Hatzikiriakos and Dealy also identified a relationship between the critical stress for the onset of macroscopic slip in binary HDPE blends and their polydispersity index. Specifically, these authors found that if the slip velocity of HDPE blends is plotted vs. normalized wall shear stress $\sigma_{Rn} = \sigma_R/(\sigma_c I^{1/4})$, where $I$ is the polydispersity index, a universal curve is obtained (54). If the critical stresses reported by Hatzikiriakos and Dealy are corrected for polydispersity using the relation $\hat{\sigma}_c = \sigma_c I^{1/4}$, the following relationship between the critical stress and weight-averaged molecular weight emerges, $\hat{\sigma}_c \sim M_w^{-0.28 \mp 0.03}$. Wang and Drada report a similar relationship between HDPE molecular weight and nominal shear stress (i.e., without the Bagley correction) at the flow transition. In particular, these authors observed that the value of the nominal shear stress $\sigma_{n,c}$ required to produce a discontinuous jump (Fig. 4.7(b)) in apparent shear rate in approximately

50% of their experiment runs decreased as, $\sigma_{n,c} \sim \overline{M}_w^{-0.5}$ (58,59). More careful consideration of the data used to deduce this trend and the unknown polydispersity of the materials suggest that a weaker relationship between $\sigma_c$ and $\overline{M}_w$ of the sort reported by Hatzikiriakos and Dealy cannot be ruled out.

Experiments by Ferguson et al. (60) using fractionated polyethylenes indicate that $\sigma_c$ is not a function of polymer molecular weight. Studies by Vinogradov et al. using narrow and broad molecular weight distribution (MWD) polybutadienes (PBD) and polyisoprenes (PI) with $\overline{M}_w > M_c$ reveal a constant $\sigma_c \approx 0.3$ MPa for PBD and $\sigma_c \approx 0.15$ MPa for PI, irrespective of molecular weight and molecular weight distribution (28). Here $M_c$ is the critical molecular weight at which entanglement effects have a noticeable influence on viscosity. This aspect of Vinogradov's observations has been confirmed recently by Yang et al. for PBD melts (61,62).

Wang and Drada used an expression analogous to Eq. (4.12) to estimate the slip length at the flow transition. Assuming that the no-slip condition is valid at shear stresses on the lower branch of the flow curve and that the flow transition is entirely because of slip, these authors substitute the ratio of the flow rates $Q_{Hc}$ and $Q_{Lc}$ at the transition for $\dot{\gamma}_{aR}/\dot{\gamma}_{a\infty}$ in Eq. (4.12) and calculate the slip length $b_c$ at the transition (58). Hay et al. (51) discussed the strengths and shortcomings of this method for quantifying slip violations. Wang and Drada reported several interesting characteristics of $\sigma_{n,c}$ and $b_c$ that may be used to isolate a molecular mechanism for slip violations in polymers. The authors, for example, show that $b_c$ for HDPE obeys the same scaling relationship with weight-averaged molecular weight as the shear viscosity, $b_c \sim \eta \sim \overline{M}_w^{3.4}$ (59). A similar finding $b_c \sim \eta \sim \overline{M}_w^{3.5}$ was also reported by Yang et al. (61) for PBD melts using the same procedure for measuring slip. Wang and Drada also report that at temperatures $T$ below 200°C, $\sigma_{n,c}/T$ increases and $b_c$ decreases with increasing temperature. At temperatures $T > 200$°C, both quantities appear to become independent of temperature. Wang and Drada (13,59) discussed these findings in terms of mesophase formation at the polymer die interface at $T < 200$°C and slip violations by disentanglement of surface attached and bulk polymer chains. Vinogradov et al. (28) reported that the critical stress for polybutadiene is independent of temperature, at temperatures well above the glass transition. They rationalize this finding in terms of the low activation energy for re-entanglement of polymer chains.

## 4.3  DIRECT METHODS FOR QUANTIFYING SLIP VIOLATIONS

### 4.3.1  Overview

In this section, we consider various methods for quantifying slip violations from measurements of velocity and/or displacement of fluid particles at solid–

liquid interfaces during flow. These methods are "direct" only in the sense that they provide quantitative information about relative motion of fluid and solid at the interface. In most cases, the fluid motion must be inferred from the response of flow tracers that are often physically larger than individual fluid elements, and always chemically different from their host. This introduces a number of fresh issues related to the compatibility of the tracers and the host fluid, and to how faithfully these tracers can be expected to track fluid motion at interfaces. None of these issues are fully understood and must be addressed on a case-by-case basis. The most common approach is to compare slip length and/or slip velocities obtained using "direct" and "indirect" methods to validate the former.

Direct velocimetry measurements are often more sensitive to slip violations than the methods discussed in Sec. 4.2.1, perhaps justifying their growing use. One advantage is that slip violations can be investigated at such low flow rates and stress levels using these measurements that interference from non-Newtonian effects and from flow instabilities can be avoided. Virtually all studies to date that have used these methods for studying slip violations at low shear stresses reveal that polymer fluids violate the no-slip condition at all stress levels, at least down to the resolution limits of the specific techniques employed.

### 4.3.2 Fluorescence Recovery of Fringe Patterns After Photobleaching

One of the most sensitive methods for studying slip violations in liquids uses fluorescent flow markers localized near the liquid–solid interface. There are currently two main approaches for quantifying slip violations using this method (22,23,63). Both methods rely on Fluorescence Recovery After Photobleaching (FRAP) for their temporal resolution and on total internal reflection (TIR) for enhancing spatial resolution near the interface. Photobleaching is the permanent quenching of fluorescence in a material by exposure to radiation. Typically, photobleaching is produced by an intense laser beam of short duration, which causes fluorophores in the material to undergo irreversible chemical change in a localized region. Transport of fluorescent molecules from the surroundings to the photobleached areas eventually restores the fluorescence, allowing the transport processes to be monitored by measuring fluorescence recovery.

If the photobleached areas are positioned near the fluid–solid interface, transport properties near that interface can be studied. This is achieved in both methods by using solid materials with higher refractive indices than the fluid and directing the laser beam to the interface via the solid, and at an angle of incidence higher than the critical angle for total internal reflection. The intensity of the evanescent field decays quickly with vertical distance $y$ from the interface, $I = I_o e^{-\delta y}$, allowing transport processes in a region of thickness $\delta$ to be probed. Migler et al. (22,23) and Miehlich and Gaub (63) used superposition of two laser beams to create a grid of photobleached and unbleached regions near the fluid–solid interface. A pattern of lower light intensity (reading grid) is used to report the time-dependent recovery of fluorescence within a region around 70 nm from the fluid–solid interface. Assuming that negligible diffusion of photobleached polymer occurs normal to the direction of the interface, the fluid under study flows with a uniform velocity $u_x e_x$, and that the grid spacing along $x$ is much smaller than along $z$, the evolution of fluorophore concentration $c(x, t)$ can be described by the one-dimensional convection–diffusion equation, which can be solved for an initial concentration profile $c_i(x, 0)$, to yield

$$c(x, t) = FT^{-1}[c_i(k)\exp(-k^2 Dt + iku_x t)] \tag{4.14}$$

where $FT^{-1}[\alpha(k)]$ is the one-dimensional inverse Fourier transform.

The fringe pattern fluorescence recovery after photobleaching (FPFRAP) is determined by integrating $c(x, t)$ over $\delta$. Miehlich and Gaub (63), for example, compared time-dependent changes in the amplitude and phase of the spatial Fourier transforms of their fluorescence images to directly determine $D$ and $u_x$. Migler et al. applied a time-dependent phase difference to the reading laser beams to produce a reading fringe pattern that oscillates with a frequency $f$ in time. Superposition of the oscillating reading pattern and the written photobleached pattern will generally yield a time-varying intensity profile with a d.c. bias, a $1f$ harmonic that oscillates with the fundamental frequency $f$, and a harmonic at $2f$. These frequency components possess the symmetries expected of the real-space analogs of the three terms in parenthesis on the right-hand side of Eq. (4.14) (22). $u_x$ is therefore determined from the oscillation frequency of the time-dependent $1f$ signal component.

In planar Couette shear flow experiments using a high-molecular weight dextrane, Miehlich and Gaub report finite velocities near the stationary wall ranging from 6 to 24 μm/sec using their approach. Migler et al. used their FPFRAP method to quantify slip violations in a high-molecular weight PDMS melt ($\overline{M}_w = 0.96 \times 10^6$ g/mol, $\overline{M}_w / \overline{M}_n = 1.27$) seeded with 5 wt.% of a lower-molecular weight, fluorescently tagged PDMS ($\overline{M}_w = 0.32 \times 10^6$ g/mol, $\overline{M}_w / \overline{M}_n = 1.18$). Shear was generated by sandwiching the mixture between two silica glass planes, one of which was functionalized with a monolayer of octadecyltrichlorosilane (OTS) to minimize physisorption of the bulk melt. A sharp transition from weak slip violations ($V_s < 0.1$ μm/sec) at low shear rates to massive levels of slip ($V_s > 10$ μm/sec) is clearly evident from the results (22). In both the weak and strong slip regimes, $V_s$ increased linearly with the true shear rate $\dot{\gamma}$, implying that the slip length $b$ in both regimes are material properties. At the highest shear rates studied, Migler et al. (21,22) measured slip velocities at the stationary OTS-coated substrate that were nearly equal to the driving plate velocity. This indicates near-complete decoupling of the motion of the polymer and glass substrate to which it adsorbs.

Extensive work by Leger et al. (24,33) using the FPFRAP technique of Migler et al. have demonstrated several important features of slip violations in entangled polymer liquids, which shed light on a possible molecular origin. Specifically, these authors studied slip violations in a high-molecular weight PDMS melt ($\overline{M}_w = 0.97 \times 10^6$ g/mol) sheared between parallel glass planes grafted with a lower-molecular weight PDMS ($\overline{M}_w = 0.96 \times 10$ g/mol). Slip measurements were performed over a wide range of surface graft densities, $v$ (24,33). For a system with $v \cong (0.0055/a^2) \cong 1.4 \times 10^{-2}$ nm$^{-2}$, for example, Leger et al. (24) reported three slip regimes similar to those observed by Migler et al. for molar mass PDMS sheared over a glass substrate grafted with OTS. Here we used $a \cong 6.2$ Å for PDMS (64).

Leger et al (24) and Durilat et al. (33) also report linear relationships between $V_s$ and $\dot{\gamma}$ at low and high shear rates. At an intermediate shear rate $\dot{\gamma}^* \cong vkT/(\eta_B \sqrt{N_e} a^3)$, a transition from modest levels of slip, $V_s \leq 10$ μm/sec and $b \leq 2$ μm, to strong slip, $V_s \geq 100$ μm/sec and $b \geq 20$ μm, is observed for values of $v$ below the threshold for surface chains to overlap. At low $v$, Durilat et al (33) report that the slip velocity $V_s^*$ on the threshold of the slip transition decreases very strongly with bulk polymer ($P$-mer) and tethered polymer ($N$-mer) molecular weight, $V_s^* \sim P^{-3.3} N^{-1}$. Here $P$ and $N$ are the degrees of polymerization of the bulk and grafted polymer molecules, respectively. The slip length $b_0$ in the weak slip regime was found to be in the range of 1–2 μm, regardless of $v$ for all materials studied. On the other hand, $\dot{\gamma}^*$ is a complicated function of graft density of $N$-mers. At low $v$, $\dot{\gamma}^*$ initially increases with surface coverage, but eventually falls off for $v \geq 3 \times 10^{-2}$ nm$^{-2}$. A mechanism based on surface exclusion of bulk chains by the tethered molecules has been put forward to explain this last observation (24,65,66).

### 4.3.3   Velocimetry Using Particle Tracers

Flow visualization using tracer particles is one of the simplest and oldest methods for determining the velocity field in liquids during continuous deformation. When the no-slip boundary condition is violated, the velocity field in a liquid sheared between stationary walls will not extrapolate to zero. Velocity measurements using small tracer particles located near the walls therefore provide another method for quantifying slip violations. Chen and Emrich (67) used smoke particles to infer slip in rarefied gases flowing in capillaries. Galt and Maxwell (16) were the first to use this method to investigate slip violations in polymer liquids. These authors photographed the near-wall distribution of ca. 50-μm silicon carbide tracer particles seeded at concentrations of ca. 0.1% by volume into LDPE melts undergoing Poiseuille flow in long tubes with circular and rectangular cross sections. Galt and Maxwell inferred the particle velocities using a stroboscopic imaging technique utilizing a camera outfitted with a high-resolution microscope objective.

At shear stresses well below those required for melt fracture, particles near the polymer–capillary interface generally manifested three types of motions during flow. Some particles remained motionless during flow, others responded to flow by drifting uniformly in the flow direction, and the vast majority moved intermittently in cycles of low and high speeds (16). While the second type of motion was only observed in a small number of the materials studied, the latter *stick–slip* motion was seen in all. No evidence of rolling or spinning was observed for particles near the capillary wall. Measurements using a LDPE with $\overline{M}_w = 3 \times 10^5$ revealed that only ca. 25% of the particles at the fluid–solid interface remained motionless during flow. About 15% of all particles at the wall were observed to drift uniformly, the remainder executed stick–slip motion. For some polymers, stick–slip motion was observed from velocimetry measurements using particles located within an annular region (thickness ca. 1/10 the capillary radius) near the capillary wall. However, in others, only particles at the wall executed stick–slip motion. The tendency of LDPE melts to manifest stick–slip flow at the wall was found to increase with pressure and weight-average molecular weight, and to decrease with polydispersity index (16).

den Otter et al. (68) used a similar method for quantifying interfacial slip in commercial HDPE, LDPE, and polydimethyl siloxane (PDMS) resins undergoing Poiseuille flow in rectangular channels. In this case, a light microscope was used to visualize motion of native dust and gel particles with diameters in the range 5–20 μm located in the bulk fluid and particles 5–10 μm in size located at the polymer wall interface. Average particle velocities were determined using a stopwatch and a 1-mm scale mark. Contrary to the observations of Galt and Maxwell, den Otter et al. (68) report no evidence of interfacial slip even at shear stresses above those where melt fracture was observed in LDPE and PDMS. The authors also found that the near-wall shear rate determined from tracer particle velocimetry agreed quite well with the shear rate predicted using a power-law fluid model with no-slip boundary conditions. den Otter et al. (68) argued that the discrepancy between their results and those reported earlier by Galt and Maxwell was a direct consequence of the larger tracer particles used in the earlier study. In particular, den Otter et al. found that near-wall velocities measured in PDMS at the onset of melt fracture decreased as the size of the tracer particles was reduced.

More recently, Muller-Mohnssen et al. (17,69) combined tracer particle velocimetry with laser microanemometry (LMA) to quantify the velocity profile in aqueous solutions of an ultra-high-molecular weight polyacrylamide ($\overline{M}_w = 1.2 \times 10^7$ g/mol) undergoing pressure-driven flow in rectangular glass channels. The authors utilized laser light scattering from 150-nm gold tracer particles to visualize their motion in flow and to quantify their velocities either directly from video clips or from a correlation analysis of the time-dependent scattered light intensity. Except at high flow rates, the first procedure was found to be more accurate. Slip violations were inferred from many point-wise measurements by first reconstructing the velocity profile in the entire channel using LMA measurements at locations starting from about 1 μm from the channel walls to the center line and then extrapolating the velocity profile to the wall. Measurements using water yield parabolic velocity profiles and profiles that extrapolate to zero velocity at the channel wall. On the other hand, measurements using polymer solutions yield finite slip velocities ranging from 5 to 70 μm/sec (17). Over the range of shear rates studied, $V_s$ was found to continuously increase with wall shear stress $\tau_w$ for all polyacrylamide concentrations and within the concentration range $0.05 < \phi < 0.4$ the constant of proportionality $k$ increased linearly with $\phi$.

To increase the near-wall resolution of their method, Muller-Mohnssen et al. used the evanescent field from a laser undergoing total internal reflection at the polymer wall interface, to excite scattering only by tracer particles within a region of about 150 μm from the wall (17,69). Surprisingly, these measurements indicated that the velocity at the wall and in an approximately 150-μm layer of fluid near the wall is zero. The relatively large displacement of the slip plane from the fluid–solid interface led the authors to propose that slip was the result of polymer depletion from the fluid layer near the wall. Calculations based on this proposal indicated that for many of the solutions studied, the fluid layer at the wall is predominately water (69)!

Archer et al. (70,71) and Mhetar and Archer (18) recently used a light microscope interfaced with a video camera to quantify the near-wall velocity field in entangled solutions of narrow molecular weight distribution (MWD) polystyrenes (PS) in diethyl phthalate and tricrecyl phosphate, and in several high-molecular weight, narrow MWD 1,4-polybutadiene (PBD) melts. Slip measurements were performed during steady shear flow in a planar Couette flow apparatus. The shear cell consisted of two rectangular glass planes supported by metal frames and aligned by a gimbal assembly (45). Polymer liquids used in the study were sandwiched between the planes to a preset thickness of 500 μm and simple shear flow generated by translating the upper plane at a fixed speed relative to the stationary lower plane. The entire shear cell was designed to fit between the collimator and objective of a fixed-stage optical/fluorescent microscope. Trace quantities of narrow size distribution silica microspheres (diameter 1.5 μm) were added to the polymer liquids used in both studies. A small number of studies also used mixtures of 1.5-μm silica microspheres and 24-nm diameter fluorescent polystyrene beads to investigate the effect of particle size on velocities measured near the polymer–solid interface (18).

A video camera and a time-lapse video recorder were used to measure the time-dependent motion of tracer particles initially resting on the stationary shear cell window. Slip velocities were deduced from this information as follows. The location of tracer particles near the polymer–wall interface was first determined by comparing the measured particle velocities obtained using very weak shear fields with the profile expected for simple shear flow with no-slip at the wall (70). Particles positioned at the polymer–wall interface are, for example, readily identified by the fact that they do not move during these measurements. When a stronger shear field is applied, particles respond in the manner illustrated in Fig. 4.8. At short times following imposition of shear, particles at the interface initially remain stationary. At long times, the tracer particle displacement increases linearly with time, allowing an average particle velocity to be determined. Because the particles used in the measurements were all initially at the polymer–wall interface, the measured velocities are the respective slip velocities $V_s$ at the apparent shear rates $\dot{\gamma}_a$ indicated.

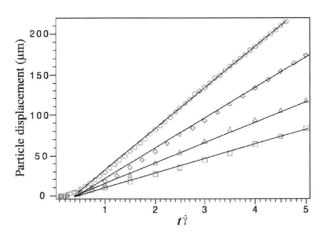

**Figure 4.8** Time-dependent displacement of 1.5-μm diameter silica tracer particles initially resting at a stationary polymer/solid interface during start-up of planar Couette shear between parallel silica glass planes. The data was obtained using a concentrated solution $\phi \approx 0.2$ of a moderate-molecular weight polystyrene $\overline{M}_w = 2.0 \times 10^6$ g/mol, $\overline{M}_w/\overline{M}_n = 1.3$ in tricrecyl phosphate (TCP). Results are for nominal shear rates $\dot{\gamma}_n = 0.01 s^{-1}$ (circles), $\dot{\gamma}_n = 0.05 s^{-1}$ (diamonds), $\dot{\gamma}_n = 0.1 s^{-1}$ (triangles) and $\dot{\gamma}_n = 0.25 s^{-1}$ (squares)(70).

From the measured slip velocity and apparent shear rate, the true shear rate $\dot{\gamma}_T$ can be determined using Eq. (4.3). The slip length $b \equiv V_s/\dot{\gamma}_T$ can also be determined from these measurements. Mhetar and Archer outfitted their planar shear cell with a shear force transducer, which facilitated direct measurements of shear stress in the polymer liquids used for the study.

Mhetar and Archer found that slip lengths obtained using a mixture of 1.5-μm and 24-nm tracer particles differed by an amount comparable to the particle diameter. This finding is consistent with the observations reported by den Otter et al. (68), and indicates that significant errors can occur in interfacial slip measurements if too large tracer particles are used. This observation also implies that tracer particle velocimetry using micron-sized particles is best suited for quantifying slip violations in systems where significant ($b > 1$ μm) levels of wall slip is present. In entangled PS/DEP solutions and PBD melts, Mhetar and Archer (18,40) observed slip violations at all shear stresses. At shear stresses in the Newtonian fluid regime, the slip velocity was observed to increase approximately linearly with shear stress $V_s \sim \sigma^{1.0}$ for PS/DEP solutions and PBD melts (Fig. 4.9). Slip behavior in this

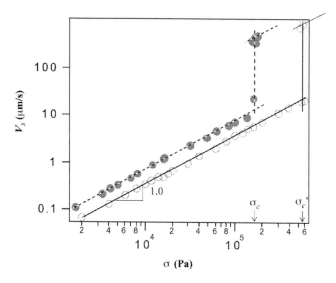

**FIGURE 4.9** Slip velocity $V_s$ vs. steady state shear stress $\sigma$ for a 1,4-polybutadiene melt $\overline{M}_w = 1.81 \times 10^5$ g/mol, $\overline{M}_w/\overline{M}_n = 1.03$. Slip measurements were performed using tracer particle velocimetry during planar Couette shear flow between bare silica glass substrates (filled symbols) and glass substrates grafted with a monolayer of a high-molecular weight PBD $\overline{M}_n \approx 9.0 \times 10^4$ g/mol (unfilled symbols). (From Ref. 34.)

regime is characterized by constant slip lengths $b_0$ ranging from 3 to 8 μm for PS/DEP solutions with $\phi \overline{M}_w$ from $2 \times 10^5$ to $6.9 \times 10^5$, and from 2 to 120 μm for PBD melts with $\overline{M}_w$ from $6.7 \times 10^4$ to $5.2 \times 10^5$. Mhetar and Archer reported that while $b_0$ for PS/DEP solutions increased linearly with $\phi \overline{M}_w$, slip lengths measured at low shear stresses increase almost quadratically with $\overline{M}_w$ for the range of PBD melts studied.

At higher shear stresses, slip velocities measured using PS/DEP solutions increase quite strongly with shear stress. In general, the first linear $V_s$ vs. $\sigma$ regime gives way to a nonlinear regime in which $V_s$ increases nearly quadratically with the $\sigma$ (18). This finding is consistent with recent observations reported in Ref. (52) and in Fig. 4.6(a) from slip measurements using gap-dependent torsional shear rheometry. At stresses above a critical value $\sigma_c$ ≈ $2.2 G_e(\phi)$, Mhetar and Archer reported a transition to large levels of interfacial slip $V_s > 10$ μm/sec for all solutions studied, where $G_e$ is the plateau modulus. A similar transition is also observed from slip measurements using gap-dependent shear flow measurements, but the ratio of the critical shear stress to the plateau modulus is lower, $\sigma_c ≈ 0.4 G_e(\phi)$ (52). The first linear $V_s$ vs. $\sigma$ regime is much broader for PBD melts than for PS/DEP solutions. For PBD melts, the linear regime also ends more abruptly. In particular, at shear stresses above a well-defined critical value, $\sigma_c ≈ 0.2 G_e ≈ 0.16$ MPa, for slip over physisorbed chains, and above $\sigma_c ≈ 0.55$ MPa for slip over a silica glass substrate grafted with long PBD chains ($\overline{M}_w = 9.1 \times 10^4$ g/mol), a dramatic transition from weak to strong slip is observed (Fig. 4.9). The critical stress for the weak-to-strong slip transition was also found to be independent of PBD molecular weight, which agrees with the observations of Vinogradov et al. (28) for a flow curve transition in capillary flow of PBD and PI melts. The critical stress reported by Vinogradov et al. for PBD melts is a factor of two larger than the $\sigma_c$ values reported by Mhetar and Archer. Finally, Mhetar and Archer observed that slip velocities in the weak slip regime could be recovered almost instantaneously by lowering the shear stress from above the critical value to below it.

In the vicinity of the weak to strong slip transition, tracer particles at the polymer–solid interface translate in cycles of intermittent fast/slow motion. This behavior is reminiscent of the stick–slip flow reported much earlier by Galt and Maxwell (16) and is accompanied by time-dependent stress oscillations. As discussed earlier, both the amplitude of stress oscillations and clarity of the stick–slip motion are enhanced when bare glass shear substrates are replaced by glass substrates modified by end-tethering a layer of high-molecular weight PBD (34). Stress oscillations are also enhanced when the glass substrates are replaced by high-energy aluminum substrates (35). Conversely, when the substrates are modified by end-tethering a single layer of heptadecafluoro-1,1,2,2-tetrahydrodecyl trichlorosilane, slip velocities in

PBD melts are enhanced by more than an order of magnitude at all shear stresses, and the weak slip flow regime transitions smoothly to the strong slip regime (34). Slip velocities in the strong slip regime are found to be insensitive to the substrate chemistry. However, slip lengths in this regime are consistently about an order of magnitude lower than expected for interfacial slip of a polymer melt over an ideal nonadsorbing substrate.

### 4.3.4  Other Direct Methods

The rate of heat transfer from a hot surface to a fluid in contact with the surface undergoing laminar flow is a function of thermal conductivity of the fluid and surface, and of the local fluid velocity. In hot film anemometry, a heated strip of material (typically metal) of known thermal conductivity is flush mounted to the wall of a flow channel through which a fluid with known thermal conductivity flows. The energy required to maintain the strip at a fixed temperature above the fluid temperature is determined using a temperature control loop and the results used to deduce the local fluid velocity. Kranynik and Schowalter (19) and Lim and Schowalter (21) used this technique to infer slip violations in aqueous polyvinyl alcohol solutions and in various narrow molecular weight distribution polybutadiene melts. In both materials, no evidence of slip was observed at low flow rates. At stresses comparable to those where flow curve discontinuity is observed, transient oscillations in pressure and heat flux from the hot film probe were seen. In polybutadiene melts, the heat flux oscillations were found to be 180° out of phase with the pressure oscillations.

Lim and Schowalter (21) interpreted this finding as evidence that a stick–slip flow near the die wall was responsible for the pressure oscillations. As discussed in Sec. 4.2.1, similar oscillations were reported previously by Kalika and Denn (31) who showed that they arise from compressibility of polymer liquid in the extruder reservoir. Lim and Schowalter, however, observed no variation in pressure or heat flux oscillation frequency with reservoir volume, which is at odds with the compressibility argument. For well-entangled polybutadiene melts, Lim and Schowalter (21) also found no evidence of a change in the slope of the flow curve or of interfacial slip from hot film measurements coinciding with macroscopic observations of sharkskin melt fracture. Furthermore, at the highest flow rates studied, the authors observed greatly attenuated heat flux oscillations, but that much higher average fluxes were needed to maintain the hot film temperature at the preset value. This result points to a transition to a state of massive, continuous slip. The critical stress for the onset of sharkskin melt fracture was found to be approximately 1 MPa, regardless of polybutadiene molecular weight, while the critical stress for the flow curve discontinuity was found to be approximately 2 MPa and to decrease weakly with polymer molecular weight, $\sigma_c \sim M_w^{-0.3}$.

Recently, Wise et al. (72) used Attenuated-Total-Reflection (ATR) Fourier-transform infrared spectroscopy to quantify time-dependent changes in concentration of deuterated polybutadiene near (within ca. 300 nm) of a polymer–solid interface. Measurements were performed at a location 1 mm upstream of the exit of a slit die during pressure-driven flow of normal (undeuterated) polybutadiene melts with the same molecular weight as the deuterated polymer. Slip was inferred from the concentration data by numerically solving the convection–diffusion equation with slip boundary conditions and fitting the predicted integrated near-surface concentration to the experimental data. At low stresses, good fits to the measured concentration data were only possible if no slip is assumed at the wall and the polymer diffusivity in the near wall region assumed to be significantly lower than in the bulk liquid. At higher wall shear stresses, a regime of surprisingly weak slip is observed. For example, at a shear stress of 0.3 MPa, the fitted slip length is less than 0.5 $\mu$m. At wall shear stresses above the spurt transition, the authors report loss of infrared signals on some runs, indicating that massive levels of convection are present in the field of view. Nonetheless, only modest slip lengths $b \approx 8\,\mu$m are found. Larger levels of slip $b \approx 30\,\mu$m were found for die surfaces modified using a fluoropolymer coating. However, these slip lengths are still more than an order of magnitude lower than reported by tracer particle velocimetry measurements using 1,4-polybutadiene melts of comparable molecular weight sheared between fluorinated parallel planes (72).

## 4.4 THEORETICAL APPROACHES FOR STUDYING SLIP VIOLATIONS

### 4.4.1 Overview

Based on the results of the previous sections, it is clear that the no-slip condition can be and is violated in polymeric liquids. Even for the limited number of systems studied, the characteristics of slip (e.g., the effect of fluid stresses on slip velocity) and location of the plane of slip appear to be complicated functions of polymer architecture, molecular weight, molecular weight distribution, solution concentration, and of the physicochemical properties of the substrate at which slip violations occur (e.g., surface energy, roughness, and tethered polymer molecular weight and graft density). To control where and how slip violations occur in polymers, it is essential that each of these effects be isolated and their fundamental role understood. Theories for slip provide a valuable means of refining the parameter space for experimentation and for clarifying the molecular-scale processes responsible for slip violations.

To begin, we will first review some basic processes by which a fluid may violate the no-slip condition. This is useful because it simultaneously introduces the key slip mechanisms that must be captured by theories and highlights the scientific richness of the subject. There are currently four accepted processes by which a polymeric fluid can violate the no-slip condition. (1) Slip can occur if the stresses in flow exceed the adhesive strength of the fluid/solid interface (73–77). (2) Cohesive failure between fluid layers near the solid–liquid interface can also give rise to slip (13,18,24,59,66,78–80). (3) In multiphase materials, slip may occur by depletion of the more viscous component from the fluid–solid interface (17,52,69). (4) Shear banding and/or constitutive instabilities in a liquid may also give rise to slip (53,81–86). Of these four slip mechanisms, wall slip by adhesive failure is the only one that actually violates the no-slip condition. Slip violations by the other three processes are collectively termed apparent slip because the fluid in molecular contact with the substrate may still satisfy the no-slip boundary condition.

To describe slip violations induced by adhesive failure at a fluid–solid interface, it is necessary to understand the nature of interactions between fluids and solid substrates, particularly how these interactions may be changed by application of a stress field (77,87). Likewise, a theory for slip by cohesive failure requires in-depth understanding of the types of forces that hold polymer molecules together in the liquid state and models for how these forces can be altered by flow (18,78–80,88). Depletion slip may arise from segregation of a lower viscosity, preferentially adsorbing component in a mixture to the fluid–solid interface or from the thermodynamic restrictions substrates impose on polymer conformations (89,90). Research into how these processes are affected by polymer architecture, molecular weight, surface chemistry, and flow fields is still in its infancy. Constitutive instabilities can produce complex flow patterns in fluids that can be mistaken for interfacial slip (83,86). These instabilities arise from multivalued relationships between shear stress and shear rate. Experimental evidence supporting the existence of constitutive instabilities is limited to Couette flow measurements using aqueous solutions of surfactants that form wormlike micelles (91,92). The conditions required to create constitutive instabilities in flexible polymers that can uncoil (stretch) in flow without breaking are poorly understood.

Much of the recent impetus for studying connections between apparent slip violations in polymers and constitutive instabilities (53,83,84,92) can be traced to the multivalued relationship between shear stress and shear rate predicted by the Doi-Edwards constitutive theory for entangled polymers (93). Recent work by several groups indicates that this prediction can be removed if the effect of convection on molecular orientation and of flow on chain stretching are properly taken into account in the model (94–98). In addition, the overwhelming majority of experiments indicate a sensitive connection between surface chemistry and roughness and interfacial slip violations (12–14,28,30–35,52,96,99,100), which is not consistent with expectations for slip by a constitutive instability. All studies that indicate a connection between slip violations and polymer depletion have been limited to dilute and semidilute polymer solutions (17,18,43,52,69,71,90), which have limited relevance to processing. In the following sections, we will therefore limit attention to theories for interfacial slip by adhesive and cohesive failure processes that are sensitive to the physicochemical properties of the polymer–solid interface.

### 4.4.2 Slip by Adhesive Failure

The simplest theory to address the effect of fluid–substrate interactions on slip violations has been traced to Tolstoi (101). Tolstoi's theory directly follows from transition state theories for liquid flow (102). In these theories, fluid flow is considered to be analogous to a chemical reaction characterized by breakage of bonds between molecules and their current neighbors and formation of bonds with new neighbors when the molecules move from their present position to adjacent holes (vacancies) in the fluid. The driving force for motion is provided by thermal energy and retarded by the flow activation energy $E_f$. The molecular mobility $\zeta^{-1}$ is then given by an Arrhenius equation

$$\zeta^{-1} = \frac{\Delta^2}{6kT\tau} \exp[-E_f/kT] \tag{4.15}$$

where $\Delta$ is the average distance between molecules, $\tau^{-1}$ is an attempt frequency, and $\zeta$ is a molecular friction coefficient. If the fluid is subjected to a uniform shear flow, shear rate $\dot{\gamma}$, the rate of energy dissipation per unit volume is $\dot{q}_M \approx \zeta(\dot{\gamma}\Delta)^2/\Delta^3$, where $\dot{\gamma}\Delta$ is the relative velocity of fluid elements. The rate of energy dissipation per unit volume may also be written in terms of the fluid viscosity, $\dot{q}_M \approx \eta\dot{\gamma}^2$, equating the two results yields, $\eta \approx \zeta/\Delta$.

Tolstoi argued that Eq. (4.15) should also be applicable for the molecular mobility at a fluid–solid interface, $\zeta_s^{-1} = (\Delta^2/(6kT\tau_s))\exp[-E_s/kT]$, where $E_s \neq E_f$ and $\tau_s^{-1} \neq \tau^{-1}$ are the flow activation energy and the attempt frequency at the surface, respectively. The fluid viscosity in the bulk liquid and at the fluid–solid interface is then related by

$$\eta/\eta_s = \frac{\tau}{\tau_s} \exp\left[\frac{E_f - E_s}{kT}\right] \tag{4.16}$$

$E_f - E_s$ is related to the difference in the work of cohesion $W_c$ of the liquid and the work of adhesion between the liquid and solid $W_a$ by, $E_f - E_s = S(W_c - W_a)$, where $S$ is the surface vacancy area occupied by the solid. If the liquid is not too volatile, $W_c - W_a = \Gamma_l(1 - \cos\theta_e)$, where $\Gamma_l$ is the liquid surface energy per unit area and $\theta_e$ is the equilibrium contact angle between liquid and solid. The connection to slip can be made by considering the velocity field developed in a layered, immiscible fluid mixture during shear between infinite parallel planes. Specifically, if one component with viscosity $\eta_s$ forms a film of thickness $\iota$ at the fluid–solid interface, and the other component with viscosity $\eta$ makes up the bulk, the velocity gradient is inhomogeneous across the gap and the mixture will appear to slip at the fluid solid surface. The slip length is easily shown to be

$$b = \iota(\eta/\eta_s - 1) \tag{4.17}$$

If it is assumed that $\tau_s = \tau$, it is evident that for $\theta_e > 0$, $\eta > \eta_s$ and the no slip condition is violated at all stress levels. It is also apparent that the largest levels of slip will occur when the fluid does not wet the substrate $\theta_e > 90°$, which is not unexpected.

In real situations, the liquid will possess some volatility and the work of adhesion will be enhanced by adsorption of vapor to vacancies on the solid (101). In addition, real substrates are rough on molecular length scales, which introduces additional modes of energy dissipation at the interface and enhances the effective surface viscosity (103,104). All of these effects reduces $b$ in Eq. (4.1), decreasing the likelihood for slip. We have so far made no mention of the role played by flow-induced stresses on adhesive failure. Several approaches have been advocated in the literature for introducing this effect. Dhori et al. (87,105), for example, invoke the entropy inequality and argue for a dependence of the contact angle on flow conditions in a bulk viscoelastic fluid. Hill proposed a simple, but elegant chemical model for polymer segment exchange kinetics at a solid–liquid interface (77). In the absence of flow, this model yields an expression similar to Eq. (4.16) for the relative rate of segment exchange $K_{sp}$ between adsorption sites on the solid surface $s$ and unbound polymer liquid $p$ near the interface.

$$K_{sp} = \exp\left[\frac{\pi a^3 \delta^2}{3W_c} \frac{(W_a - W_c)}{kT}\right] \tag{4.18}$$

Here $\delta$ is the solubility parameter and $a$ is a characteristic monomer size.

When the polymer liquid is subjected to flow, bound segments can be thought of as the conduits through which momentum is transported from the bulk liquid to the solid. Because only a small number of segments per chain may be bound, Hill points out that these bound segments will be under tremendous tension, which favors desorption. To account for this effect, the interfacial free energy is modified to include the flow contribution to the binding free energy $W_{II}$, yielding,

$$\tilde{K}_{sp} = \exp\left[\frac{\pi a^3 \delta^2}{3W_c} \frac{(W_a - W_c)}{kT} - \frac{W_{II}}{kT}\right] \tag{4.19}$$

For an arbitrary stress field, $W_{\Pi}$ would be expected to depend on all invariants of the extra stress tensor $\Pi$. Hill argues for a dependence on the first invariant, $W_{\Pi} = tr(\Pi)/\alpha\rho_e$, where $\alpha$ is the segment surface coverage, $\rho_e = \rho/(N_e a^3)$ is the entanglement density, $\rho$ is the bulk density, $N_e = M_e/m_0$ is the number of segments between entanglements, and $M_e$ is the entanglement molecular weight. For a homogeneous shear flow, $\tilde{K}_{sp}$ is related to the slip velocity $V_s$ by

$$V_s = \frac{a}{12\eta_s} \tilde{K}_{sp}^{-1/2} \sigma \tag{4.20}$$

where $\eta_s$ is the monomeric viscosity at the fluid–solid interface and $\sigma$ is the shear stress at the wall. Equation (4.20) has several features that are qualitatively consistent with observations, particularly for slip in pressure-driven flows. First, for weak shear fields $W_{\Pi}/kT$ is small and apart from numerical prefactors the slip length is of similar form to that predicted by Eq. (4.17). Thus depending on the relative magnitudes of the work of adhesion and cohesion, slip violations can be observed at all shear stresses. Second, for strong shear fields, Eq. (4.20) admits a nonlinear relationship between $V_s$ and $\sigma$, as well as between $V_s$ and normal stresses developed in flow. As discussed earlier, such a relationship has been advocated on the basis of experiments (55). Third, Eq. (4.20) admits a sudden increase in slip velocity above a critical stress that is set by balancing the free energy contributions in $\tilde{K}_{sp}$ and instability of $V_s$ near the critical condition. A regime of stable flow is predicted to be recovered for constant stress flow at stresses above the critical value. Finally, the equation indicates that while the temperature dependence of $V_s$ is complex, the temperature dependence of the slip length $b = \eta V_s/\sigma$ is set by $\tilde{K}_{sp}$ and hence provides a useful test for the model.

A comprehensive comparison of the predictions of Eq. (4.20) with experimental results, mainly from capillary flow experiments using commercial polyethylenes and metallic substrates, is provided by Hill (77). Many features observed experimentally, including the absence of a lower critical stress for slip violations and onset of strong slip above a critical stress of order 0.4 MPa for LLDPE, and the dependence of slip velocities measured in extrusion on die length, are correctly captured by Hills theory for slip by adhesive failure.

### 4.4.3 Slip by Cohesive Failure

Slip violations in entangled polymer liquids can also arise from cohesive failure near the polymer solid interface. Most theories follow from the original works by Brochard-Wyart and de Gennes (78,79) who recognized that dynamic dissimilarities between entangled polymer molecules tethered to a surface and polymer chains in a bulk liquid can lead to premature disentanglement, and hence apparent slip. In this case, the slip plane is located roughly one molecular diameter from the polymer–solid interface. As discussed below, the number of polymer molecules adsorbed per unit area $v$ is a key variable in models for slip by interfacial disentanglement. This gives a natural mechanism for introducing processes such as described in Sec. 4.3.2, and therefore provides a promising framework for a complete model for interfacial slip in polymers.

To understand the nature of the slip transition, consider a bulk polymer liquid subject to a laminar shear flow between parallel planes. Molecular fluid layers in the bulk will slide by each other with a relative velocity proportional to the layer separation $\Delta$ and shear stress $\sigma$ and inversely proportional to the bulk fluid viscosity $\eta$. If this picture is extended to the interface between the bulk liquid and a molecularly thin film of strongly adsorbed molecules, we anticipate a relative interfacial velocity $V_{\text{rel}}$ given by

$$V_{\text{rel}} = \frac{\sigma \overline{R}}{\xi} \tag{4.21}$$

where $\overline{R} \approx \Delta$ is the average molecular size and $\xi$ is the resistance the adsorbed layer provides to relative motion, and has dimensions of viscosity. If $\xi \neq \eta$, the layer will appear to slip with velocity speed, $V_s = (\sigma \overline{R}/\eta)[(\eta/\xi) - 1]$. The slip length is then given by an expression analogous to Eq. (4.17)

$$b = \overline{R}\left[\frac{\eta}{\xi} - 1\right] \tag{4.22}$$

If adsorbed molecules are assumed to resist shear by the same mechanisms as in the bulk liquid, $\xi \approx v\overline{R}^2\eta$ to a first approximation. For $v$ small this yields

$$b = b_0 \approx \left(\frac{1}{\overline{R}v}\right) \text{(78)} \tag{4.23}$$

In entangled polymer liquids, relative motion between fluid planes is resisted by friction between monomers on adjacent sides of the interface and by intermolecular entanglements spanning the interface. The friction coefficient per molecule is

$$\zeta = N\zeta_m + \zeta_p \tag{4.24}$$

where $\zeta_m$ is a monomeric friction coefficient, $\zeta_p$ is the extra friction provided by entanglements, and $N$ is the average degree of polymerization. For well-entangled polymer liquids $\zeta_p \gg N\zeta_m$, so most of the frictional resistance arises from the presence of entanglements. In the extreme case where adsorbed polymer chains become so highly oriented by shear that their effective entanglement constraints with the adjacent layer is lost, $\zeta$ would fall precipitously to $N\zeta_m$ per tethered chain. In this case, we may write

$$\xi = \xi_\infty \approx \overline{R}^{-1}(N\zeta_m + v\overline{R}^2 N\zeta_m) \tag{4.25}$$

which on substitution in Eq. (4.22) yields

$$b = b_\infty \approx \frac{\overline{R}^2 \eta}{(N\zeta_m + v\overline{R}^2 N\zeta_m)} \tag{4.26}$$

In situations where $v \ll 1/\overline{R}^2$, the first term in the denominator dominates and $b_\infty \approx a(\eta/(\zeta_m/a)) = a(\eta/\eta_m)$, where $\eta_m$ is the monomeric viscosity. It is therefore apparent that cohesive failure by disentanglement of surface adsorbed and bulk polymer chains yields a dramatic increase in the slip length. Brochard-Wyart and de Gennes computed the shear stress required for the disentanglement transition to be $\sigma = \sigma_c = vkT/\sqrt{N_e}a$, which apart from a possible dependence of $v$ on $N$, is not a function of polymer molecular weight (78).

At shear stresses intermediate between these two extremes, the degree of intermolecular entanglement changes with shear stress and the slip velocity gradually increases. If we assume the slip length in this regime to be larger than the molecular size, the bulk melt slides by the surface adsorbed molecules with a uniform velocity $V$. To determine how the structure and hence friction coefficient of the adsorbed layer changes in flow, Brochard-Wyart and de Gennes considered the equivalent problem of determining the structure and friction coefficient of a single entangled polymer molecule pulled by one end with a uniform velocity $V$ through a stationary entangled melt of similar molecules (Fig. 4.10(a)). The authors contend that at all pulling speeds, the equilibrium structure of the pulled molecule is set by a balance between an elastic restoring force, $F_e \approx kT/D(V)$, where $D(V)$ is the cross-sectional diameter of the molecule, and the viscous drag force, which the authors assumed to be given by Stokes law at both low ($L$) and high ($H$) pulling speeds; $F_L(V) \approx \eta\overline{R}V$ and $F_H(V) \approx \eta\overline{L}V$. Here $\overline{L}$ is the average contour length and $\overline{R}$ the average coil size at equilibrium. To interpolate between the high and low speed results, they proposed a simple formula for the overall drag force, $F(V) = \eta V(\overline{R}^2 + \overline{L}^2)^{1/2}$.

Enforcing this balance consistently above and below the disentanglement transition yields the following expression for the evolution of molecular contour length, $V = (kT/\eta\overline{R}^2)L/(\overline{R}^2 + \overline{L}^2)^{1/2}$, which finally yields a relationship between the slip length and slip velocity

$$b(V_s) = \frac{\overline{R}^2 \eta}{N\zeta_m + vkT(\overline{L}(V_s)/V_s)} \tag{4.27}$$

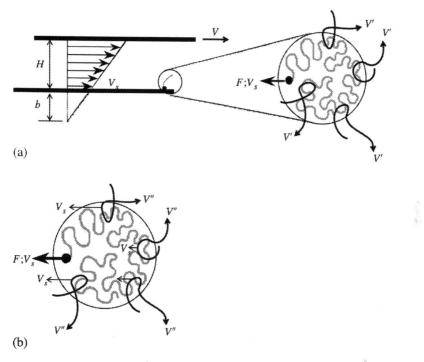

(a)

(b)

**FIGURE 4.10** (a) When the slip length is comparable to the size of polymer molecules tethered at the fluid–solid interface, these molecules experience the bulk as a viscous entangled liquid moving by at velocity $V_s$. This frictional resistance $F$ offered by each tethered molecule can be determined by a thought experiment where a molecule tethered at one end is pulled with a constant speed $V_s$ relative to a stationary entangled melt. In this case, the surrounding chains reptate with a speed $V' \gg V_s$, so that the pulled chain preserves its Gaussian coil structure (78,80). (b) Schematic of a single tethered molecule pulled by one end, at a fixed speed $V_s$, through a melt of entangled polymer chains. The pulled molecule moves by dragging chains entangled with it at a speed $V_s$. The dragged molecules reptate with a speed $V''$ through their surrounding tubes and thereby release entanglements with the tethered molecule. (From Ref. 18.)

and between the slip velocity and shear stress

$$V_s = \left\{ \frac{\overline{R}^2 \eta}{N\zeta_m + vkT(\overline{L}(V_s)/V_s)} \right\} \sigma \tag{4.28}$$

Equation (4.28) predicts that the slip velocity should increase roughly as the square root of $\sigma$, and saturate near the critical stress. Beyond a critical velocity $V_s^* \cong kT/(\eta \overline{R}^2) \cong (a/\tau_m)[P(P/N_e)^2 N]^{-1}$, the cross-section of the pulled chain falls below the entanglement threshold (tube diameter) and the chain disentangles from the surrounding fluid, causing macroscopic slip. At shear stresses $\sigma > \sigma_c$, $V_s$ is large, $V_s = V_s^\dagger = va^3/(\tau_m \sqrt{N_e})$, and the first term in the denominator of Eq. (4.28) dominates. $V_s$ is therefore predicted to increase linearly with shear stress above the slip transition. More refined calculations by Ajdari et al. (80) and by Brochard-Wyart et al. (79) yield similar results for $\sigma \geq \sigma_c$, but different predictions at lower stress. To compute the friction coefficient between surface adsorbed and bulk chains, these authors consider a situation where slip occurs between a layer of long $N$-mer molecules end-tethered to the substrate and a bulk melt of entangled $P$-mer molecules. Again, the problem can be solved by studying how the friction coefficient of a single $N$-mer changes as it is pulled through a stationary $P$-mer melt at a fixed velocity $V = V_s$.

At low pulling speeds, the pulled $N$-mer preserves its Gaussian coil (ball) structure, but the friction coefficient is larger than for Stokes-drag on a sphere of diameter $\overline{R}(N)$ through a melt of entangled $P$-mers. Specifically, the authors argue that the $N$-mer can only preserve its ball structure if the surrounding $P$-mers reptate away at a faster rate $V_P = (P/N_e)V$ from each entanglement with the $N$-mer (Fig. 4.10(a)). This produces a friction coefficient $\zeta_{CR}$ and a linear stress–velocity relationship (assuming noninteracting $N$-mers)

$$\sigma \approx v \zeta_{CR} V_s \qquad (4.29)$$

where $\zeta_{CR} = \zeta_m(P^3/N_e^2)(N/N_e)^{1/2}$ for $N < P$, and $\zeta_{CR} = \zeta_m(P^{2.5}/N_e^{1.5})(N/N_e)$ for $N > P$. This leads to two expressions for the slip length at low stresses

$$b = b_0 \approx \frac{\eta(P)}{v\zeta_{CR}} \approx \frac{(N/N_e)^{-1/2}}{va}, \text{ for } N < P \qquad (4.30)$$

and

$$b_0 \approx \frac{(P/N_e)^{1/2}}{(N/N_e)va}, \text{ for } N > P \qquad (4.31)$$

If multiple entanglements between a given $N$-mer and the same $P$-mer are considered, $b_0$ is given by

$$b_0 \approx \frac{(N/N_e)^{-1}}{va}, \text{ for } N < N_e^2 \qquad (4.32)$$

At pulling speeds greater than the rate of $N$-mer relaxation, $V > V_1 \approx \overline{R}(N)^2/\tau_{CR}$, the $N$-mer begins to deform, beginning at the pulling point. Here

$\tau_{CR} = \overline{R}(N)^2/(kT/\zeta_{CR})$ is the constraint release relaxation time of the $N$-mer. In this regime, the cross-section of the pulled chain is not uniform along the contour length as assumed by Brochard-Wyart and de Gennes, rather faster relaxing sections of the chain near the ends remain ball-like, while slower relaxing sections near the pulling point are highly deformed. The overall structure can therefore be visualized as a sequence of spherical blobs with increasing diameters with distance from the pulling point. The stress velocity relationship for this trumpet-shaped structure can be shown to be

$$\sigma \approx \frac{vkT}{a}\left(\frac{V_s}{A}\right)^{1/2} \exp\left[\frac{V_s}{V_1} - 1\right], \text{ for } V_s > V_1 \tag{4.33}$$

where $A = kT(N_e/N)^3/(\sqrt{N_e}a\zeta_m)$. The slip velocity therefore increases very slowly with shear stress.

When the shear stress reach a value

$$\sigma_c \approx \frac{vkT}{\sqrt{N_e}a} \approx v\sqrt{N_e}a^2 G_e \tag{4.34}$$

the pulled $N$-mer disentangles from the surrounding melt. Here $G_e = kT/N_e a^3$ is the plateau modulus of the entangled polymer liquid. When this occurs, the $N$-mer partially relaxes and re-entangles. If the pulling speed is maintained at the critical value indefinitely, the $N$-mer will go through cycles of disentanglement/re-entanglement, yielding asymmetric stress oscillations similar to the slip–stick processes observed experimentally. If the velocity is increased above a critical value $V_s^*$ given by

$$V_s^* \approx \frac{aN_e^{5/2}}{\tau_m P^3(N - N')}\left(1 + \frac{1}{2}\ln\left(\frac{aN_e^{3/2}}{\tau_m P^3 V_s^*}\right)\right) \tag{4.35}$$

with, $N' = (8/15)NN_e \ln(P^3/(N^2 N_e))$, the pulled chain will disentangle from the bulk and the friction coefficient will drop precipitously. For the actual situation of interest, disentanglement will lead to a sudden jump in the slip velocity to $V_s = V_s^{\dagger} \approx va^3/(\tau_m\sqrt{N_e})$ and an increase in the slip length to $b = b_\infty \cong \eta(P)/(vNN_m a) \approx (P/vNa)(P/N_e)^2$.

Mhetar and Archer (18) proposed a model for slip violations in polymers that is in many ways similar to the models of Brochard and de Gennes (78) and of Ajdari et al. (80). For example, this model uses the paradigm of a single tethered $N$-mer chain pulled at constant speed $V = V_s$ through a melt of entangled $P$-mer molecules to determine the friction coefficient of the $N$-mer and, for the independent *mushroom* case, the stress–velocity diagram. As in the model of Ajdari et al., this model also assumes that the balance between the rate of pulling and relaxation rate of subsections of the pulled $N$-mer determines the cross-section it presents for entanglements with $P$-mers.

Thus at pulling speeds $V < V_1 = \overline{R}(N)/\tau_{CR}(N)$, the $N$-mer relaxes faster than it is perturbed by pulling. In this regime, the $N$-mer retains a globular (ball) structure as it translates through the $P$-mer melt. To compute the drag on the $N$-mer, Mhetar and Archer followed an approach suggested long ago by Buche (106), namely that the tethered $N$-mer translates at the pulling speed $V$ by dragging the $N/N_e$ $P$-mers entangled with it (Fig. 4.10(b)). These molecules in turn drag the approximately $P/N_e$ $P$-mers entangled with them a distance of order the reptation tube diameter before the second-generation $P$-mers release their constraints with the first generation dragged $P$-mers. The drag force per $N$-mer entanglement is therefore $f_p \approx V(P/N_e)P\zeta_m$ and the total drag force on the $N$-mer is $F_{ball} = V(N/N_e)((P/N_e)P\zeta_m)$. If there are $v$ grafted $N$-mer chains per unit area of surface and the surface chains are independent this implies that in the ball regime

$$\sigma \approx v\eta_p a \frac{N}{P} V_s \text{ for } V < V_1 \tag{4.36}$$

The slip length is therefore given by

$$b = b_0 \approx \frac{P}{vNa} \tag{4.37}$$

For $V > V_1$, the pulled $N$-mer begins to deform, starting near the location of greatest constraint (the pulling point). This yields a molecular cross section that is a combination of a ball-like section near the free end and an oriented cylindrical section near the pulling point. The friction coefficient of the $N$-mer is a combination of the ball and cylinder friction coefficients. As the pulling speed is progressively increased, the number of monomers belonging to the cylinder section grows and the number in the ball falls, the friction coefficient therefore decreases as the pulling rate goes up. In this regime, the stress–velocity relationship is more complicated

$$\sigma = av\eta_m V_s \left[ \frac{(N-K)^2}{N_e} + K\left(\frac{P}{N_e}\right)^2 \right] \tag{4.38}$$

where $K \approx \left[ a/(\tau_m V)(N_e/P)^2 \right]^{2/3}$. Eventually, at a pulling velocity greater than the Rouse retraction rate $\overline{L}/\tau_R$, the cross-section of the $N$-mer falls below the $P$-mer tube diameter and the chain disentangles. The critical shear stress at which the disentanglement occurs is, in this case, given by

$$\sigma_c \cong v\frac{N}{\sqrt{N_e}b}\frac{kT}{N_e} = \frac{vNa^2}{\sqrt{N_e}}G_e \tag{4.39}$$

Thus in the limiting case $v \cong (Na^2)^{-1}$, the critical stress is related in a simple way to the plateau modulus and molecular weight between entanglements, both material properties that are independent of polymer molecular weight,

$$\sigma_c \cong \left. \frac{G_e}{\sqrt{N_e}} \right|_{v \cong (Na^2)^{-1}} \tag{4.40}$$

Again, following disentanglement the $N$-mer will partially relax (as dictated by the relative rates of pulling and relaxation) and re-entangle. If the pulling speed is maintained at the critical value for disentanglement, the $N$-mer will execute cycles of disentanglement/re-entanglement, also resulting in slip–stick stress dynamics. If the stress is instead increased slightly above $\sigma_c$, the slip velocity is predicted to jump from $V_s^* \approx a/(N N_e^{1/2} \tau_m)$ to a very large value, $V_s^{\dagger} \approx a/(N_e^{3/2} \tau_m)$, independent of polymer molecular weight. The slip length in this strong slip regime is a constant $b = b_{\infty} \cong \eta(P)/(a N v \eta_m)$ (18).

Extrusion experiments by Legrand et al. (107) and, more recently, by Barone et al. (108) indicate that fluorescent molecules preadsorbed to the surface of extruder dies remain adsorbed even after the stick–slip transition. These findings are not consistent with expectations for slip by adhesive failure, and have been interpreted as providing proof of the disentanglement slip mechanism (108). Most parameters needed to determine slip levels by disentanglement processes are available from rheological experiments. Other parameters, for example, the areal density and molecular weight of tethered chains, have been measured for a few polymer systems. It is therefore possible to provide some comparisons between predicted and measured levels of slip. In Tables 4.1 and 4.2, for example, we compare values of $b_0$, $V_s^*$, $\sigma_c$, $V_s^{\dagger}$, and $b_{\infty}$ predicted using the three theories introduced in the previous section, with experimental results. Experimental results in Table 4.1 are from FPFRAP measurements using a highly entangled PDMS melt ($\overline{M}_w = 0.97 \times 10^6$ g/mol,

**TABLE 4.1**  Slip Parameters for a PDMS Melt from FPFRAP Velocimetry Measurements Compared with Theoretical Predictions

|  | $b_0$ [μm] | $V_s^*$ [μm/sec] | $\sigma_c$ [MPa] | $V_s^{\dagger}$ [mm/sec] | $b_{\infty}$ [mm] |
|---|---|---|---|---|---|
| Experiment | $12.5 \pm 3.3$ | 8.0 | 0.29 | 0.5 | $0.4 \pm 0.05$ |
| Brochard et al. [78] | 0.045 | $3.4 \times 10^{-5}$ | 0.002 | 0.199 | 26.2 |
| Ajdari et al. [80] | 0.26 | $1.8 \times 10^{-6}$ | 0.002 | 0.199 | 26.2 |
| Mhetar and Archer [18] | 2.6 | 198.6 | 0.17 | 20.0 | 26.2 |

**TABLE 4.2** Slip Parameters for a PBD Melt from Tracer Particle Velocimetry Measurements Compared with Theoretical Predictions

|  | $b_0$ [μm] | $V_s^*$ [μm/sec] | $\sigma_c$ [MPa] | $V_s^\dagger$ [mm/sec] | $b_\infty$ [mm] |
|---|---|---|---|---|---|
| Experiment | 1.8 | 10 | 0.2 | 0.2 | 0.03 |
| Brochard et al. [78] | 0.003 | $2 \times 10^{-5}$ | 0.015 | 0.93 | 50 |
| Ajdari et al. [80] | 0.0004 | $1.5 \times 10^{-6}$ | 0.015 | 0.93 | 6.7 |
| Mhetar and Archer [18] | 1.1 | 124 | 0.12 | 0.97 | 6.7 |

$P/N_e \cong 78$). While the data in Table 4.2 are from tracer particle velocimetry measurements using a high-molecular weight 1,4-polybutadiene melt $\bar{M}_w = 1.83 \times 10^5$ g/mol, $\bar{M}_w/\bar{M}_n = 1.04$, $P/N_e \cong 101$. To quantify slip violations in PDMS, the high-molecular weight polymer was sheared between glass substrates to which a lower-molecular weight PDMS $\bar{M}_w = 0.96 \times 10^5$ g/mol is grafted with known density, $v = 1.4 \times 10^{-2}$ nm$^{-2}$.

Published values for the plateau modulus $G_e \approx 0.2$ MPa, entanglement molecular weight $M_e = 1.23 \times 10^4$ g/mol, statistical segment length $a = 6.2$ Å, and segmental friction coefficient $\zeta_m \cong 10^{-8.05}$ dyn · sec/cm for PDMS (64,109) were used in the comparisons. The segmental hoping time $\tau_m$ for PDMS was estimated from $\zeta_m$ using the relation, $\tau_m \cong a^2 \zeta_m/3kT \cong 0.3$ nsec. The zero-shear viscosity of the bulk PDMS melt was estimated to be $\eta_0 = \eta_0(P) \cong 5.3 \times 10^4$ Pa·sec from published viscosity data for a lower molecular weight PDMS ($\bar{M}_w = 0.33 \times 10^6$ g/mol, $\eta_0 = 2 \times 10^3$ Pa·sec) (110) using the relation $\eta_0 = P(P/N_e)^2 \eta_m$ (93). Using this viscosity value, the shear stress at the weak/strong slip transition was determined to be $\sigma_c \cong \dot{\gamma}^* \eta(P) \cong 0.29$ MPa. This value is about one-half the critical shear stress reported by Piau and El Kissi (29) for the onset of flow curve discontinuity in well-entangled PDMS melts subject to capillary flow. The graft density and adsorbed polymer molecular weight for the 1,4-PBD melt are unknown. To facilitate rough comparisons, Mhetar and Archer assumed $P \cong N$ and $v \cong (Na^2)^{-1}$ (18,40). The plateau modulus $G_e \cong 1$ MPa, entanglement molecular weight $M_e = 1822$, and segmental relaxation time $\tau_m = 0.2$ nsec for the 1,4-polybutadiene melt were recently determined using mechanical rheometry (111). The statistical segment length of 1,4-polybutadiene at 25°C is given by Ferry to be $b = 6.75$ Å (64).

Results in Tables 4.1 and 4.2 indicate that while all three models predict a transition from weak to strong slip above a critical shear stress, the predicted slip parameters and value of the critical stress are quite different from those observed experimentally. Although differences are perhaps expected for scaling models of the sort considered here, some are large enough that the underlying physical assumptions in the model should be evaluated. The most significant difference between the slip model of Mhetar and Archer (18) and that of Ajdari et al. (80) is the assumption in the former that the pulled $N$-mer chain instinctively drags its entangled neighbors. That the predictions provided by the former model are closer to the experimental results for both materials considered here, as well as others discussed elsewhere (40), suggests that the dragging assumption may be closer to reality, at least in the case of the disentanglement slip transition. If the plateau modulus and entanglement molecular weight of polyethylene are taken to be $G_e \cong 2.6$ MPa and $M_e \cong 828$ g/(mol) (109), Eq. (4.40) yields $\sigma_c = 0.48$ MPa, which is slightly larger than the critical stress reported for linear polyethylenes (see Sec. 4.2.1).

All of the above theories for slip by disentanglement assume that the contributions of adsorbed chains to the shear stress at the fluid–solid interface is simply additive. The experiments by Leger et al. using surfaces modified by end-tethering PDMS chains at high surface coverages (see Sec. 4.2.2) indicate that this cannot be true at high coverages. Two recent studies have attempted to determine how grafted chains at high surface coverages influence slip by disentanglement processes (67,79). The main effect seems to be that at high surface coverages, molecules in the bulk melt are expelled from the polymer–solid interface. This reduces the degree of interpenetration between bulk and interfacial chains, enhancing the likelihood for slip violations by a disentanglement mechanism (24,66).

## ACKNOWLEDGMENTS

Support from the National Science Foundation Tribology program (grant no. CMS0004525) and from the Department of Energy Nanoscience program (grant no. DE-FG02-02ER4600) are gratefully acknowledged.

## REFERENCES

1.  Goldstein, S. *Modern Developments in Fluid Dynamics*; Oxford University Press: London, 1938; Vol. 2, 676–680.
2.  Silliman, W.J.; Scriven, L.E. Separating flow near a static contact line-slip at a wall and shape of a free surface. J. Comput. Phys. 1980, *34*, 287–313.
3.  Dussan, E.B. Spreading of liquids on solid surfaces—static and dynamic contact lines. Annu. Rev. Fluid Mech. 1979, *11*, 371–400.
4.  Schnell, E. Slippage of water over non-wettable surfaces. J. Appl. Phys. 1956, *27*, 1149–1152.

5. Churaev, N.V.; Soboloev, V.D.; Somov, A.N. Slippage of liquids over lyophobic solid surfaces. J. Colloid Interface Sci. 1984, *97*, 574–581.

6. Richardson, S. On the no-slip boundary condition. J. Fluid Mech. 1973, *59*, 707–719.

7. Zhu, Y.X.; Granick, S. Limits of the hydrodynamic no-slip boundary condition. Phys. Rev. Lett. 2002, *88*, 106102-1–106102-4.

8. Zhu, Y.X.; Granick, S. Apparent slip of Newtonian fluids past adsorbed polymer layers. Macromolecules 2002, *35*, 4658–4663.

9. Reiter, G.; Demirel, A.L.; Granick, S. From static to kinetic friction in confined films. Science 1994, *263*, 1741–1744.

10. Benbow, J.J.; Lamb, P. New aspects of melt fracture. S.P.E. Trans. 1963, *3*, 7–17.

11. Denn, M.M. Issues in viscoelastic fluid mechanics. Annu. Rev. Fluid Mech. 1990, *22*, 13–34.

12. El Kissi, N.; Piau, J.M. The different capillary flow regimes of entangled polydimethyl siloxane polymer—macroscopic slip at the wall, hysteresis, and cork flow. J. Non-Newtonian Fluid Mech. 1990, *37*, 55–94.

13. Wang, S.-Q. Molecular transitions and dynamics at polymer/wall interfaces: origins of flow instabilities and wall slip. Adv. Polym. Sci. 1999, *138*, 227–275.

14. Joshi, Y.M.; Lele, A.K.; Mashelkar, R.A. Slipping fluids: a unified transient network model. J. Non-Newtonian Fluid Mech. 2000, *89*, 303–335.

15. Denn, M.M. Extrusion instabilities and wall slip. Annu. Rev. Fluid Mech. 2001, *33*, 265–287.

16. Galt, J.; Maxwell, B. Velocity profiles for polyethylene melts. Mod. Plast. December; 1964, *42*, 115–189.

17. Muller-Mohnssen, H.; Lobl, H.P.; Schauerte, W. Direct determination of apparent slip for a ducted flow of polyacrylamide solutions. J. Rheol. 1987, *31*, 323–336.

18. Mhetar, V.R.; Archer, L.A. Slip in entangled polymer solutions. Macromolecules 1998, *31*, 6639–6649.

19. Kraynik, A.M.; Schowalter, W.R. Slip at the wall and extrudate roughness with aqueous solution of polyvinyl alcohol and sodium borate. J. Rheol. 1981, *25*, 95–114.

20. Schowalter, W.R. The behavior of complex fluids near solid boundaries. J. Non-Newtonian Fluid Mech. 1988, *29*, 25–36.

21. Lim, F.J.; Schowalter, W.R. Wall slip of narrow molecular weight distribution polybutadienes. J. Rheol. 1989, *33*, 1359–1382.

22. Migler, K.B.; Hervet, H.; Leger, L. Slip transition of a polymer melt under shear stress. Phys. Rev. Lett. 1993, *70*, 287–290.

23. Migler, K.B.; Massey, G.; Hervet, H.; Leger, L. The slip transition at polymer–solid interface. J. Phys. Condens. Matter 1994, *6*, A301–A304.

24. Leger, L.; Raphael, E.; Hervet, H. Surface-anchored polymer chains: Their role in adhesion and friction. Adv. Polym. Sci. 1999, *138*, 185–225.

25. Mooney, M. Explicit formulas for slip and fluidity. J. Rheol. 1931, *2*, 210–222.

26. Hatzikiriakos, S.G.; Dealy, J.M. Wall slip of molten high-density polyethylene. I. Sliding plate rheometer studies. J. Rheol. 1996, *35*, 497–523.

27. Henson, J.D.; Mackay, M.E. Effect of gap on the viscosity of monodisperse polystyrene melts: slip effects. J. Rheol. 1995, *34*, 359–373.

28. Vinogradov, G.V.; Protasov, V.P.; Dreval, V.E. The rheological behavior of flexible-chain polymers in the region of high shear rates and stresses, the critical process of spurting, and supercritical conditions of their movement at $T > T_g$. Rheol. Acta 1984, *23*, 46–61.

29. Piau, J.M.; El Kissi, N. Measurement and modeling of friction in polymer melts during macroscopic slip at the wall. J. Non-Newtonian Fluid Mech. 1994, *54*, 121–142.

30. Ramamurthy, A.V. Wall slip in viscous fluids and influence of materials of construction. J. Rheol. 1986, *30*, 337–357.

31. Kalika, D.S.; Denn, M.M. Wall slip and extrudate distortion in linear low density polyethylene. J. Rheol. 1987, *31*, 815–834.

32. Person, T.J.; Denn, M.M. The effect of die materials and pressure-dependent slip on extrusion of linear low density polyethylene. J. Rheol. 1997, *41*, 249–265.

33. Durilat, E.; Hervet, H.; Leger, L. Influence of grafting density on wall slip of polymer melt on a polymer brush. Europhys. Lett. 1997, *38*, 383–388.

34. Mhetar, V.R.; Archer, L.A. Slip in entangled polymer melts. 2. Effect of surface treatment. Macromolecules 1998, *31*, 8617–8622.

35. Dao, T.T.; Archer, L.A. Stick–slip dynamics of entangled polymer liquids. Langmuir 2002, *18*, 2616–2624.

36. Black, W.B.; Graham, M.D. Wall-slip and polymer-melt flow instability. Phys. Rev. Lett. 1996, *77*, 956–959.

37. Reimers, M.J.; Dealy, J.M. Sliding plate rheometer studies of concentrated polystyrene solutions: large amplitude oscillatory shear of a very high molecular weight polymer in diethyl phthalate. J. Rheol. 1996, *40*, 167–186.

38. Dhori, P.K.; Giacomin, A.J.; Slattery, J.C. Common line motion. 2. Sliding plate rheometry. J. Non-Newtonian Fluid Mech. 1997, *71*, 215–229.

39. Leger, L.; Hervet, H.; Charitat, T.; Koutsos, V. The stick–slip transition in highly entangled poly(styrene–butadiene) melts. Adv. Colloid Interface Sci. 2001, *94*, 39–52.

40. Mhetar, V.R.; Archer, L.A. Slip in entangled polymer melts. 1. General features. Macromolecules 1998, *31*, 8607–8616.

41. Mhetar, V.R.; Archer, L.A. Secondary flow of entangled polymer fluids in plane Couette shear. J. Rheol. 1996, *40*, 549–572.

42. Koran, F. M. Eng Dissertation. Chemical Engineering; McGill University: Montreal, 1994.

43. Dao, T.T.; Archer, L.A. Relaxation dynamics of entangled polymer liquids near solid substrates. Langmuir 2001, *17*, 4042–4049.

44. Islam, M.T.; Archer, L.A. Nonlinear rheology of highly entangled polymer solutions in start-up and steady shear flow. J. Polym. Sci. B Polym. Phys. 2001, *39*, 2275–2289.

45. Larson, R.G.; Mead, D. Development of orientation and texture during shearing of liquid crystalline polymers. Liq. Cryst. 1992, *12*, 751–768.

46. Macosko, C.W. Rheology Principles Measurements and Applications; Wiley-VCH: New York, 1994. Chapter 5.

47. Yoshimura, A.; Prud'homme, R.K. Wall slip corrections for Couette and parallel disk viscometers. J. Rheol. 1988, 32, 53–67.

48. Brunn, P.O.; Ryssel, E. Lamb's slip hypothesis—revisited for torsional and cone-plate flow. Z. Angew. Math. Mech. 1999, 79, 485–491.

49. Dealy, J.M. Comments on slip of complex fluids in viscometry by P. Brunn, S. Muller, S.B. Schorer. Rheol. Acta 1998, 37, 195.

50. Henson, J.D.; Mackay, M.E. The effect of molar mass and temperature on the slip of polystyrene melts at low stress levels. J. Rheol. 1998, 42, 1505–1517.

51. Hay, G.; Mackay, M.E.; McGlashan, S.A.; Park, Y. Comparison of shear stress and wall slip measurement techniques on a liner low density polyethylene. J. Non-Newtonian Fluid Mech. 2000, 92, 187–201.

52. Sanchez-Reyes, J.; Archer, L.A. Interfacial slip violations in polymer solutions: Role of microscale surface roughness. Langmuir, ASAP, 2003, 19, 3304–3312.

53. Larson, R.G. Instabilities in viscoelastic flows. Rheol. Acta 1992, 31, 213–263.

54. Bagley, E.B. End corrections in the capillary flow of polyethylene. J. Appl. Phys. 1957, 28, 624–627.

55. Hatzikiriakos, S.G.; Dealy, J.M. Wall slip of molten high-density polyethylene. II. Capillary rheometer studies. J. Rheol. 1992, 36, 703–741.

56. Hay, G.; Mackay, M.E.; Awati, K.M. Pressure and temperature effects in slit rheometry. J. Rheol. 1999, 43, 1099–1116.

57. Awati, K.M.; Park, Y.; Weisser, E.; Mackay, M.E. Wall slip and shear stresses of polymer melts at high shear rates without pressure and viscous heating effects. J. Non-Newtonian Fluid Mech. 2000, 89, 117–131.

58. Wang, S.-Q.; Drada, P.A. Superfluid like stick–slip transition in capillary flow of linear polyethylene melt: 1. General features. Macromolecules 1996, 29, 2627–2632.

59. Wang, S.-Q.; Drada, P.A. Stick–slip transition in capillary flow of polyethylene. 2. Molecular weight dependence and low-temperature anomaly. Macromolecules 1996, 29, 4115–4119.

60. Ferguson, J.; Haward, R.N.; Wright, B. Flow properties of polyethylene whole polymers and fractions. J. Appl. Chem. USSR 1964, 14, 53–65.

61. Yang, X.; Ishida, H.; Wang, S.-Q. Wall slip and absence of interfacial flow instabilities in capillary flow of various polymer melts. J. Rheol. 1998, 42, 63–80.

62. Yang, X.; Halasa, A.; Ishida, H.; Wang, S.-Q. Experimental study of interfacial and constitutive phenomena in fast flow. 1. Interfacial instabilities of monodisperse polybutadiene. Rheol. Acta 1998, 37, 415–423.

63. Miehlich, R.; Gaub, H.E. Holographic pattern photo bleaching apparatus for the measurement of lateral transport processes at interfaces—design and performance. Rev. Sci. Instrum. 1993, 64, 2632–2638.

64. Ferry, J.D. Viscoelastic Properties of Polymers; John Wiley & Sons: New York, 1980. Chapter 17.

65. Aubouy, M.; Raphael, E. Structure of an irreversibly adsorbed polymer layer in a solution of mobile chains. Macromolecules 1994, 27, 5182–5186.

66. Joshi, Y.M.; Lele, A.K. Dynamics of end-tethered chains at high surface coverage. J. Rheol. 2002, 46, 427–453.

67. Chen, C.J.; Emrich, R.J. Investigation of the shock-tube boundary layer by a tracer method. Phys. Fluids 1963, 6, 1–9.

68. den Otter, J.L.; Wales, J.L.S.; Schijf, J. The velocity profiles of molten polymers during laminar flow. Rheol. Acta 1967, 6, 205–209.

69. Muller-Mohnssen, H.; Weiss, D.; Tippe, A. Concentration dependent changes of apparent slip in polymer solution flow. J. Rheol. 1990, 34, 223–244.

70. Archer, L.A.; Larson, R.G.; Chen, Y.L. Direct measurements of slip in sheared polymer solutions. J. Fluid Mech. 1995, 301, 133–151.

71. Archer, L.A.; Chen, Y.L.; Larson, R.G. Delayed slip after step strains of highly entangled polystyrene solutions. J. Rheol. 1995, 39, 519–525.

72. Wise, G.M.; Denn, M.M. Surface mobility and slip of polybutadiene melts in shear flow. J. Rheol. 2000, 44, 549–567.

73. Vinogradov, G.V.; Dreval, V.E.; Protasov, V.P. The static electrification of linear polymers extruded through ducts above the glass transition temperature. Proc. R. Soc. Lond. A 1987, 409, 249–270.

74. Lau, H.C.; Schowalter, W.R. A model for adhesive failure of viscoelastic fluids during flow. J. Rheol. 1986, 30, 193–206.

75. Anastasiadis, S.H.; Hatzikiriakos, S.G. The work of adhesion of polymer/wall interfaces and its association with the onset of wall slip. J. Rheol. 1998, 42, 795–812.

76. Hill, D.A.; Hasegawa, T.; Denn, M.M. On the apparent relation between adhesive failure and melt fracture. J. Rheol. 1990, 34, 891–918.

77. Hill, D.A. Wall slip in polymer melts: a pseudo-chemical model. J. Rheol. 1998, 42, 581–601.

78. Brochard, F.; de Gennes, P.G. Shear-dependent slippage at a polymer/solid interface. Langmuir 1992, 8, 3033–3037.

79. Brochard-Wyart, F.; Gay, C.; de Gennes, P.G. Slippage of polymer melts on grafted surfaces. Macromolecules 1996, 29, 377–382.

80. Ajdari, A.; Brochard-Wyart, F.; de Gennes, P.G.; Leibler, L.; Viovy, J.L.; Rubinstein, M. Slippage of an entangled polymer melt on a grafted surface. Physica A 1994, 204, 17–39.

81. Huseby, T.W. Hypothesis on a certain flow instability in polymer melts. Trans. Soc. Rheol. 1966, 10, 181–190.

82. Vinogradov, G.V.; Ivanova, L.I. Wall slippage and elastic turbulence of polymers in the rubbery state. Rheol. Acta 1968, 7, 243–254.

83. McLeish, T.C.B. Stability of the interface between two dynamic phases in capillary flow of linear polymer melts. J. Polym. Sci. B Polym. Phys. 1987, 25, 2253–2264.

84. McLeish, T.C.B.; Ball, R.C. A molecular approach to the spurt effect in polymer melt flow. J. Polym. Sci. B Polym. Phys. 1986, 24, 1735–1745.

85. Kolakka, R.W.; Malkus, D.S.; Hansen, M.G.; Ierley, G.R.; Worthing, R.A. Spurt phenomena of the Johnsen-Segalman fluid and related models. J. Non-Newtonian Fluid. Mech. 1988, 29, 303–335.

86. Adewale, K.P.; Leonov, A.I. Modeling spurt and stress oscillations in flows of molten polymers. Rheol. Acta 1997, *36*, 110–127.
87. Dhori, P.K.; Slattery, J.C. Common line motion. 1. Implications of entropy inequality. J. Non-Newtonian Fluid Mech. 1997, *71*, 197–213.
88. Joshi, Y.M.; Lele, A.K.; Mashelkar, R.A. Molecular model of wall slip: role of convective constraint release. Macromolecules 2001, *34*, 3412–3420.
89. Lee, L.T.; Guiselin, O.; Lapp, A.; Farnoux, B.; Penfold, J. Direct measurements of polymer depletion layers by neutron reflectivity. Phys. Rev. Lett. 1991, *67*, 2838–2841.
90. Cohen, Y.; Metzner, A.B. Apparent slip-flow of polymer solutions. J. Rheol. 1985, *29*, 67–102.
91. Rehage, H.; Hoffmann, H. Viscoelastic surfactant solutions—model systems for rheological research. Mol. Phys. 1991, *74*, 933–973.
92. Spenley, N.A.; Cates, M.E.; McLeish, T.C.B. Nonlinear rheology of wormlike micelles. Phys. Rev. Lett. 1993, *71*, 939–942.
93. Doi, M.; Edwards, S.F. *The Theory of Polymer Dynamics*; Oxford University Press: Oxford, 1984.
94. Ianniruberto, G.; Marrucci, G. On compatibility of the Cox–Merz rule with the model of Doi and Edwards. J. Non-Newtonian Fluid Mech. 1996, *65*, 241–246.
95. Mead, D.W.; Larson, R.G.; Doi, M. A molecular theory for fast flows of entangled polymers. Macromolecules 1998, *31*, 7895–7914.
96. Mhetar, V.R.; Archer, L.A. A new proposal for polymer dynamics in steady shearing flows. J. Polym. Sci. B Polym. Phys. 2000, *38*, 222–233.
97. Milner, S.T.; McLeish, T.C.B.; Lichtman, A.E. Microscopic theory of convective constraint release. J. Rheol. 2000, *45*, 539–563.
98. Islam, M.T.; Archer, L.A. Nonlinear rheology of highly entangled polymer solutions in start-up and steady shear flow. J. Polym. Sci. B Polym. Phys. 2001, *39*, 2275–2289.
99. El Kissi, N.; Leger, L.; Piau, J.M.; Mezghani, A. Effect of surface properties on polymer melt slip and extrusion defects. 1994, 52, 249–261.
100. Wang, S.Q.; Drada, P.A. Stick–slip transition in capillary flow of linear polyethylene. 3. Surface conditions. Rheol. Acta 1997, *36*, 128–134.
101. Blake, T.D. Slip between a liquid and a solid: DM Tolstoi's (1952) theory reconsidered. Colloids Surf. 1990, *47*, 135–145.
102. Hirshfelder, J.O.; Curtiss, C.F.; Bird, R.B. *Molecular Theory of Gases and Liquids*; John Wiley and Sons: New York, 1954.
103. Pearson, J.R.A.; Petrie, C.J.S. In *Polymer Systems: Deformation and Flow*; Wetton, R.E., Whorlow, R.W., Eds.; Macmillan: London, 1968.
104. Richardson, S. No-slip boundary condition. J. Fluid Mech. 1973, *59*, 707–719.
105. Dhori, P.K.; Giacomin, A.J.; Slattery, J.C. Common line motion. 3. Implications in polymer extrusion. J. non-Newtonian Fluid. Mech. 1997, *71*, 231–243.
106. Bueche, F. Viscosity, self-diffusion, and allied effects in solid polymers. J. Chem. Phys. 1952, *20*, 1959–1964.
107. Legrand, F.; Piau, J.M.; Hervet, H. Wall slip of a polydimethyl siloxane

extruded through a slit die with rough steel surfaces: Micrometric measurement at the wall with fluorescent-labeled chains. J. Rheol. 1998, *42*, 1389–1402.

108. Barone, J.R.; Wang, S.Q.; Farinha, J.P.S.; Winnik, M.A. Polyethylene melt adsorption and desorption during flow on high-energy surfaces: characterization of postextrusion die wall by laser scanning confocal fluorescence microscopy. Langmuir 2000, *16*, 7038–7043.

109. Fetters, L.J.; Lohse, D.J.; Richter, D.; Witten, T.A.; Zirkel, A. Connection between polymer molecular weight, density, chain dimensions, and melt viscoelastic properties. Macromolecules 1994, *27*, 4639–4647.

110. Leger, L.; Hervet, H.; Auroy, P.; Boucher, E.; Massey, G. The reptation model: Tests through diffusion measurements in linear polymer melts. In *Rheology of Polymer Melt Processing*; Piau, J.-M., Agassant J.-F., Eds; Elsevier: Amsterdam, 1996; 1–16.

111. Juliani; Archer, L.A. Linear and nonlinear rheology of bidisperse polymer blends. J. Rheol. 2001, *45*, 691–708.

# 5

## Sharkskin Instability in Extrusion

**Kalman B. Migler**
National Institute of Standards and Technology, Gaithersburg,
Maryland, U.S.A.

### 5.1 INTRODUCTION

The instabilities that occur in the pressure-driven extrusion of molten polymers are fascinating from the scientific perspective but troublesome and sometimes catastrophic from the industrial one. Over the past 50 years, there has been a sustained interest in the understanding and control of these instabilities. They can occur in common extrusion operations such as the manufacture of polymeric rods, tubes, sheets, and wire coating. Over the years, processors have learned to work around these processing defects by a variety of means: slowing the manufacturing rate, using materials that process better but have reduced final mechanical properties, or addition of processing additives (Chapters 8–9). From the scientific point of view, the fundamental understanding of these instabilities remains as a major challenge despite much progress in the past decade. The primary motivation for continued research is the promise of a more rational design of materials and processes to enhance manufacturing efficiency.

Several reviews have been written to summarize the state of knowledge at various points in time (1–4). In the following set of chapters, we provide a much more in-depth discussion of these instabilities, breaking them down according to their common classification, which is primarily derived from the extrudate appearance, the flow curve, and the site initiation of the flow instability.

### 5.1.1  The Flow Curve

The industrial extrusion process is typically carried out in a single- or twin-screw extruder. In the case of thermoplastics, solid pellets are fed via a metering screw into the extruder. In the first section of the extruder, the pellets are melted under the combined effects of temperature and mechanical shear/grinding forces. In the second section (if required), mixing occurs, either between two different polymers or between a polymer and a solid. In a typical third section, the screw pressurizes the molten polymer. In the fourth section, the pressurized polymer is forced to pass through a narrow constriction where it is extruded into the atmosphere (or into a water bath).

The instabilities discussed in this and the following four chapters are those that derive from the fourth section, where the pressurized polymer passes through the narrow orifice and the combined effects of polymer elasticity, fracture, and wall slip become important. To simplify the study of these instabilities, a capillary rheometer is often employed, as it typically allows for experiments over a wider range of shear rates as well as greater control over either the flow rate or the pressure.

As we shall see, there are several common themes that emerge in the various extrusion instabilities and consequent techniques for their elimination. For example, wall slip plays a prominent role in sharkskin, stick–slip, and in the use of additives to provide stable flow. Another theme is the issue of the fracture of viscoelastic melts that may have a critical role in sharkskin and gross melt fracture.

The geometry of a typical extrusion experiment is shown in Fig. 5.1. It consists of two main chambers: the barrel (reservoir) and the capillary. In a capillary rheometry experiment, the reservoir is circular in cross-section, but the capillary cross-section can be either circular or rectangular. In industrial extrusion, the geometry may be more complex, but these two regions still exist. In the case shown, the transition from the reservoir to the capillary is abrupt (with a 180° entrance angle), but in industrial extrusion, the angle is generally smaller, such as 90° to lessen dead spots due to recirculation flows (see Chapter 3) that develop in the corners.

Capillary rheometer experiments are typically conducted in one of two modes: controlled pressure or controlled flow rate. In a controlled flow rate mode, the piston is moved at a constant velocity and the pressure is measured, whereas in a controlled pressure experiments, the pressure on the piston is controlled (often through application of a pressurized gas) and the velocity of the piston (or the mass of the extrudate per time) is measured. In stable flow,

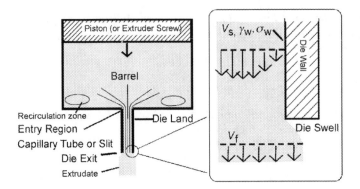

**FIGURE 5.1**  Geometry of a typical capillary extrusion experiment.

these two methods should give coincident curves but during unstable flow, they may be quite different. In industrial extrusion where a gear pump is used as the final stage, the system should be considered as controlled flow rate, but in the case where a screw is the final pressurizing element, then the conditions are neither controlled flow nor controlled pressure.

The flow curve that is derived from a capillary rheometer experiment is fundamental to our understanding of processing instabilities and thus will be independently discussed in the following four chapters. It relates the volumetric flow rate to the pressure required to generate that flow. In the case of flow through a tube, the starting point for the flow curve is the fundamental Hagen–Poiseuille equation that describes a Newtonian fluid without wall slip:

$$Q = \frac{(P_d - P_e)R^4}{8\eta L} \tag{5.1}$$

where $Q$ is the volumetric flow rate, $R$ and $L$ are the radius and length of the tube, $P_d - P_e$ is the pressure drop along the tube, and $\eta$ is the viscosity of the fluid. In terms of the flow curve, this equation can be rewritten as:

$$\sigma_w = \eta\dot{\gamma}_a \tag{5.2}$$

where the stress at the wall of the tube is given by

$$\sigma_w = \frac{P_d - P_{end}}{4L/D} \tag{5.3}$$

and the apparent shear rate at the wall is given by

$$\dot{\gamma}_a \equiv \frac{32Q}{\pi D^3} \tag{5.4}$$

and we have used $2R = D$. Thus, for a Newtonian fluid, the plot of $\sigma_w$ vs. $\dot{\gamma}_a$ is simply a straight line whose slope gives the viscosity. For such a fluid, the velocity distribution is parabolic with a maximum at the center:

$$v(r) = \frac{2v_f}{R^2}\left(R^2 - r^2\right) \tag{5.5}$$

and the velocity of the final extrudate (in the absence of die swell) is:

$$v_f = \frac{Q}{\pi R^2} = \frac{R\dot{\gamma}}{4} \tag{5.6}$$

For further details on the capillary rheometry, the reader is referred to two recent textbooks (5,6). The case of slip and shear-thinning is discussed in Chapters 3 and 6.

A typical flow curve for a linear polyethylene (PE) of relatively narrow molecular mass distribution [also known as molecular weight distribution (MWD) in the literature] is shown in Fig. 5.2. The first regime is stable flow and the surface of the extrudate is smooth. Usually (although not in Fig. 5.2) the curve is linear at sufficiently low shear stress, reflecting Newtonian flow or flow at a constant level of shear thinning. The second regime is the sharkskin instability that is discussed in detail in the remainder of this chapter. Sometimes (but not always), there is a change in the slope of the curve at the onset of sharkskin. The third regime is the stick–slip flow characterized by oscillations in the pressure and flow fields and the dramatic transitions in the

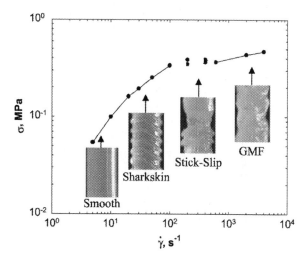

**FIGURE 5.2** Typical flow curve relating the apparent wall shear stress to the apparent wall shear rate.

extrudate (see Chapter 6). In a controlled pressure experiment, the analogous instability is known as spurt and has a different signature. The fourth regime is known as superextrusion. While not observed in the case of Fig. 5.2, it is characterized by smooth flow at a relatively high extrusion rate. The final regime is gross melt fracture (GMF), characterized by a chaotic appearance of the extrudate of large-scale amplitude (Chapter 7).

### 5.1.2 Introduction to Sharkskin

In this chapter, we focus on the sharkskin instability. Sharkskin is most commonly observed in polyethylenes of sufficiently narrow molecular weight. However, it has been observed in a variety of other materials such as polypropylene (PP), polydimethylsiloxane (PDMS), and polyisobutylene (PIB). One reason why sharkskin is of great importance is that as the extrusion flow rate increases, it is the first instability to occur and thus it is the first problem that must be solved, before one needs to worry about the other instabilities. The past decade has seen a renewed urgency in the problem of sharkskin because of metallocene-based polyethylenes. These materials allow a much greater control over the polymerization process and a much narrower molecular weight distribution can be achieved than with traditional methods. This leads to enhanced and tailored physical properties. However, because of this narrow MWD, they suffer from the sharkskin instability.

The cause of sharkskin has long been a subject of great debate within the processing and rheological communities. Debates have focused on many aspects: the boundary condition between the polymer and wall (i.e., slip, stick, or slip–stick), the reason for the rough surface, the parameters defining the onset of the instability, the reasons why sharkskin is observed in some polymers and not others. Many ideas have been advanced as to the cause of the instability. While the subject is still not resolved, much progress has been made over the past decade in reconciling the various aspects of this instability. One reason for the unresolved issues is that experiments are carried out using widely varying sets of materials (both the polymers and the wall), geometries, material conditions, etc. and results obtained in one experiment may not carry over to a different set of conditions. A second reason is the inherent complexity of the phenomena, involving concepts in polymer physics that are poorly understood, such as polymer–wall interactions under stress and transient elongational flows. In this chapter, we summarize our current understanding by focusing on aspects that appear to be universal. To the extent possible, we defer a direct discussion of the controversial issue of the actual cause of sharkskin to the end of the chapter, so that we may focus attention on the large body of experimental (and some numerical) work that has been carried out.

## 5.2 CHARACTERIZATION OF SHARKSKIN

We start our discussion by describing some of the salient features of sharkskin, beginning with several images of sharkskin, as its appearance is ultimately its most important characteristic. In Fig. 5.3 , we show a sequence of images of a polyethylene that has been extruded at $T = 177°C$ through a 1.6-mm sapphire die (7). At shear rates below those shown here, the extrudate is smooth. These optical micrographs were taken of the cold extrudate, and reflect the effects of increasing throughputs. Figure. 5.3A shows sharkskin at throughputs slightly above that where it is first detectable by eye. It is a small-scale distortion; that is, the amplitude and the wavelength are small with respect to the diameter of the extrudate. As the throughput increases, both the wavelength and amplitude increase until we arrive at Fig. 5.3E in which the amplitude of the distortions are a significant fraction of the diameter of the tube. It is not possible to tell by visual appearance alone whether this last image corresponds to sharkskin, to GMF, or to both (see Chapter 7). Additional evidence from the flow curves (see below) and possibly

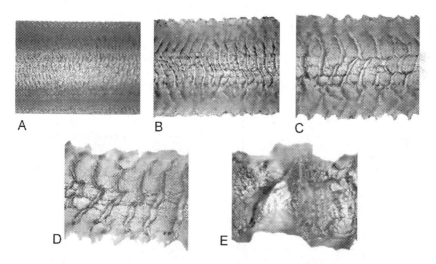

**FIGURE 5.3** Optical micrographs of cold postextrusion micrographs as a function of flow rate. The processing conditions were $T = 160°C$ and no polymer processing additive (PPA). Each image is actually a composite of two micrographs in which the side and top are focus. Relative error in throughputs is 0.05. (A) $Q = 1.0$ g/min. (B) $Q = 2.2$ g/min. (C) $Q = 3.8$ g/min. (D) $Q = 6.3$ g/min. (E) $Q = 11$ g/min. The width of each image corresponds to 3 mm. (From Ref. 7.)

flow visualization or selective coloration experiments would be necessary to make the determination.

In Fig. 5.4, we show two images of an extruded polyethylene that is imaged by scanning electron microscopy (SEM) (8). The interesting feature about these images is that the sharkskin texture appears to be comprised of microtears or microcracks in the surface of the polymer. These tears grow in wavelength and amplitude as the throughput increases. As we shall discuss in some detail below, one influential view of sharkskin is that it is caused by a tearing of the surface as the polymer exits the die. For PDMS, Piau and El Kissi have published detailed sets of sharkskin images as a function of molecular weight, die geometry, and throughput (9–12). These images are taken on-line because PDMS is a room temperature melt and the sharkskin structure will relax over time. They emphasize that as the throughput increases, sharkskin is first observed at specific locations along the extrudate. These are likely areas where there is an enhanced roughness of the die at the exit. Because the extrudate is continuously removed from the die exit, the sharkskin regions show up as "thin lines" in the flow direction. As the throughput increases, both the amplitude and the density of the lines increase. Such effects are also observed in polyethylene; in that case, it occurs when the amplitude is so small that it is difficult to observe by eye.

An obvious feature from the images is that as the throughput increases, the amplitude and wavelength increase. The quantification of this trend by Venet and Vergnes (13) is shown in Fig. 5.5 via the technique of profilometry, although the data is complemented by studies that use optical microscopy and a cutting technique to view the cross-sections. Interestingly, we see that there

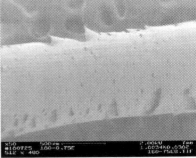

**FIGURE 5.4** Scanning electron micrographs of cold postextrusion micrographs of polyethylene as a function of flow rate. These images highlight how the sharkskin texture is comprised of microtears in the surface of the polymer. (From Ref. 8.)

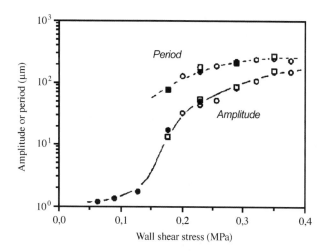

**FIGURE 5.5** Evolution of the amplitude of sharkskin and its spatial period (wavelength) as a function of wall shear stress. Comparison of three methods. Open circles = optical microscopy; closed circles = profilometry; squares = observation of cross sections. (From Ref. 13.)

is a sharp upturn in the amplitude of sharkskin at approximately 0.15 MPa. For lower shear stresses, there is still sharkskin with an amplitude of approximately 1 μm as revealed by profilometry, and their equipment does not reveal a regime in which there is no sharkskin. This is an interesting finding with respect to much sharkskin literature, which usually reports the onset of sharkskin as measured via optical microscopy. Also notable in this plot is the value of 0.15 MPa as the wall shear stress at which sharkskin is first observed. Values near 0.1 MPa as the stress onset have been observed over a range of geometries and a range of polymers. We return to this point later.

A second fundamental quantity that naturally falls out of these measurements is the frequency of the sharkskin instability; that is, how many sharkskin ridges leave the die per unit time. This quantity, $f$, can be related to the wavelength, $\lambda$, by the relation $f = BV_f/\lambda$, where $B$ is a dimensionless factor accounting for die-swell. Because $V_f$ is proportional to the apparent shear rate $\dot{\gamma}_a$, we have $f \sim \dot{\gamma}_a/\lambda$. As both these quantities increase with shear stress, it is not clear how $f$ will change. Figure 5.6 shows that in fact there is a weak dependence of frequency with the increasing shear stress (they plot frequency vs. amplitude) (13). A different behavior was found by Inn (14) who showed that for a polybutadiene sample, the frequency first increases with shear stress, reaches a peak and then decreases; however, similar to Venet and Vergnes, the wavelength continuously increases.

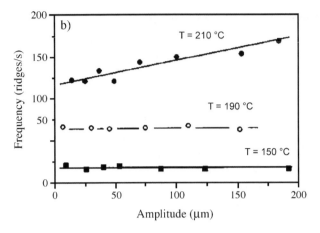

**FIGURE 5.6** The change of the frequency of the sharkskin instability with the amplitude of the defect. (From Ref. 13.)

Another observation from the various photographs of the sharkskin extrudate in this chapter (Figs. 5.3, 5.4, 5.9, 5.15, and 5.16) is that the structure is rather complex. While there is a dominant wavelength in the images, there is also structure at a smaller length scale. An interesting attempt to characterize this topographic complexity was carried out by Tzoganakis et al. (15), whereby they carried out a fractal analysis of the topography. Fractal objects possess a similarity of scale, whereby if one were to magnify a portion of the surface, it would look similar to the original portion, except for the change in length scale. Essentially, they measured the fractal dimension, $D$, of the surface that is a measure of how the surface fills space because of its irregular, multilength scale nature. While there is not a physical reason advanced as to why this may work for sharkskin, the surface topography can fit to the power law characterizing a fractal geometry of the form $r = N^{-1/D}$, where $D = 1$ would indicate a smooth structure and $D > 1$ indicates sharkskin. A significant increase in $N$ was observed at $\sigma \approx 0.15$ MPa.

While the above phenomenology related to the sharkskin instability is well established (16), the underlying cause of sharkskin, and hence the best practices for its elimination, are still the area of active debate. The debates have revolved around the following five interrelated issues, which must be resolved before we have a full understanding of the phenomena: the flow boundary condition, the site of initiation, the cause of the surface roughness, the kinetics of the instability, and the effects of molecular parameters. We sequentially focus on these issues in the following sections.

## 5.3 FLOW BOUNDARY CONDITION

### 5.3.1 High-Energy Walls

A central theme in sharkskin research is the proper understanding of the relationship between the flow at the die wall and the instability. Does the polymer stick, slip, or undergo an alternating stick–slip behavior while sharkskin is occurring? As Shaw remarked, "For this and other reasons, the slip at the wall of polymeric melt has become one of the most important and controversial issues in rheology (17)." Understanding this question is crucial in attempts to suppress sharkskin by various means, such as through additives, lubricants, treated wall surfaces, and others.

As discussed in Chapter 4, there are two primary methods to assess slippage in capillary flow: those based on the measurement of pressure and throughputs during capillary extrusion and those based on direct measurements of polymer velocity. The concept that the polymer does slip during sharkskin was first suggested by Ramamurthy (18) in an influential paper in which he further asserted that sharkskin is initiated by this slippage. The experimental evidence presented in favor of slippage during sharkskin is twofold. First, it was reported that at the onset of sharkskin, there is a "kink" in the flow curve, a sudden decrease in the slope of the curve as throughput is increased. This kink was interpreted by Kalika and Denn (19) as an indication of slippage; for a given wall shear stress, the overall polymer flow is greater because of the slippage.

The second evidence comes from quantitatively modeling the slip by a Mooney analysis by studying the effects of $L/D$ on the flow curve. As discussed by Archer (Chapter 4), this evidence supports the idea of strong slippage in the sharkskin regime. Again, the conclusion is the slippage in the die wall coincides with sharkskin, and further that the onset of slippage occurs at slightly lower shear stress than the onset of sharkskin.

However, there are several researchers who have drawn the opposite conclusion from the rheological data—that the polymer does not slip in the die land during sharkskin. The data is drawn from the same two features as before: the question of the existence of a kink and also the results of a Mooney analysis. El Kissi and Piau (9) carried out rheological measurements on a PDMS sample as well as a PE sample. In the PE sample, the data is corrected for both the entrance pressure effect, as well as for the effect of the change in polymer viscosity as a function of pressure (in this model, the polymer viscosity decreases with increasing pressure). In their work, they then plot the "true" wall shear stress as a function of apparent wall shear rate. They observe sharkskin at a wall shear stress of approximately 0.17 MPa. The authors draw two conclusions from this plot: first, that there is no clear kink in the data; second, the data is better described as having a continuous change in slope as shear stress is increased. This is then opposite to the conclusion drawn by Ramamurthy (18) and Kalika and Denn (19). Other authors have also presented data in which there is no obvious change in slope during sharkskin (14). Note that Fig. 5.2 shows a gradual change in slope near the onset of sharkskin.

The second conclusion drawn by El Kissi and Piau (9) is that the shear stress values as a function of apparent shear rate from different values of $L/D$ superpose on the same curve. The implication from the perspective of the Mooney analysis is clear—that within the experimental uncertainty, there is no slippage. Wang and et al. (20) have also presented data in which stress curves from different $L/D$ systems do coincide and thus indicate that there is no slippage. Thus, the kink that has appeared in some experiments does not appear to be a universal feature of the flow curve. Further, no satisfactory understanding of the disagreement regarding the existence of slippage in the mechanical measurements exists at this point.

As the rheological measurements are inconclusive on the question of whether or not sharkskin occurs in the presence of slippage, several researchers have carried out studies using flow visualization. Experimentally, these studies have utilized both laser Doppler velocimetry (LDV) as well as direct particle tracking velocimetry. The conclusion of most of these studies is that in high surface energy materials (steel, glass, sapphire), the polymer sticks at the wall during sharkskin melt fracture. As described below, a recent example is provided by Migler et al. (21) in which the flow of a metallocene linear low-density polyethylene (LLDPE) sample was extruded through a sapphire capillary tube (see Chapter 8). Other direct measurements have also reported that sharkskin occurs during a stick boundary condition in the die land (12,22,23). However, there has been a recent preliminary report of a measurement via LDV in which a LLDPE slips during conditions of sharkskin (24).

There have also been suggestions that the boundary condition at the exit of the die is different from that in the die land. For example, the polymer may be sticking upstream but slipping near the exit (25), or it could be sticking upstream and undergoing a stick–slip oscillation at the exit (26,27). Thus experiments were carried out to examine the velocity field near the exit. It was found that when the velocimetry is conducted within 20 μm of the exit, there is still a stick boundary condition (7).

There have been two experiments to examine the behavior of tracer particles right at the wall as close to the exit as possible. The first report is by Inn et al. (23) who used polybutadiene and found smooth motion of particles near the exit, although the distance from the exit was not specified. Figure 5.7 by Migler et al. (7) in PE shows the motion of particles as a function of time was tracked to within 20 μm of the exit during conditions of sharkskin. In this case, the velocity profile at the wall was first obtained, and the radial distance of each particle from the wall was determined from a fit of the velocity profile.

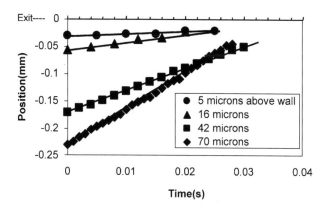

**FIGURE 5.7** Position–time curves for single particles as a function of the axial position in the die land but near the exit. The different curves correspond to particles of different radial position. The time lapse from the first measurement at $t = 50$ sec to the last one at $t' = 25$ msec corresponds to approximately one cycle of sharkskin. The height of the particle from the wall is determined by flow curve fitting. Conditions: $Q = 52.2$ g/min. (From Ref. 7.)

The primary result is that there is no disturbance of the velocity field found to within 20 μm of the exit. Thus the idea that there is a stick–slip of the polymer at the wall was not observed under these conditions. Thus it shows that neither slip nor stick–slip boundary is a necessary condition for sharkskin.

However, as the temperature was lowered (which favors slippage), an oscillation of the flow field in the exit region was indeed observed. Figure 5.8 shows position vs. time plots for particles near the wall at several flow rates. Strong sharkskin prevails in the three flow rates shown. The particles are inside the tube and the position $x = 0$ corresponds to the exit of the tube— similar to Fig. 5.7. At a flow rate of 4.4 g/min, a slight dip in the curve was observed at $x = -0.75$ μm. At higher flow rates, two or three inflections in the flow curve are seen. By simultaneously observing the position of the particle inside the die and the sharkskin kinetics outside the die, it was found that the period between the inflections corresponds exactly to the period of sharkskin. Note that the severity of the inflections increases as the particle approaches the exit.

## 5.3.2 Wall Material, Surface Energy, and Slippage

Thus far, we have discussed the extrusion of polymers through wall materials with a high surface energy, but it is well established that changing the wall material to one with lower surface energy can reduce or eliminate sharkskin.

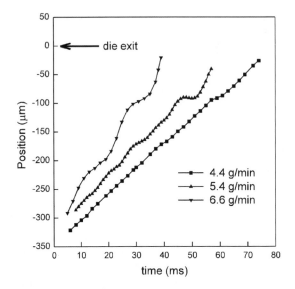

FIGURE 5.8 Position–time curves for single particles as a function of the axial position in the die land but near the exit. The three curves correspond to different flow rates. The period of the oscillation corresponds to that of the resulting sharkskin. (From Ref. 7.)

This occurs when a wall material allows slippage of the polymer against it. If everything else is held constant and one wall material exhibits less shear stress for a given throughput, then that material is allowing slip at its surface. This idea has been extensively used in understanding sharkskin.

The most widely used example of this is when a processable fluoropolymer is added to a polyethylene resin at low mass fraction. The fluoropolymer coats the wall and causes the polyethylene to slip against the fluoropolymer, and sharkskin is eliminated (21) (Chapter 8). A related example concerns the use of a die with a poly(tetrafluoroethylene) (PTFE) coating where it was shown by velocimetry that it allows massive slip (in the case of PB) (12). Additionally, there is a reduction in wall shear stress. In terms of the effort to understand sharkskin at a fundamental level, the ability to change the flow boundary condition while keeping everything else the same provides opportunities to test various theories on the origin of sharkskin that will be discussed later.

We now return to the question of whether slippage in the die land occurs during sharkskin by examining the behavior of low energy surfaces. There are several examples in the literature in which the polymer/wall interface is modified to allow slippage, and the onset of sharkskin is not eliminated but rather its onset is delayed to higher shear rates. The importance of this finding has not been emphasized in the literature. It shows that sharkskin can also occur under conditions of slippage, as well as under conditions of stick.

One example is from the work of El Kissi et al. (10), where a low-energy single layer coating was applied to the surface. At a given throughput, it was found that the treated surface causes an approximately 10% reduction in shear stress. Significantly, the treated surface allowed an increase in throughput by a factor of 2 before the onset of sharkskin. The second example is from Wise et al. (28), who showed differences in the onset of sharkskin between copper steel and Teflon. The Teflon showed that the critical throughput for the onset of sharkskin is increased by a factor of about 10 over the metallic walls.

Migler et al. conducted a direct measurement of slippage during extrusion on a low surface energy wall (7) where the slippage of the polymer at the wall at the exit was directly measured along with the occurrence of sharkskin. In that case, weak sharkskin was observed at their highest reported throughputs. The results presented in this chapter on slippage now show that sharkskin has been observed under conditions of stick, of slip, and of stick–slip depending on the wall material, the throughput, and the temperature.

There is a second example in which changing the wall material was seen to delay or eliminate sharkskin. Returning to the influential paper by Ramamurthy (18), in addition to the claim of slippage during sharkskin, he tested numerous ferrous and nonferrous metals in a blown-film line and a capillary rheometer for their susceptibility to sharkskin. In the case of a blown-film line, the first observation is that the appearance of the polymer extrudate showed time-dependent behavior (with the exception of chrome-plated steel). It took as long as 6 hr for the extrudate to reach its steady state appearance. In the case of copper and beryllium copper, the metal showed signs of corrosion. Remarkably, in the case of $\alpha$-brass, it was observed that the steady state extrudate was free of sharkskin after an induction period in which the brass changed its color. This result was confirmed by Ghanta et al. (29) who also showed that there is a reduction of shear stress for a given throughput. This last finding strongly indicates that the brass causes the polymer to slip after an induction period and similar to what is known about the fluoropolymer coatings; this slippage leads to the disappearance of sharkskin. In Ramamurthy's work (18), he conjectured the opposite—that the brass causes the polymer to stick at the wall. However, based on the results of Ghanta et al., this conjecture can now be said to be incorrect. In fairness to Ramamurthy, he did not do the shear stress measurements on the blown-film line; he may have altered his conjecture had he seen those results.

## 5.4   DEPENDENCE OF SHARKSKIN ON ASSORTED PARAMETERS

To understand and control sharkskin, there are numerous variables that the experimentalist can control. We have already discussed the effects of material throughput (i.e., apparent wall shear rate), but here we review what has been learned about several others, such as temperature, die geometry, molecular mass and its distribution, and branching content.

### 5.4.1   Die Geometry

The dependence on the die geometry can be quite important for the onset of sharkskin and has been examined by several groups. Whereas for the purposes of a Mooney analysis one tends to work at constant $L/D$, for the case of sharkskin, it makes sense to independently vary $L$ while keeping $D$ fixed, and vice versa. Howells and Benbow (30) observed that as die radius increases, the apparent shear rate at the onset of sharkskin decreases according to $\dot{\gamma}R \approx A$, where $A$ is a constant. This was called the "linear velocity criterion" because according to Eq. 5.6, $\dot{\gamma}R \approx 4V_f$. Their results were obtained over a somewhat limited range from $R = 0.5$ to $1.5$ mm, although this is the typical range for polymer processing. They also found no systematic dependence on the die entry geometry. Another finding was that in the case of PDMS extrusion, the sharkskin wavelength increased with increasing die radius at constant $A$.

Upon varying $L$ for constant $D$, it was found by Venet and Vergnes (13) that the severity of sharkskin markedly increased upon going from an orifice die, to a capillary die of finite length. However, they found little systematic effect when changing $L$ from 5.5 to 22 mm at constant value of $D$. Fujiyama and Inata (31,32) also found little variation at constant $D$ ($D = 1$ mm) with $L$ ranging from 5 to 20 mm. This result contrasts with that of Moynihan (33), who found that there is a decrease in the severity of sharkskin as the length increases. Perhaps these two results can be reconciled by noting that in the experiments where no effect of length was found, $L/D$ ranges from 4 to 20, whereas in the case of Moynihan, the $L/D$ values vary from 12 to 75. It is generally considered that fully developed flow occurs when $L/D > 10$, thus perhaps the difference between the two cases is that there is fully developed flow in the latter.

In the case of a series of dies for which $L/D$ is held constant, both $L$ and $D$ vary simultaneously; because increasing $D$ increases the severity of sharkskin and increasing $L$ decreases it, these two effects may cancel each other out. Fujiyama and Inata found no effect on the onset of sharkskin when $L/D = 10$ for $D$ ranging from 0.5 to 2.0 mm. In the case of Wang et al., for a die of $L/D = 15$, it was found that the sharkskin amplitude and wavelength scaled with $D$. From the work of Venet, one can conclude that the effects of $L$ are weak at this $L/D$ value; thus we expect the effects of $D$ to dominate, as observed by Wang. Previously, it was mentioned that the critical stress for the onset of sharkskin is near 0.1 MPa. From these results, we see even for one sample, the onset of sharkskin can vary as a function of geometry; thus this value of 0.1MPa should be thought of as an order of magnitude estimate rather than an absolute number. There are other ways of varying the geometry; for example, it was found by Rutgers et al. (34) that employing a rounded exit reduces the severity of sharkskin by approximately a factor of 2.

## 5.4.2  Temperature

It was reported by Howells and Benbow (30) that there are two potential easy cures for sharkskin that involve manipulation of temperature fields. For a given throughput, one may either increase or decrease the temperature! They pointed out that it might only be necessary to change the boundary conditions at the exit. Clearly, the temperature dependence merits further consideration.

An important contribution to this area was made by Wang et al. (35) who considered the effect of Mw and temperature on the frequency of sharkskin and found that the frequency scales in the same fashion as the molecular relaxation time, $\tau$. As this time is determined from classic linear viscoelastic measurements, this affords an important connection between rheology and the sharkskin. It is well known that has a strong dependence on Mw, $\tau \sim M^{3.4}$, and this is reflected in the reported time scales for sharkskin.

On the other hand, it is well known that the amplitude of sharkskin depends strongly on the Mw. As discussed by Shaw (14), the relation between $f$ and $\tau$ helps understand the differences between PIB, which has a character-istic frequencies of 1 $sec^{-1}$ as compared to PE, which has a characteristic frequency closer to 50 $msec^{-1}$. Returning to the case of PDMS in which the sharkskin cycle occurs over a relatively long time scale, these low-frequency instabilities are caused when the Mw of PDMS becomes extremely high. El Kissi and Piau examined the sharkskin behavior in a wide variety of PDMS samples. While a specific relation between $f$ and $\tau$ is not discussed, it is clear that the frequency of sharkskin rapidly increases with increasing Mw (9–12).

The same relation between $f$ and $\tau$ also holds when considering the effects of polymer temperature (36). It was found that both scale in a WLF-like fashion with temperature, over a reasonably broad range from 160°C to 240°C (20). There are two points to note. First, that there is a linear relationship between the time scale determined by $f$ and $\tau$ and second that they are both of the same order of magnitude.

If we examine the effect of sharkskin amplitude with temperature, then it is clear that at higher temperatures, the amplitude decreases. At higher temperatures, this is reasonable as the viscosity and elasticity of the melts decrease. Clearly, in fact, this has been suggested as a potential method industrially to eliminate sharkskin. Cogswell had mentioned that die tip heating is employed as a method to diminish sharkskin. While the shear rate at the onset of sharkskin increases with temperature, the effect on the shear stress on temperature is much weaker, and some have reported that it is independent. The following chapter contains a more extensive discussion of the effects of temperature on the onset of sharkskin, as its temperature dependence and that of the stick–slick instability are similar.

The existence of smooth extrudates at low temperatures has been explored by several groups and remains somewhat mysterious. Kolnaar and Keller (37) found a narrow temperature window from 150°C to 152°C in which the flow is smooth and a corresponding drop in the measured pressure (at constant flow rate). Pudjijanto and Denn (38) found a window for LLDPE from 139°C to 146°C in which the stick–slip instability disappears (the stick part of the cycle had sharkskin) and enhanced processing were found over a wider temperature range by Perez-González and Denn (39). Their interpretation of these results is of a flow-induced mesophase, whereby the PE adopts a liquid crystalline-like structure. Recently, Santamaria et al. (40) examined the behavior of a large number of single site metallocene polyethylenes. At temperatures above the melt temperature, but below that typical of industrial processing, they were able to obtain smooth extrudates at reasonably high shear rates for all their samples, including those suspected of having long-chain branching. At higher shear, gross melt fracture or crystallization occurred. They determined the upper temperature limit of this phenomenon and found that short side chain branches promoted sharkskin. Cogswell also mentions that tie tip cooling is a patented method to reduce sharkskin.

### 5.4.3 Molecular Structure

Howells and Benbow (30) compared the susceptibility of a fractionated narrow MWD polyethylene with its parent material, in the case where both materials had approximately equal viscosities. They found that the material with the narrow MWD had significantly more severe sharkskin. The narrow MWD sample is expected to exhibit less shear thinning and less elasticity than the broad MWD parent. They also compared two silicone rubbers, and found that the one which exhibited stronger sharkskin had a flow curve that was more Newtonian, and exhibited less elastic recovery.

Kazatchkov et al. (41) analyzed two detailed series of LLDPEs; one was a series that kept the Mw relatively constant while varying the MWD by a wide range; and the second series did the opposite; it held MWD constant (at ~3.8) while varying Mw. Examining the rheology, they found, for example, that the greater the MWD, the greater the relaxation time, the zero shear viscosity, and the shear thinning. The linear viscoelastic moduli of the polymers were used to determine their linear viscoelastic spectrum. Through the fitting process, one determines the relaxation spectrum, how the modulus decreases with time. They found that at shorter times, the greater the MWD, the *less* the elastic modulus. Thus these materials are actually less elastic at short times, and it is the short time behavior that may be critical because of the sudden extensional pulse that occurs to the material as it exits the die (discussed below). They also find that increasing MWD increases the shear rate at which sharkskin occurs, with a dramatic improvement in processability at MWD = 9.

A second example concerns polypropylene, which typically has a broad molecular weight distribution and sharkskin is generally not reported. However, Fujiyama and Inata recently prepared narrow distribution polypropylenes and did indeed observe sharkskin for their two narrow samples (Mw/Mn = 2.1 and 3.1), but not for the two broader polypropylenes (Mw/Mn = 7.0 and 6.6), which only showed GMF. The data shows enhanced shear thinning in the broad MWD. Further details on the effects of MWD on sharkskin and a comparison to the stick–slip instability are offered in the next chapter.

Another approach is to systematically vary a material property, and explore its effect on sharkskin. Yamaguchi et al. (8) recently reported experiments on a series of metallocene copolymers of similar MWD, but that vary in their content of copolymer (1-Hexene). The measured rubbery plateau modulus (and entanglement density) is a decreasing function of the 1-Hexene content. The stress at the onset of sharkskin is reported to then be an increasing function of entanglement density.

## 5.5  SITE OF INITIATION

Different models of sharkskin may posit different physical locations where the flow first becomes unsteady (die entrance, die land, or die exit). Thus an important piece of the puzzle is to know the site of initiation of the instability. It is now widely accepted that sharkskin initiates in the exit region of the die (1). In fact, it has been suggested that the definition of sharkskin should include the fact that it is initiated at the exit. Perhaps the most dramatic evidence comes from experiments where modifications of flow boundary condition at the capillary exit tube were performed. Moynihan et al. (33)

modified the last 1 mm of a capillary die by dissolving a thin fluoropolymer layer onto it. It was found that this was able to eliminate sharkskin in a LLDPE. In the case of polybutadiene, Inn et al. (14) modified the exit by applying a soap solution with a cotton swab to the polymer–wall–air interface during extrusion as shown in Fig. 5.9. Remarkably, this was able to eliminate sharkskin, although the degree of back penetration of the solution into the capillary tube must have been extremely small. This observation points to the importance of the flow/stress exactly at the exit corner.

The opposite experiment was performed by Dhori et al. (42). In this case, a capillary tube is fully coated with a fluoropolymer layer, which results in the disappearance of sharkskin, as expected. Then, they damage the coating at the very exit to the capillary by machining off the edge of the tube by a fraction of a millimeter. In this case, the polymer slips everywhere except the exit. They find that sharkskin returns, although its magnitude as a function of throughput remains small. In some cases, they report that the sharkskin from the damaged die eventually reappears, presumably because of downstream migration of the undamaged fluoropolymer into the exit region. Thus if the

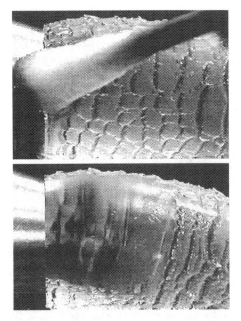

**FIGURE 5.9** Elimination of sharkskin in polybutadiene extrusion via application of a soap solution with a cotton swab to the extrudate–wall–air interface. (From Ref. 14.)

die land is slippery and the exit is sticky, one finds sharkskin; conversely, if the land is sticky and the exit is slippery, one finds a smooth extrudate.

## 5.6   MEASUREMENT OF LOCAL STRESS FIELDS VIA BIREFRINGENCE

As many explanations for sharkskin center on the stress fields in the exit region, flow birefringence measurements are a potentially powerful tool to map out these fields. Because of its spatial and temporal specificity, it has been applied by a number of researchers (12,22,43,44). It has been particularly useful in confirming the location of regions of high stress, as well as in determining the temporal stability of the flow fields.

Flow birefringence is based on the well-known phenomena in which the molecules orient in a stress field, making the sample locally birefringent. The key to using birefringence is the stress optical law:

$$\Delta n = C\sigma \tag{5.7}$$

where $\Delta n$ is the birefringence and $C$ is stress optical coefficient, which is available for different polymers in the literature. Thus one can map and visualize the stress field throughout the sample. Here the setup involves a monochromatic lamp, a polarizer, and analyzer that are placed in conjunction with a quarter wave plate on both sides of the sample, and the transmitted light is recorded with a camera. In the resulting images, one sees a series of bright and dark fringes. An isocontour indicates a continuous region under the same level of birefringence. Through the stress optical law, moving from a dark to light region indicates an increase in stress by a fixed amount. Figure 5.10 by Vinogradov et al. (26) shows the extrusion of a polybutadiene sample, showing the concentration of the stress lines in both the entry and exit regions.

Several researchers have utilized birefringence to examine the temporal stability of the stress field during conditions of sharkskin and to compare the stability at the exit with that upstream (in the die land). In a study by Barone and Wang (44), the laser beam was passed through the sample in the flow gradient direction, thus the collected signal is integrated from the high stress region near the wall to the low shear stress region in the center of the slit. They observed large oscillations in time of the recorded intensity at the exit region whereas the flow upstream is stable. The frequency of these oscillations matched the time period of the sharkskin instability.

A similar conclusion was drawn by Legrand and Piau (22), where they implemented techniques pioneered by G.G. Fuller in which the measurements are carried out using a laser beam that is modified by a photoelastic modulator. If a focused laser traverses the rectangular direction along the wide axis perpendicular to flow, then in principle, the state of stress is con-

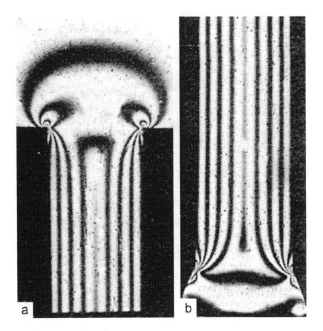

**FIGURE 5.10** Measure of the stress field via the birefringence technique during flow through a slit die under conditions of sharkskin. The left figure is the upstream region and the right figure is the exit region. The stress lines are stable with time in the die and flow is downward. (From Ref. 26.)

stant along the path of the beam and one can measure the local shear stress and first normal stress of the material at different locations. For stresses slightly above that encountered in sharkskin, they reported stress oscillations to occur only at the exit; both the velocity fields and the stress fields were temporally constant upstream of it. While they reported shear stress measurements at the wall near the exit, they indicated that these measurements were unreliable.

In contrast to the above, Inn et al. (23) found that the stress field was smooth in time in the immediate vicinity of the exit in the case of a polybutadiene sample exhibiting sharkskin. Perhaps these contradictory results can be reconciled by noting that Migler et al. (7) found that the velocity field can be either smooth or oscillating in the exit region depending on the relative stress and temperature.

Upon modification of the flow boundary condition to a slip boundary condition by using a PTFE die, Piau et al. (12) found that the stress at the exit was reduced; however, there was still a concentration of stress lines at the exit lip. Barone and Wang (43) showed that the integrated stress at the exit was reduced in the presence of a slip boundary condition for a given level of apparent wall stress as measured through birefringence. However, it was found that with the slippery boundary condition, one could extrude at a higher throughput rate; in fact, for a sufficiently high rate, the integrated exit stress under conditions of no sharkskin can exceed that in the no-slip condition where sharkskin occurs.

## 5.7  NUMERICAL SIMULATION OF STRESS AND VELOCITY FIELDS

Numerical simulations augment the experimental measurements because they are able to simultaneously track the stress and velocity fields. At present, the work does not attempt to simulate the kinetics of sharkskin, but rather to simulate these fields below the stress levels of sharkskin when the flow is still steady. By analyzing the resulting stress fields, progress can be made by determining the magnitude and shape of the stress fields, particularly in the exit region. The ultimate goal of the simulations is to predict the onset of the sharkskin and its dependence on material properties, die geometry, etc. The simulations are subject to the difficulties outlined in Chapter 2.

There are two groups that have recently carried out numerical simulations that are correlated to experimental results. In both cases, the polymer is a LLDPE and the simulations are based on finite element calculations. When the simulations use a no-slip boundary condition at the wall, there is the problem of a stress singularity at the exit lip. Thus they cannot compute the stresses at the wall but they compute it along a streamline that is a small distance from the wall (roughly 50 μm for a 1-mm diameter die). Naturally, an important parameter in the simulations is the size of the computational mesh near the wall.

In the case of Rutgers and Mackley (34,45,46), a constitutive model (BKZ) is used to interpret the rheological data so that a discrete relaxation spectrum (eight relaxation times are employed) and the damping factor can be determined (see Chapter 2). These relaxation times and corresponding moduli are input into a commercial software package. The boundary conditions used in the simulations are fully developed flow 25-mm upstream from the slit wall; no-slip boundary condition (for a steel wall); and an 80-mm-long free surface past the die exit. Confidence in this methodology is based on satisfactory predictions of the simulation with experimentally measured quantities such as the principal stress difference in the centerline (from birefringence measurements), pressure difference, end pressure losses, die swell, and centerline velocities.

Figure 5.11 shows the results of a simulation of the extrusion of a LLDPE through a slit die (46). A number of critical fields are simultaneously tracked for a streamline that is located 50 μm from the wall. The simulation covers both the entry region (at 28 mm) and the exit region (at $x = 36$ mm). An important result from these simulations is the existence of sharp peaks in all stress fields at the die exit: $N_1$, the principal stress difference, and the shear stress. Correspondingly, there are sharp peaks in the flow field as seen by examining the shear and the extension rate. Examining the shear rate, for example, we see a constant value in the die land $28 < x < 36$ but at its peak near the exit, it is nearly double the value in the land. The extension rate is zero

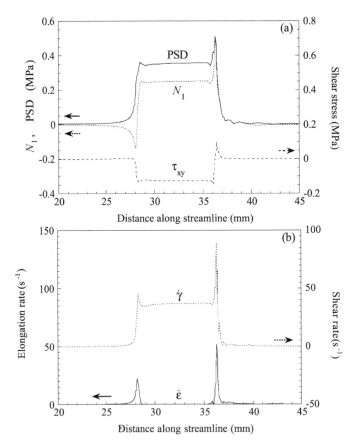

**FIGURE 5.11**   The tracking of stresses (a) and rates (b) along the streamline at 50 μm from the wall. $N_1$ = first normal stress; PSD = principal stress differences. (From Ref. 46.)

in the die land because the velocity at the wall is constant, but briefly jumps to a value nearly equal to the shear rate in the die land as it exits the die. Under conditions that first create sharkskin, they find a principal stress difference at the exit of approximately 0.6 MPa. When they compare the stress fields that are produced in the no-slip vs. the slip boundary condition, they find that the slip condition still produces a peak in stress at the exit, but its amplitude is about six times smaller than in the no-slip condition.

In the work by Venet and Vergnes (47), the Phan-Thien and Tanner model is used with five relaxation times. Qualitatively similar findings to those mentioned in the preceding paragraph were found, but the size of the jumps at the exit is relatively smaller. For example, they find a peak tangential stress of approximately 0.15 MPa at the exit under conditions that first produce sharkskin. They also find a "traction zone," a region where the polymer undergoes rapid acceleration as it exits the die. As the flow rate is increased, the traction zone remains limited (from 30 to 175 μm for a 1.39-mm die). To make the simulations relevant to the sharkskin problem, one must determine what causes sharkskin, and what determines its onset.

## 5.8  WHAT DETERMINES THE ONSET?

It is intuitively clear that there is a critical value of a physical parameter involved with capillary extrusion that becomes exceeded at the point in the flow curve at which sharkskin is first observed; however, it is uncertain which physical parameter. We explore several candidates such as wall shear stress and rate, exit extensional stress, and exit extensional rate, and a recently proposed deformation rate.

It has been previously noted that sharkskin occurs when the wall shear stress is in the range of 0.5 to 2 MPa. Because of this observation, the wall shear stress has been proposed as the parameter that coincides with the onset of sharkskin. However, while the wall shear stress certainly correlates with the onset, there are several reasons that it is not the determining parameter. Intuitively, as the wall shear stress is significantly lower in the land region than near the exit, the value of the wall shear stress itself is not directly relevant to a stress that causes the sharkskin. Second, as reported earlier, some (but not all) studies have shown the wall shear stress at the onset of sharkskin varies as a function of the tube's $L$ and $D$. Third, Migler et al. (21) reported experiments in which the wall shear rate was directly measured through velocimetry for two cases; one in which there was a stick boundary condition and one in which there was slip (via a fluoropolymer coated surface) condition. In the case of stick, there was sharkskin at all reported shear rates, whereas for slip, there was no sharkskin. Because one can have sharkskin in the stick case at weak shear rate and no sharkskin in the case of slip, although the shear rate is much higher, then wall shear rate is not the determining parameter. Because the materials and geometry are the same (the fluoropolymer wall thickness is less than 1 μm), this result is also true for wall shear stress.

A promising route is to measure the relevant parameters in the exit region where according to the birefringence, numerical simulations, and theory, there are peaks in the extensional stress and strain. Again, we compare the slip and no slip conditions. As pointed out by Cogswell (48), an important parameter to consider is the amount of material stretching at the exit. In the no-slip case, the material near the wall must accelerate as it exits, from a velocity of

$$V_r = \frac{4V_f \delta R}{R} \tag{5.8}$$

inside the die (where $\delta R$ is the distance from the wall), to the final extrudate velocity $V_f$ over a distance $\Delta x$. Thus the material is stretched as it accelerates, so that if one were to paint a small rectangle in the polymer at the wall just inside the die, it would deform upon exiting. The amount of deformation is dependent on the width of the rectangle (in the radial direction): The thinner it is, the more it becomes stretched, and for a stick boundary condition the answer diverges in the limit of an infinitely thin rectangle. We consider the ratio $T = V_f/V_\delta$, where $V_\delta$ is the velocity of a region 20 μm from the wall. Figure 5.12 shows a plot of $T$ as a function of throughput for the cases with and without PPA (21). The remarkable result is that the presence of slip causes a reduction in the total material deformation by an order of magnitude. This result shows why the shear stress as determined by birefringence is lower upon use of a slippery boundary condition. The deformation is relatively insensitive to the total material throughput. Also, in the case of sharkskin, $T$ represents an "apparent deformation" because the true deformation is reduced by the

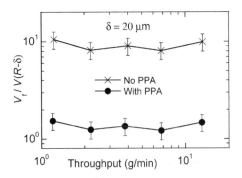

**FIGURE 5.12** Deformation with and without PPA during sharkskin.

sharkskin. Note that the above quantity $T$ neither take into account the rapidity of the deformation nor assess the relative importance of viscous and elastic effects.

A second fundamental parameter describing the exit flow is the peak extensional strain rate. This was previously discussed (Fig. 5.11) in regard to numerical modeling. Through particle tracking velocimetry, the velocity at the wall (inside the die) and the velocity at the air–polymer interface just outside the die were compared (7). The extensional strain rate was estimated as $\dot{\varepsilon} = \Delta v / \Delta x$, where $\Delta v$ is the difference between the polymer velocity at the air–polymer interface just outside the die to that at a point just inside the die (but near the wall). The peak extensional rate exceeds the apparent wall shear rates. As shown in Fig. 5.13, there is a result similar to that obtained in the case of wall shear rate. One can have sharkskin in the stick case at weak shear rate and no sharkskin in the case of slip, although the shear rate is much higher. This is reasonable because the extensional strain rate should be a linear function of wall shear rate (in the Newtonian case).

Motivated by Figs. 5.12 and 5.13 and the fact that extensional stress is a function of both the strain rate and the time of deformation, Migler et al. (7) defined a phenomenological parameter that incorporates both the total material deformation and the extensional strain rate. Because sharkskin initiates in the exit region, the measured value:

$$T = \frac{V_+}{V_-} \tag{5.9}$$

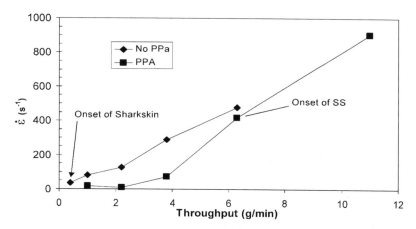

**FIGURE 5.13** Extensional strain rate at the polymer–air interface as the material exits the tube. We show cases with and without PPA. The onset of sharkskin is denoted by the arrows. (From Ref. 7.)

is used, which is the deformation in the immediate vicinity of the exit. If we assume that the particle undergoes uniform acceleration as it goes from $x_-$ to $x_+$, then the time it takes the particle to move between these two points is $\Delta t = 2\Delta x / (V_+ V_-)$. Then the deformation rate is defined by:

$$\dot{T} = \frac{T-1}{\Delta t} \tag{5.10}$$

$T - 1$, rather than $T$ is utilized in the expression for $\dot{D}$, so that the deformation rate would be zero in the case of a pure plug flow through the exit. Using Eqs. (5.9) and (5.10), and the time for a particle to move by $\Delta x$, one has:

$$\dot{T} = \dot{\varepsilon}_{max}(T+1) \tag{5.11}$$

In the case of a stick boundary condition, at the onset of sharkskin, one is concerned with the case $\delta R \ll R$ so that $T - 1 \approx T$ and then $\dot{T} = \dot{\varepsilon}_{max}T$ so that the deformation rate is the product of the extensional strain rate and the extensional deformation.

In Fig. 5.14, we compare the values of $\dot{T}$ for the cases with and without PPA. Note that the addition of PPA causes a dramatic reduction in $\dot{T}$ for a given mass throughput. Significantly, the onset of sharkskin as noted in this plot occurs for similar values of $\dot{T}$. This is the first measurable quantity that we have found, which is comparable at the onset of sharkskin as the boundary condition is changed from stick to slippage.

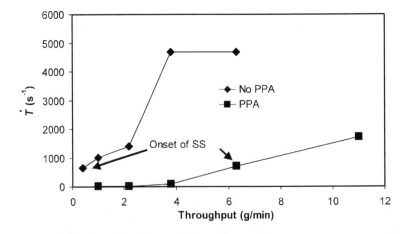

**FIGURE 5.14** Reconfiguration rate as a function of the flow rate for cases with and without PPA. The onset of sharkskin is denoted by the arrows. (From Ref. 7.)

## 5.9  KINETICS OF SHARKSKIN

It is fascinating to visually observe the sharkskin kinetics that occurs just outside the extrusion die. More importantly, any description for the cause of sharkskin must be consistent with the kinetics as observed visually. For the case of polymers with a longer relaxation time, this can be accomplished in real time with a simple close-up lens (14). For the case of shorter relaxation times, such as polyethylene, one must employ a high-speed camera because the typical time of the sharkskin cycle is approximately 20 msec (see Fig. 5.15). Despite the different time and length scales, there is a distinct similarity in the reported descriptions of sharkskin between the systems of polyethylene, polyisobutylene, and PDMS. All descriptions describe some kind of splitting (or peeling) of the polymer into two layers near the exit lip.

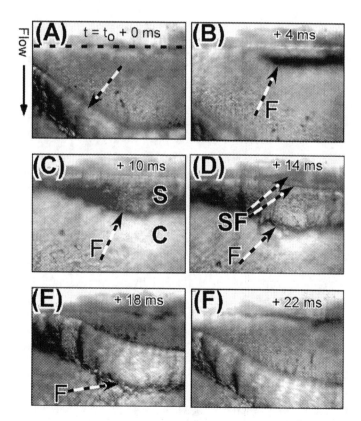

**FIGURE 5.15**  High-speed photographs of the sharkskin instability. (From Ref. 7.)

For the case of PE, Migler et al. (7) has given a detailed description of one cycle of the instability. As will be shown below, an interesting feature is that there are two material failures during each sharkskin cycle. In Fig. 5.15, the extrudate is shown as it exits capillary tube. The horizontal dashed line shows the location of the edge of the capillary tube, which is the same for all six video micrographs. The flow direction is from top to bottom as shown by the arrow to the left of Fig. 5.15A. At the bottom of Fig. 5.15A, the surface disturbance from the previous cycle of sharkskin is visible—see the dashed arrow. In Fig. 5.15B, $t = (t_0 + 14)$ msec, we see the failure line labeled "F" occurring at the air–polymer–tube contact line. In subsequent images, this "fracture line"—always labeled as F—moves downstream until in Fig. 5.15C, where it is below the field of view. In Fig. 5.15C, we see that the fracture surface divides the material into two regions, a surface layer—labeled S—and a core region—labeled C. The surface layer bulges upward. In Fig. 5.15C, the velocity of the surface region in the immediate vicinity of the exit lip $\sim 20$ μm is near zero; thus we say that the surface region sticks to the tube at this point in the cycle. The material in the growing surface region is composed of polymer that had been adjacent to the capillary wall while inside the capillary tube and the material in the core region is composed of polymer that had been at a smaller radius.

In Fig. 15B–C, the fracture line F is a point of material discontinuity—the velocity of the core region is significantly greater than that of the surface region. Once the material has detached from the tube, fracture line F in Fig. 5.15D is no longer a point of material discontinuity—the velocity of the core and surface at this line become identical. Figure 15E–F shows this failure line flowing. Meanwhile, back at the exit lip area, once the failure line is not active, the extensional stress again starts to build up. Figure 15F is essentially the same point in the sharkskin cycle as Fig. 5.15A.

The arrow in Fig. 5.15B shows the initiation of fracture. To the right of this point, the fracture line is visible, while to the left, fracture has not yet occurred. In Fig. 5.15C, the initiation point is at the leftmost part of the image. The initiation point propagates circumferentially around the capillary tube; it could equally well propagate from left to right. It seems reasonable to assume that the stress to continue the propagation of an existing fracture line is less than the stress required to propagate a new fracture line. When the initiation point travels undisturbed around the circumference for several cycles of sharkskin, the sharkskin ridges spiral around the extrudate.

We note that we observe a dark region that cyclically appears with the same frequency as sharkskin in the first 20 μm upstream of the exit. This coincides with the surface failure that occurs when the surface layer detaches from the exit region (Fig. 5.15D). This dark ring could either be associated with a detachment from the inner surface of the capillary tube or could be caused by a breaking of the material right near the wall.

Howells and Benbow (30) were the first to report results from PDMS where the kinetics is more readily observed. They described a "failure of the melt caused by the body of the extrudate moving away form a relatively stationary bank of polymer held near the die exit plane." This is different language for the same phenomena described above for the LLDPE.

It was pointed out by Howells and Benbow (30) that a structure similar to that of sharkskin could be produced in a film of PDMS by pulling a blade across its surface. Like the case of sharkskin, a stress singularity occurs at the air–polymer–metal contact line. Recently, a very detailed set of images and velocimetry measurements were carried out by Mizunuma and Takagi (49). It is interesting to note that their images look quite similar to those shown in Fig. 5.15. Also, they observe a dark region that appears cyclically inside of the exit die, in a fashion similar to the observation reported above. Because of the increased length scale of the PDMS, the dark region was much more pronounced.

## 5.10  MECHANISM OF SHARKSKIN

The mechanism of sharkskin has long been debated in the literature. In this section, we will focus on the two most popular ideas; the *exit stick–slip* and the rupture hypotheses. While most authors have focused on these two ideas, there have certainly been other suggestions; we first briefly mention them here but give them less space in this review in part because they have thus far received less discussion in the literature.

In a numerical model, Tremblay (50) predicted the stress profile of a fluid in a capillary die. He found that the stress could become negative near the exit region for the fluid near the wall–air–polymer three-phase line. He conjectured that the negative stress gives rise to cavitation, which is then the cause of the sharkskin surface melt fracture, previously described. However, there has been little experimental support for this idea. Recently, Son and Migler (51) did observe cavitation that initiated in the die near the wall just before the material exits the capillary tube. However, cavitation was not observed in the sharkskin regime—only in the gross melt fracture regime.

Many of the models are concerned with the effects of the potential complexity of the flow boundary condition. For example, in work by Black and Graham (52), it is theoretically shown that in a model incorporating slip and viscoelasticity, the viscoelastic shear flow is unstable to short wavelength perturbations during slip. To the extent that we define sharkskin as an instability that occurs at the exit, and that some experiments show sharkskin in the absence of a slip (or very weak slip), then this model may be more relevant to a different instability.

Joseph (53) has postulated ideas based on the similarities in appearance between the sharkskin extrudate and that which occurs in core/annular flow when the annular region is of lower viscosity than the core, e.g., oil/water. In that case, the instability occurs in the pipe itself and not at the exit. It does not seem that this idea can be reconciled with the experimental observations of the smooth and stable stress lines seen in the die during sharkskin via birefringence and velocimetry observations.

### 5.10.1  Exit Stick–Slip Hypothesis

One theory for the cause of sharkskin is based on the idea of an unstable boundary condition at the exit lip; the flow boundary oscillates in time between that of stick and slip. We refer to this as *exit stick–slip*, to distinguish it from that known as stick–slip, which is the topic of the next chapter (but is also briefly discussed below). The *exit stick–slip* hypothesis is compelling because it offers a simple explanation for the sharkskin phenomena and offers an explanation as to why the extrudate surface has ridges and valleys. But as described below, there are several experimental observations that seem to contradict the *exit stick–slip* theory as described in its simplest form; thus it should be used with caution. However, elements of the idea are correct; it does seem clear from kinetics measurements that a periodic failure (the term detachment was used in Sec. 5.9) at the boundary is occurring; thus the concept of a periodic buildup of stress that causes a local "failure" at the wall–polymer interface does need to be retained.

The simple version of the *exit stick–slip* hypothesis was put forth by Vinogradov (26) and rests on three well-established facts. First, as we have pointed out several times in this chapter, the stress field at the exit lip is significantly greater than the wall shear stress acting upstream. This was established through birefringence measurements as well as numerical modeling. Second, many of the systems that exhibit sharkskin also exhibit the phenomena of *exit stick–slip*. Third, it is well known that die swell at the exit in polymeric materials can be much larger than for Newtonian fluids as it is caused by a release in the built-up elastic energy during the capillary flow. As demonstrated explicitly by Barone et al. (27), the amount of die swell in a stick boundary condition is greater than that during a slip boundary condition. This is reasonable because it is the high shear rate near the wall that generates much of the elasticity; additionally, the central region of the flow must slow down as the material exits the tube, and this generates a compressive extensional flow. In contrast, the more slip, the more plug-like the flow through the tube and the less elastic energy is built up in the polymer.

Putting the three elements together, the *exit stick–slip* hypothesis states that as the throughput in the capillary die is increased, the stress at the wall may be in the stick condition, but the stress at the exit lip is even larger. Eventually, the exit lip would reach the condition where the shear stress exceeds that required to cause the transition. The entire capillary wall does not slip, only the exit region. As the polymer slips at the exit, the die swell decreases, which causes the valley in the extrudate. As the flow boundary condition at the exit is one of slip, the extra stress is relieved, which then causes the stress at the exit to decrease until the slip boundary condition returns. Once there is slip, the stress and elasticity starts to increase again, leading to increased die swell. So the system jumps back and forth between these two boundary conditions and the extrudate exhibits increased and decrease levels of die swell. More recently, Wang (36) has utilized a molecular theory by Brochard and de Gennes, which is based on a proposed relationship between a coil–stretch transition at the wall and a stick to slip transition.

However, there is experimental evidence against this picture. As discussed earlier, there have been two experiments in search of a stick–slip behavior during sharkskin, and they both found that one can achieve sharkskin without a stick–slip behavior (7). Other experiments then showed that weak sharkskin could occur under a slip boundary condition, and that under very strong sharkskin, an oscillation in the velocity at the exit was observed, although definitive evidence of exit stick–slip was not made. These results indicate that sharkskin is not dependent on the occurrence of an exit stick–slip boundary condition, or if it were, it would have to be within the last 10 μm of the exit.

A second reason against the *exit stick–slip* hypothesis concerns the issue of die-swell. Figure 5.16 shows the extrudate in the global stick–slip regime, where it undergoes the transition from global slip (smooth—left third of extrudate) to the stick boundary condition (sharkskin—right portion of extrudate). As this sharkskin occurs in the global stick–slip part of the flow curve (see Fig. 5.2), and the amplitude of the ridges is ~ 35% of $R$, we classify this as strong sharkskin. According to the *exit stick–slip* theory, the valley is the slip part and the ridges are the stick. However, the diameter of the sharkskin valleys are less than the diameter of the global slip extrudate, in fact, they are less than the diameter of the capillary die. If anything, one expects the diameter of the sharkskin valleys to be larger than that during global slip because for the case of global slip, there is much less elastic energy built up during the extrusion. Further, there is no explanation for the fact that the valleys are of lower diameter than the capillary tube within the context of the *exit stick–slip* hypothesis. Another discrepancy is that the part corresponding to the global stick slip is very smooth whereas that in the valley is quite rough; evidence that they were formed under different processes.

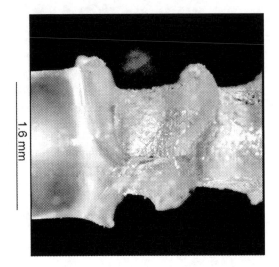

1.6 mm

FIGURE 5.16 Photograph of the stick–slip instability showing a section of the extrudate where the transition from slip (left side) to stick (right side) is shown.

Finally, the reader is referred to a colorful exchange on this topic that appeared in the *Journal of Rheology* between Cogswell and Wang (54) to "determine the mechanism of sharkskin." The capillary was first filled with a PE containing a small amount of carbon black, which was followed by extrusion in the sharkskin regime where the barrel is then filled with uncolored PE. First, the extrudate is completely black, but as the uncolored material is extruded, it comes through the PE in the middle surrounded by the black PE. This black ring continues to thin down in time. The question is what should happen when the thickness of the ring becomes comparable to the thickness of sharkskin?

According to the *exit stick–slip* model, the slip state (valleys) should appear black because if the outer ring is black in the capillary and then it comes out as a plug, it should remain black. The stick state (ridges) should also be black but perhaps less so—the thickness of the black ring in the capillary would be greater than that in the extrudate because of the velocity rearrangements upon exiting the die. In the experiment, as the black ring becomes sufficiently thin, the ridges are black and the valleys are white, a result that seemingly would not agree with the *exit stick–slip* model, unless that model was increased in complexity from that put forward by Vinogradov. As described below, it is consistent with the Cogswell model where it is taken as a proof of a material discontinuity. However, as described below, there is certainly reason to believe that elements of the exit stick–slip model should be combined with elements of the Cogswell model. In this modified idea, detachment may be a better word than slip for describing the behavior.

### 5.10.2  Cogswell Model

In 1977, Cogswell (48) proposed a model for sharkskin that focuses on the rapid acceleration or stretching that the polymer near the wall experiences as it exits the die. In the context of a Newtonian fluid with a stick boundary condition, the material near the wall must accelerate as described in Sec. 5.8. Concomitantly, the material in the center undergoes compression as it decelerates from $2V_f$ to $V_f$. The idea is that when the material near the lip is sufficiently stretched, the extensional stress exceeds a critical material-dependent value and the molten polymer fractures. According to Cogswell, the maximum tensile stress that a polymeric liquid can sustain is approximately 1 MPa. Thus when the extensional stress at the exit exceeds 1MPa, one can anticipate sharkskin.

This idea is then explored with a model material "silly putty," chosen because its viscosity is independent of shear rate, it exhibits negligible die swell, it can be modeled by a simple one-element Maxwell model, and its extensional viscosity could be approximated as being independent of stress. It also showed an elastic extensibility of 0.2 units of Hencky strain. Reasonable agreement was found between the calculated stretching and the stretch rate at fracture.

The analysis was furthered by Inn et al. (14) who extended Cogswell's analysis to include the effects of a power law shear thinning as well as die swell. They explicitly showed the existence of the layer near the wall, which becomes stretched under flow, i.e., the "traction zone" previously described by numerical methods (47). This provides a rational argument to explain why the narrow part of the extrudate in the sharkskin regime can be less than the diameter of the die—the rupture that first occurs at the air–polymer interface can propagate radially inward to relieve the extensional stress inside the polymer. Further, the authors provide a force balance between the extensional and shear stresses acting on the surface layer and core with the tensile force acting on the surface layer.

Thus, when considering a fluid element that exits the capillary from the near wall region, it undergoes a rapid extensional stress over a finite duration, i.e., an extensional pulse. The Cogswell model then provides a framework for understanding the effect of molecular weight distribution. It is known that upon application of extensional strain rate, the resulting stress is an increasing function of time. Comparing polymers of the same viscosity, a more elastic polymer will have weaker moduli and longer relaxation times. The more elastic polymer will take longer to buildup stress. Thus, one may imagine that the stress in the more elastic polymer does not build up to the critical level for fracture in the time scale of the extensional pulse. The polymers with greater MWD exhibit greater shear thinning and are more elastic than their narrow MWD cousins, so this may be an explanation for the fact that sharkskin is absent or weaker in broad MWD polymers (41,48).

A side-view cartoon of the kinetics from the perspective of the Cogswell model is shown in Fig. 5.17. In Fig. 5.17A, the polymer's jump in velocity from the polymer–tube interface to the polymer–air interface is depicted. The extensional stress at the air–polymer–sapphire contact line builds up over time until the material fractures (Fig. 5.17B). The fracture is initially located at the air–polymer surface. In Fig. 5.17C, the material near arrow 1 goes toward the surface region of the extrudate while the material at arrow 2 goes toward the core region. The extensional strain rate as the material flows from arrow 2 to the core region is much reduced from what occurs at the surface in Fig. 5.17A. The surface region grows in size and it eventually peels off the

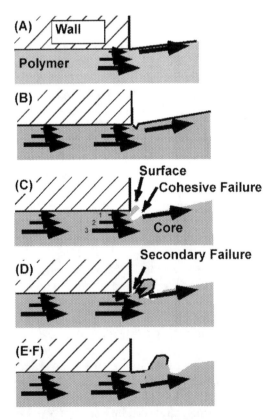

FIGURE 5.17 Sketch of sharkskin kinetics. (From Ref. 7.)

sapphire (Fig. 5.17D); this is the surface failure. Once it peels off, the fracture line heals as the surface and core regions move off.

While many observations are consistent with this rudimentary model and there is no experimental data that contradicts it, definitive proof has not been obtained yet. The primary issue in directly confirming the Cogswell approach is the difficulty in measuring the stress at the exit, coupled with the difficulty in measuring the extensional stress at fracture for various viscoelastic polymer melts.

The Cogswell model does not address the complexity of the kinetics. Indeed, based on the kinetics visualization, it was mentioned that there was a part of the cycle in which the surface region detaches from the wall. This failure then may be related to the stick–slip hypothesis mentioned above. While "slip" may not be the best word to use, there seem to be one part of the cycle where the material is sticking (causing rupture) and one part of the cycle where it is detaching (slipping).

A promising approach by Rutgers and Mackley (46) is to utilize flow modeling to estimate the stress in the exit region (described earlier) coupled with separate experiments to measure the stress at breakage, obtained by extrusion of the polymer through a capillary, followed by stretching of the molten extrudate until it breaks. Order-of-magnitude agreement is found between the exit stress at the onset of sharkskin and the stress required for breakage.

## 5.11 CONCLUSION

While a great deal has been learned about the sharkskin instability, we still lack understanding at a deep level. The precise understanding and control of the sharkskin instability will require advances in several key areas of fundamental polymer physics. For example, to better estimate the extensional stresses at the die exit, a methodology for measuring the extensional viscosity of molten polymers at high strain rates must be developed. A concomitant improvement in numerical modeling is needed to better predict these rapid, nonlinear deformations. The field of fracture of molten polymers must be further developed and new test methods are needed. Finally, the relationship between molecular architecture, rheology, and sharkskin must be further advanced. Progress in these areas will enable the development of new materials and processes that will allow more efficient manufacturing.

## ACKNOWLEDGMENTS

I wish to acknowledge S. Kharchenko for assistance and providing Figs. 5.2 and 5.16.

## REFERENCES

1. Denn, M.M. Extrusion instabilities and wall slip. Annu. Rev. Fluid Mech. 2001, *33*, 265–287.
2. Denn, M.M. Issues in viscoelastic fluid-mechanics. Annu. Rev. Fluid Mech. 1990, *22*, 13–34.
3. Larson, R.G. Instabilities in viscoelastic flows. Rheol. Acta 1992, *31*, 213–263.
4. Piau, J.M.; Agassant, J.F. *Rheology for Polymer Melt Processing*; Elsevier: Amsterdam, 1996.
5. Macosko, C.W. *Rheology Principles, Measurements, and Applications*; VCH: New York, 1994.
6. Morrison, F.A. *Understanding Rheology*; Oxford University Press: New York, 2001.
7. Migler, K.B.; Son, Y.; Qiao, F.; Flynn, K. Extensional deformation, cohesive failure, and boundary conditions during sharkskin melt fracture. J. Rheol. 2002, *46*, 383–400.
8. Yamaguchi, M.; Miyata, H.; Tan, V.; Gogos, C.G. Relation between molecular structure and flow instability for ethylene/alpha-olefin copolymers. Polymer 2002, *43*, 5249–5255.
9. El Kissi, N.; Piau, J.M. Adhesion of linear low-density polyethylene for flow regimes with sharkskin. J. Rheol. 1994, *38*, 1447–1463.
10. El Kissi, N.; Leger, L.; Piau, J.M.; Mezghani, A. Effect of surface-properties on polymer melt slip and extrusion defects. J. Non-Newton. Fluid Mech. 1994, *52*, 249–261.
11. El Kissi, N.; Piau, J.M.; Toussaint, F. Sharkskin and cracking of polymer melt extrudates. J. Non-Newton. Fluid Mech. 1997, *68*, 271–290.
12. Piau, J.M.; El Kissi, N.; Mezghani, A. Slip-flow of polybutadiene through fluorinated dies. J. Non-Newton. Fluid Mech. 1995, *59*, 11–30.
13. Venet, C.; Vergnes, B. Experimental characterization of sharkskin in polyethylenes. J. Rheol. 1997, *41*, 873–892.
14. Inn, Y.W.; Fischer, R.J.; Shaw, M.T. Visual observation of development of sharkskin melt fracture in polybutadiene extrusion. Rheol. Acta 1998, *37*, 573–582.
15. Tzoganakis, C.; Price, B.C.; Hatzikiriakos, S.G. Fractal analysis of the sharkskin phenomenon in polymer melt extrusion. J. Rheol. 1993, *37*, 355–366.
16. Denn, M.M. Extrusion instabilities and wall slip. Annu. Rev. Fluid Mech. 2001, *33*, 265–297.
17. Shaw, M.T.; Wang, L. Sharkskin melt fracture: recent findings using model geometries. International Conference on Rheology, 2000.
18. Ramamurthy, A.V. Wall slip in viscous fluids and influence of materials of construction. J. Rheol. 1986, *30*, 337–357.
19. Kalika, D.S.; Denn, M.M. Wall slip and extrudate distortion in linear low-density polyethylene. J. Rheol. 1987, *31*, 815–834.
20. Wang, S.Q.; Drda, P.A.; Inn, Y.W. Exploring molecular origins of sharkskin, partial slip, and slope change in flow curves of linear low density polyethylene. J. Rheol. 1996, *40*, 875–898.

21. Migler, K.B.; Lavallée, C.; Dillon, M.P.; Woods, S.S.; Gettinger, C.L. Visualizing the elimination of sharkskin through fluoropolymer additives: coating and polymer–polymer slippage. J. Rheol. 2001, 45, 565–581.

22. Legrand, F.; Piau, J.M. Spatially resolved stress birefringence and flow visualization in the flow instabilities of a polydimethylsiloxane extruded through a slit die. J. Non-Newton. Fluid Mech. 1998, 77, 123–150.

23. Inn, Y.W.; Wang, L.S.; Shaw, M.T. Efforts to find stick–slip flow in the land of a die under sharkskin melt fracture conditions: polybutadiene. Macromol. Symp. 2000, 158, 65–75.

24. Merten, A.; Seidel, C.; Munstedt, H. Comparison of the sharkskin defect of two linear polyethylenes and the influence of a slip additive as investigated by laser-doppler velocimetry. Abstract of the First Annual European Conference on Rheology, 2003.

25. Hatzikiriakos, S.G. The onset of wall slip and sharkskin melt fracture in capillary-flow. Polym. Eng. Sci. 1994, 34, 1441–1449.

26. Vinogradov, G.V.; Insarova, N.I.; Boiko, B.B.; Borisenkova, E.K. Critical regimes of shear in linear polymers. Polym. Eng. Sci. 1972, 12, 323–334.

27. Barone, J.R.; Plucktaveesak, N.; Wang, S.Q. Interfacial molecular instability mechanism for sharkskin phenomenon in capillary extrusion of linear polyethylenes. J. Rheol. 1998, 42, 813–832.

28. Wise, G.M.; Denn, M.M.; Bell, A.T. Surface mobility and slip of polybutadiene melts in shear flow. J. Rheol. 2000, 44, 549–567.

29. Ghanta, V.G.; Riise, B.L.; Denn, M.M. Disappearance of extrusion instabilities in brass capillary dies. J. Rheol. 1999, 43, 435–442.

30. Howells, E.R.; Benbow, J. Flow defects in polymer melts. Trans. Plastics Inst. 1962, 30, 240–253.

31. Fujiyama, M.; Inata, H. Melt fracture behavior of polypropylene-type resins with narrow molecular weight distribution. 1. Temperature dependence. J. Appl. Polym. Sci. 2002, 84, 2111–2119.

32. Fujiyama, M.; Inata, H. Melt fracture behavior of polypropylene-type resins with narrow molecular weight distribution. II. Suppression of sharkskin by addition of adhesive resins. J. Appl. Polym. Sci. 2002, 84, 2120–2127.

33. Moynihan, R.H.; Baird, D.G.; Ramanathan, R. Additional observations on the surface melt fracture-behavior of linear low-density polyethylene. J. Non-Newton. Fluid Mech. 1990, 36, 255–263.

34. Rutgers, R.P.G.; Mackley, M.R. The effect of channel geometry and wall boundary conditions on the formation of extrusion surface instabilities for LLDPE. J. Non-Newton. Fluid Mech. 2001, 98, 185–199.

35. Deeprasertkul, C.; Rosenblatt, C.; Wang, S.Q. Molecular character of sharkskin phenomenon in metallocene linear low density polyethylenes. Macromol. Chem. Phys. 1998, 199, 2113–2118.

36. Wang, S.Q. Molecular transitions and dynamics at polymer/wall interfaces: origins of flow instabilities and wall slip. Adv. Polym. Sci. 1999, 138, 227–275.

37. Kolnaar, J.W.H.; Keller, A. A singularity in the melt flow of polyethylene with wider implications for polymer melt flow rheology. J. Non-Newton. Fluid Mech. 1997, 69, 71–98.

38. Pudjijanto, S.; Denn, M.M. A stable island in the slip–stick region of linear low-density polyethylene. J. Rheol. 1994, 38, 1735–1744.
39. Perez-Gonzalez, J.; Denn, M.M. Flow enhancement in the continuous extrusion of linear low-density polyethylene. Ind. Eng. Chem. Res. 2001, 40, 4309–4316.
40. Santamaria, A.; Fernandez, M.; Sanz, E.; Lafuente, P.; Munoz-Escalona, A. Postponing sharkskin of metallocene polyethylenes at low temperatures: the effect of molecular parameters. Polymer 2003, 44, 2473–2480.
41. Kazatchkov, I.B.; Bohnet, N.; Goyal, S.K.; Hatzikiriakos, S.G. Influence of molecular structure on the rheological and processing behavior of polyethylene resins. Polym. Eng. Sci. 1999, 39, 804–815.
42. Dhori, P.K.; Jeyaseelan, R.S.; Giacomin, A.J.; Slattery, J.C. Common line motion. 3. Implications in polymer extrusion. J. Non-Newton. Fluid Mech. 1997, 71, 231–243.
43. Barone, J.R.; Wang, S.Q. Rheo-optical observations of sharkskin formation in slit-die extrusion. J. Rheol. 2001, 45, 49–60.
44. Barone, J.; Wang, S.Q. Flow birefringence study of sharkskin and stress relaxation in polybutadiene melts. Rheol. Acta 1999, 38, 404–414.
45. Mackley, M.R.; Rutgers, R.P.G.; Gilbert, D.G. Surface instabilities during the extrusion of linear low density polyethylene. J. Non-Newton. Fluid Mech. 1998, 76, 281–297.
46. Rutgers, R.; Mackley, M. The correlation of experimental surface extrusion instabilities with numerically predicted exit surface stress concentrations and melt strength for linear low density polyethylene. J. Rheol. 2000, 44, 1319–1334.
47. Venet, C.; Vergnes, B. Stress distribution around capillary die exit: an interpretation of the onset of sharkskin defect. J. Non-Newton. Fluid Mech. 2000, 93, 117–132.
48. Cogswell, F.N. Stretching flow instabilities at the exits of extrusion dies. J. Non-Newton. Fluid Mech. 1977, 2, 37–47.
49. Mizunuma, H.; Takagi, H. Cyclic generation of wall slip at the exit of plane couette flow. J. Rheol. 2003, 47, 737–757.
50. Tremblay, B. Sharkskin defects of polymer melts—the role of cohesion and adhesion. J. Rheol. 1991, 35, 985–998.
51. Son, Y.; Migler, K.B. Cavitation of polyethylene during extrusion processing instabilities. J. Polym. Sci. B Polym. Phys. 2002, 40, 2791–2799.
52. Black, W.B.; Graham, M.D. Wall-slip and polymer-melt flow instability. Phys. Rev. Lett. 1996, 77, 956–959.
53. Joseph, D.D. Steep wave fronts on extrudates of polymer melts and solutions: lubrication layers and boundary lubrication. J. Non-Newton. Fluid Mech. 1997, 70, 187–203.
54. Cogswell, F.N.; Barone, J.R.; Plucktaveesak, N.; Wang, S.Q. Letter to the editor: the mystery of the mechanism of sharkskin. J. Rheol. 1999, 43, 245–252.

# 6

## Stick–Slip Instability

**Georgios Georgiou**
University of Cyprus, Nicosia, Cyprus

### 6.1 INTRODUCTION

Among the class of polymer extrusion instabilities, collectively known as *melt fracture* (1), the *stick–slip instability* is the only one that is associated with pressure and extrudate flow rate oscillations, observed while the throughput is controlled. These oscillations result in extrudates that are characterized by alternating rough and relatively smooth regions.

The stick–slip instability has been the subject of experimental studies since the late 1950s and has been given many different names by researchers studying polymer extrusion instabilities. The first reports of the phenomenon are those of Tordella (1) and Bagley et al. (2) who studied the capillary flow of high-density polyethylenes (HDPEs). More systematic observations on HDPEs have been reported by Lupton and Regester (3) and Myerholtz (4) in the 1960s. Then, Vinogradov et al. (5–8) thoroughly investigated the flow of narrow molecular weight distribution (MWD) polyisoprenes (PIs) and poly-butadienes (PBs) in both pressure- and flow rate-controlled experiments, and introduced the term *spurt flow* for the stick–slip instability. Weill (9), who proposed a phenomenological model based on the combination of a double-branched pressure/flow rate diagram with the compressibility of the polymer melt in the reservoir to describe the experimental pressure and flow rate oscillations, used the term *main flow instability*.

The term *slip–stick instability* is also very often used. For example, this has been used by Lin (10,11) who experimentally studied the capillary extrusion of PI, polystyrene (PS), and PE samples and associated the onset of the instability with the maximum exhibited by a nonmonotonic constitutive equation. Two frequently cited contributions from the 1980s are the articles by Ramamurthy (12) and Kalika and Denn (13) who studied the flow of linear low-density polyethylenes (LLDPEs) and the effects of different die materials. More recently, Piau and El Kissi et al. (14–21) have extensively studied the extrusion instabilities for moderately and highly entangled polydimethylsiloxanes (PDMS), LLDPEs, and PBs, and employed the term *cork flow* for the stick–slip instability. In the early 1990s, Hatzikiriakos and Dealy (22–24) studied the stick–slip instability of HDPEs for which they used the term *cyclic melt fracture*. Other names used for the instability are *oscillating melt fracture* and *bamboo fracture*.

Important contributions have also been made by Wang et al. (25–29) who systematically studied the extrusion of linear PE melts and the molecular origins of the stick–slip instability. Recent work on the subject concerns the use of different die materials (30,31) and processing aids (32), with which the instabilities are delayed or eliminated, and direct velocity measurements that provide useful information for the instability (33–35). Additional information and references are provided in the following sections, as well as in other chapters of this book (see also reviews of extrusion instabilities in Refs. (19,29,36–39)).

Generally speaking, the stick–slip instability occurs at apparent shear rates above the *sharkskin instability* regime, which is discussed in Chapter 5 (present volume, by Migler), and below the *gross melt fracture* regime, which is discussed in Chapter 7 (present volume, by Dealy). In certain cases, the latter instability may be observed before the occurrence of the stick–slip instability. The sequence of all these instabilities is discussed in Sec. 6.2 by looking at the different regimes of the *flow curve*; that is, the plot of the wall shear stress vs. the apparent shear rate. Sec. 6.3 focuses on the stick–slip instability regime. In Sec. 6.3.1, the experimental flow curves, the range of the stick–slip instability regime, and the corresponding critical shear stresses are reviewed, and the effects of the capillary diameter, the capillary length-to-diameter ratio, temperature, molecular weight, and MWD are discussed. Emphasis is also given on the capillary diameter effects from which wall slip velocities can be inferred using the Mooney or analogous techniques. In Sec. 6.3.2, the effects of the geometry and the operating conditions on the pressure and extrudate flow rate oscillations are presented. In Sec. 6.3.3, experimental observations concerning the presence of a secondary oscillation regime at higher shear rates are reviewed. Sec. 6.3.4 is concerned with the appearance of the extrudates obtained in the stick–slip instability regime, and Sec. 6.3.5 is devoted to the role of wall slip and to the effects of the die material and processing aids on the onset of extrusion instabilities.

In Sec. 6.4, one-dimensional phenomenological relaxation/oscillation models describing the experimental pressure and flow rate oscillations are reviewed. By combining the melt compressibility in the reservoir with a multivalued flow curve, these models predict oscillatory flow that follows, more or less, the hysteresis loop defined by the flow curve. Sec. 6.5 reviews some mechanisms proposed for explaining the stick–slip instability. In addition to the compressibility/slip mechanism, which is consistent with the experimental observations and leads to pressure and flow rate oscillations similar to the experimental ones, two other mechanisms, based on the combination of nonlinear slip with elasticity and the nonmonotonicity of the constitutive equation, respectively, and their limitations are briefly discussed. Finally, concluding remarks are provided in Sec. 6.6.

## 6.2 REGIMES OF EXTRUSION INSTABILITIES

### 6.2.1 Flow Curves in General

To relate the stick–slip instability to the other extrusion instabilities and facilitate our discussion, we recall here the notion of the capillary flow curve. This is the plot of wall shear stress, $\sigma_w$, vs. the apparent shear rate, $\dot{\gamma}_A$, which is usually obtained by means of a capillary rheometer, under either constant flow rate (i.e., constant piston speed), or constant pressure drop operation. The apparent wall shear rate is calculated from the volumetric flow rate $Q$ as follows:

$$\dot{\gamma}_A \equiv \frac{32Q}{\pi D^3} \tag{6.1}$$

where $D$ is the diameter of the capillary. The wall shear stress is calculated by

$$\sigma_w = \frac{(P_d - P_{end})}{4L/D} \tag{6.2}$$

where $P_d$ is the driving pressure determined for the force on the piston, $P_{end}$ is the Bagley end correction for the pressure drop, and $L$ is the length of the capillary. The plot of $\sigma_w$ vs. $\dot{\gamma}_A$ is referred to as the *apparent flow curve*. It is clear from Eqs. (6.1) and (6.2) that the apparent flow curve can also be viewed as the plot of the pressure drop across the capillary vs. the volumetric flow rate. If the end correction is ignored, then the calculated shear stress

$$\sigma_A = \frac{P_d D}{4L} \tag{6.3}$$

is an "apparent" wall shear stress.

The wall shear stress, $\sigma_w$, is typically an increasing function of the shear rate, $\dot{\gamma}_A$; that is, the flow curve consists of a positive-slope branch. Slip along the capillary wall and/or shear thinning result in the reduction of $\sigma_w$ corresponding to a certain value of $\dot{\gamma}_A$; that is, they shift the flow curve to the right. They may also cause reduction of the slope of the flow curve. This is easily demonstrated by considering the flow of a power-law fluid with uniform slip along the wall. For any slip equation, the following expression is obtained for the wall shear rate, $\dot{\gamma}_w$, in steady capillary (Poiseuille) flow of a power-law fluid:

$$\dot{\gamma}_w = \frac{4n}{1+3n}\left(\frac{\sigma_w}{K}\right)^{\frac{1}{n}} + \frac{8u_s}{D} \tag{6.4}$$

where $u_s$ is the wall slip velocity, $K$ is the consistency index, and $n$ is the power-law exponent. In the case of a Newtonian fluid, $n = 1$ and $K$ is replaced by the viscosity, $\eta$:

$$\dot{\gamma}_w = \frac{1}{\eta}\sigma_w + \frac{8u_s}{D} \tag{6.5}$$

In experiments, when slip is present, the apparent shear rate is calculated by

$$\dot{\gamma}_A = \frac{32Q}{\pi D^3} + \frac{8u_s}{D} \tag{6.6}$$

The plot of $\dot{\gamma}_A$ ($\sigma_w$) is the Mooney plot, which is used to calculate the slip velocity.

The effect of slip on the Newtonian flow curve is illustrated in Fig. 6.1, where we employed a power-law slip equation

$$\sigma_w = k_s u_s^m \tag{6.7}$$

where $k_s$ and $m$ are parameters depending on the polymer and the capillary wall properties. Linear slip ($m = 1$) reduces the shear stress so that the flow curve is shifted to the right while the slope remains constant. With nonlinear slip (i.e., power-law slip with $m < 1$), the flow curve is shifted to the right and its slope is reduced. The effect of shear thinning in the absence of slip is shown in Fig. 6.2. Shear thinning shifts the flow curve to the right and reduces its slope. The flow curve of a fluid obeying the Maxwell or the Oldroyd-B constitutive equation coincides with the Newtonian flow curve, which indicates that the effect of elasticity is not as important as that of shear thinning.

### 6.2.1.1. Multivalued Flow Curves

In theory, the flow curve may exhibit maxima and minima. As illustrated in Fig. 6.1, a nonmonotonic slip equation results in a nonmonotonic flow curve.

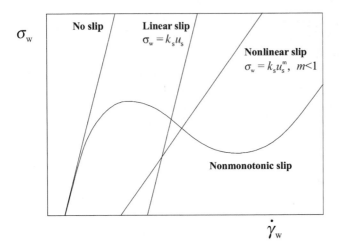

**FIGURE 6.1** Effect of slip on the Newtonian flow curve.

However, the negative-slope branch of the flow curve is not observed in practice because the corresponding steady state capillary flow solutions are unstable (40,41). Nonmonotonic macroscopic slip equations for polymer melts, exhibiting one maximum and one minimum, have been proposed by El Kissi and Piau (14) and Leonov (42,43). Piau and El Kissi (18) also proposed a slip equation with two maxima and two minima for LLDPEs and PBs.

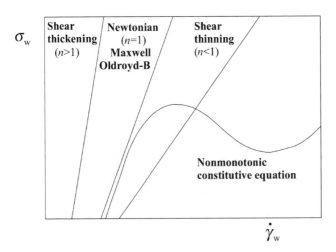

**FIGURE 6.2** Effect of shear thinning and elasticity on the flow curve, in the absence of slip.

As shown in Fig. 6.2, the flow curve is also nonmonotonic when the constitutive equation is nonmonotonic. Nonmonotonic constitutive models include the three-constant Oldroyd model, the Doi–Edwards model with a Rouse relaxation mode, the Johnson–Segalman model with an added Newtonian viscosity, the Giesekus model, and the KBKZ model with an extraviscous term. Nonmonotonicity of the constitutive equation is predicted by reptation theories for highly entangled polymer melts and can be physically viewed as the separation of flow into two dynamic regimes with histories at low and high deformation rates, away from or near to the solid boundary, respectively, which implies the existence of two characteristic times of the viscoelastic fluids (44). Nevertheless, steady state solutions corresponding to the negative-slope branch of the flow curve are again unstable. Only certain shear-banded steady state solutions in this regime are stable (see Ref. 45 and references therein). However, shear banding has been experimentally observed only with wormlike surfactant semidilute solutions, which exhibit a narrow spectrum of relaxation times, and not with polymer melts (46).

### 6.2.2 Flow Curves in Extrusion and Regimes of Instability

During the extrusion of polymers from capillary or slit dies under constant flow rate (i.e., fixed piston speed) operation, a variety of distortions are observed on the extrudate surface and core, when the apparent shear rate exceeds a critical value (12,13,15). A similar range of extrudate defects is also obtained in pressure-controlled experiments above a critical wall shear stress (2,8). *Melt fracture* is a generic term used for all these different extrudate distortions, the size and the severity of which generally increase with the volumetric flow rate or the pressure.

Before discussing the various regimes of extrudate distortion, it is instructive to first consider the typical apparent flow curve for a linear polymer melt, such as HDPE or LLDPE, which is shown in Fig. 6.3. This consists of two positive-slope, stable branches separated by an unstable zone, and contains several interesting features that cannot be explained on the basis of shear thinning and/or linear slip.

In a capillary experiment in which the piston speed (i.e., the shear rate) is gradually increased, the shear stress follows the left positive-slope branch until it reaches the upper point A (Fig. 6.3). Beyond this point, oscillations of the pressure and of the extrudate flow rate appear and continue until the apparent shear rate exceeds the value corresponding to the lower point C of the right positive-slope branch (3,13,23). Beyond this shear rate, the shear stress follows the right branch of the flow curve. If the piston speed is now gradually reduced, the shear stress will again follow the right branch and descend to its lower point C, after which oscillations are again observed until

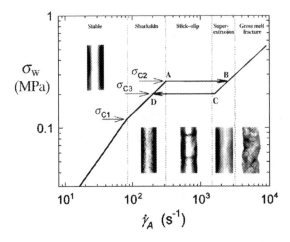

**FIGURE 6.3** A typical apparent flow curve and regimes of instability for a linear polymer melt (i.e., HDPE or LLDPE).

the apparent shear rate reaches the value corresponding to the upper point A of the left branch. Below this shear rate, oscillations cease and the shear stress nicely follows the left branch.

In an experiment in which the pressure (i.e., the wall shear stress) is gradually increased, the shear rate increases following the left branch of the flow curve. When the wall shear stress reaches the critical value $\sigma_{c2}$ that corresponds to point A, the flow rate suddenly jumps from A to the point B of the right branch (Fig. 6.3). This sudden dramatic (tens to thousands of times) increase in the flow rate was first observed by Bagley et al. (2) and was later called as *spurt effect* by Vinogradov et al. (7). If the pressure is further increased, the shear rate follows the right branch. If the pressure is now decreased, the shear rate continuously decreases until it reaches the critical shear stress $\sigma_{c3}$ that corresponds to point C, at which the flow rate jumps suddenly to point D of the left branch. This effect, referred to as *flow curve hysteresis* (1,2,7), is a consequence of the multivaluedness of the flow curve: For shear stresses between $\sigma_{c3}$ and $\sigma_{c2}$, two stable shear rates are possible. In this experiment, where the pressure is controlled, the shear rate range between points A and C is never reached.

As far as the appearance of the extrudate is concerned, five distinct flow regimes are, in general, observed during the extrusion of linear, high-molecular weight, and narrowly distributed polymers (e.g., HDPE or LLDPE) from slit or axisymmetric dies at a controlled flow rate: the *stable*, the *sharkskin*, the *stick–slip*, the *superextrusion*, and the *gross melt fracture* regimes. As shown in Fig. 6.3, the stick–slip instability regime separates the left branch of the flow curve (which corresponds to the stable and sharkskin regimes) from the right branch (which corresponds to the superflow and gross melt fracture regimes).

In the stable regime, i.e., at small shear rates, the viscosity of the polymer follows a power-law relation, and the extrudate is smooth, glossy, and often transparent. The stable regime persists up to a shear rate corresponding to a critical wall shear stress, $\sigma_{c1}$, at which the extrudate starts gradually losing its glossiness and transparency and develops a small-amplitude, short-wavelength periodic disturbance on its surface, which is commonly called sharkskin or surface melt fracture. Sharkskin is observed up to the end point of the left branch of the flow curve, i.e., until the upper critical stress $\sigma_{c2}$, and increases in intensity as the shear rate increases (12,24,47).

Some researchers associate sharkskin with a small departure from the no-slip boundary condition, which is sometimes accompanied by a sharp change in the slope of the flow curve (12,13,23,48). However, others presented smooth flow curves with continuous slowly changing slope over the stable and sharkskin regimes, which show that sharkskin occurs without macroscopic

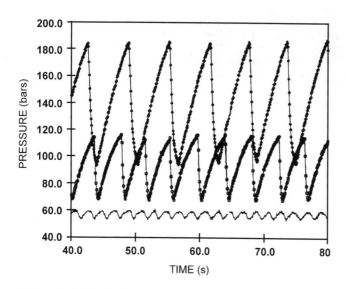

**Figure 6.4** Typical pressure oscillations in the stick–slip instability regime obtained with a linear HDPE, $D = 1.3$ mm and $L/D = 16$ (large amplitude, $\dot{\gamma}_A = 177$ sec$^{-1}$), $L/D = 8$ (medium amplitude, $\dot{\gamma}_A = 117$ sec$^{-1}$), and $L/D = 4$ (small amplitude, $\dot{\gamma}_A = 140$ sec$^{-1}$). (From Ref. 51.).

slip (19,29,49). It is well established that sharkskin is initiated in the die exit region. The proposed origins of sharkskin, which include cracking of the fluid because of the high tensile stresses at the die exit (14,49,50) and local stick–slip because of polymer molecule disentanglement (29), are reviewed in Chapter 5 (present volume, by Migler).

In the stick–slip instability regime that succeeds the sharkskin regime, the flow becomes unstable. As already mentioned, oscillations of the pressure and the instantaneous flow rate are observed although the mean flow rate is kept constant (3,4,24,51). The extrudate emerges from the capillary in bursts, and is characterized by alternating rough and relatively smooth zones. Typical pressure oscillations and extrudate shapes obtained in the stick–slip instability regime are shown in Figs. 6.4 and 6.5, respectively. As discussed below, these oscillations are attributed to the compressibility-induced peri-

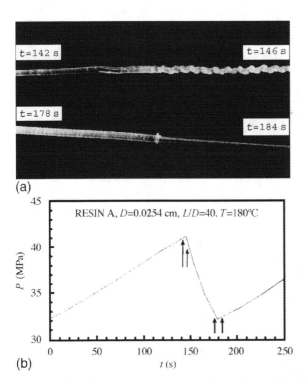

**FIGURE 6.5** (a) Extrudate shapes of a HDPE melt in the stick–slip instability regime. (b) Pressure oscillation. The arrows indicate the times at which the two segments are produced; $\dot{\gamma}_A = 1482 \ sec^{-1}$. (From Ref. 24.)

odic transitions between the two stable branches of the flow curve, from a weak to a strong wall slip and vice versa. In contrast to the sharkskin instability, which is a surface distortion initiated in the die exit region, the stick–slip instability is a volume distortion initiated in the die land region.

Above the stick–slip instability regime, i.e., along the right branch of the flow curve, the pressure oscillations cease and stable solutions are again obtained. Along this branch, the velocity profile in the capillary is nearly plug (33,52,53), which implies the occurrence of strong slip. For a certain range of shear rates, mostly with long dies, a second stable regime may exist, where the extrudate is again smooth and transparent. This is known as the *super-extrusion* or *superflow* regime. Given that it permits stable extrusion at flow rates much higher than those of the first stable region, this regime has been exploited for the extrusion of certain polymers.

Finally, in the gross melt fracture (or *wavy fracture*) regime, the extrudate is grossly distorted. In the early stages, the extrudate is often relatively smooth and wavy, but at higher shear rates it becomes irregular and chaotic. Gross melt fracture is associated with instabilities in the die entry region (7,15,52). For more details, the reader is referred to Chapter 7 (present volume, by Dealy).

The various instabilities mentioned above may not all appear, depending on the polymer used. Moreover, the order of their appearance may not agree with the scenario of Fig. 6.3. For example, most commercial long-chain branched polymers, such as low-density polyethylene (LDPE) and branched polysiloxane, do not exhibit sharkskin and stick–slip instability, but they do exhibit gross melt fracture (39). On the other hand, in certain cases before the appearance of sharkskin, superficial scratching (or matting) may appear in the form of stripes laid longitudinally along the axis and usually spaced equally around the free surface of the extrudate (24,54). Moreover, at high shear rates, as the sharkskin intensity increases, the extrudate distortions may become quite large, which indicates that sharkskin is not always a small-amplitude, short-wavelength instability (17). With HDPEs, it is also possible to observe the occurrence of gross melt fracture, associated with an abrupt change in the slope of the flow curve, before the appearance of the stick–slip instability (54). In other cases, the stick–slip instability is observed before sharkskin. With some polymers, a second distinct oscillation region has been reported at higher shear rates (19,55). Finally, when the pressure is controlled so that the shear rate jumps to the right branch, the extrudate can exhibit a regular helical shape instead of gross melt fracture (24).

Four possible shapes of experimental flow curves are shown in Fig. 6.6. The first flow curve (Fig. 6.6a) is continuous and corresponds to the case where no stick–slip regime is observed. This is mostly encountered with hardly to moderately entangled polymers, such as LDPE, polypropylene (PP),

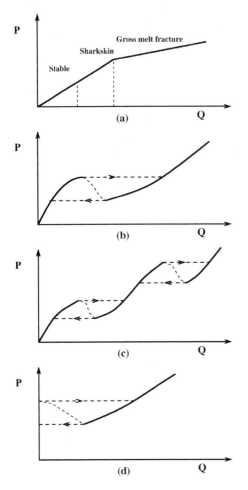

FIGURE **6.6** Shapes of experimental flow curves with no (a), one (b), and two (c) oscillation regimes, and with an oscillation regime with no left branch (d). The latter shape is observed with certain viscoplastic materials, such as food products.

polymethyl-methacrylate (PMMA), nylon (PA 66), polyvinylchloride (PVC), and PS, and is characteristic of extrusion with adherence at the wall. The onset of gross melt fracture coincides with a slope discontinuity of the flow curve, which is attributed to the triggering of a hydrodynamic instability upstream of the contraction region (54). The second flow curve (Fig. 6.6b) is the typical discontinuous flow curve with a stick–slip region. As already mentioned, this is obtained with linear low-molecular weight, narrowly distributed polymers that are sufficiently entangled to slip along the wall. In the third flow curve (Fig. 6.6c), a second oscillation regime appears at higher shear rates. This is mostly observed with highly entangled polymers. Finally, the fourth flow curve (Fig. 6.6d) shows an extreme case in which there is no left stable branch and the oscillation regime starts at zero volumetric flow rate. This situation arises with certain viscoplastic materials (i.e., materials exhibiting yield stress), such as food products (56). In a pressure-controlled experiment, no flow is observed below the critical shear stress at which the system jumps to the stable positive-slope branch of the flow curve. In a speed-controlled experiment, the pressure and the extrudate flow rate oscillate with the same period. What is interesting in this case is that for certain periods of time, no fluid emerges from the capillary, although the piston in the reservoir steadily advances at constant speed (56).

## 6.3  EXPERIMENTAL OBSERVATIONS

In this section, we review experimental observations in the stick–slip insta-bility regime, with emphasis on experiments with speed-controlled rheome-ters. Whenever we refer to "flow rate oscillations," these concern the instantaneous flow rate in the capillary; that is, the flow rate of the extrudate.

An example of an experimental flow curve obtained by means of capillary rheometer experiments on a HDPE (24) is shown in Fig. 6.7, where the squares correspond to steady state data and the triangles correspond to averaged transient data obtained in the stick–slip regime. The nearly instan-taneous jumps from point A to B and from point C to D occur at the critical wall shear stresses $\sigma_{c2}$ (onset of the stick–slip instability at the end-point A of the left branch) and $\sigma_{c3}$ (initial point C of the right branch), respectively. Thus these two critical stresses define the limiting values between which the wall shear stress oscillates, the amplitude of the oscillation being thus equal to the shear stress difference ($\sigma_{c2} - \sigma_{c3}$). Similarly, the amplitudes of the sudden apparent shear rate increase and decrease are determined by the shear rate differences between points A and B and points C and D, respectively (24,51,57–59). Hence for a given capillary (i.e., given $D$ and $L/D$), the onset of the stick–slip instability (i.e., the critical shear stress $\sigma_{c2}$) and the amplitudes of the oscillations are completely determined by the hysteresis loop defined by the flow curve. Some rare exceptions to this general observation have been reported: In experiments with PBs, Vinogradov et al. (7) observed a reduction of the amplitude of the pressure oscillation with flow rate (in the unstable regime), whereas Durand (60) observed a systematic amplitude growth (to both sides of the oscillation) on a HDPE.

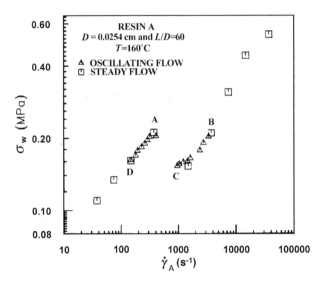

**FIGURE 6.7** The apparent flow curve of a HDPE. No steady flow is possible between the two branches. (From Ref. 24.)

As already mentioned, the stick–slip instability is not observed with most commercial long-chain branched polymers, e.g., LDPE and branched polysiloxane (39). It is mostly observed with linear high-molecular weight and narrow-distributed polymers, such as HDPE (see, e.g., Refs. 2–4,12,24, 51,53,55,61–64), LLDPE (12,13,30,54,55,64,65), PDMS (15,16,18), PI (7,10,11), and PB (7,19,48,49,54,59). However, it should be added that the presence of branching may result only to the attenuation and not to the complete suppression of the stick–slip instability. This is, for example, the case with long-chain branched metallocene PEs (66). The stick–slip instability has also been observed with Teflon fluoropolymer resins [such as polytetrafluoroethylene, PTFE (67); tetrafluoroethylene-hexafluoro-propylene copolymer, TFE–HFP (1,68,69); and tetrafluoroethylene hexafluoropropylene-perfluoro(alkyl vinyl ether), TFE–HFP–PAVE (69)], with various copolymers [such as ethylene-propylene (4,5,8), ethylene-butylene (4), and styrene–butadiene (70) copolymers], with ethylene propylene diene monomer (EPDM) compounds (61), and with other systems, such as highly entangled PB solutions (71), aqueous solutions of polyvinyl alcohol and sodium borate (72), wormlike micellar solutions (46,73), pastes (74), and chocolate (56). There are also some scarce and rather old reports of the instability with PP (6,38), polyisobutylene, PIB (7,75), and PS (10,11) melts.

### 6.3.1 Flow Curves and Onset of the Stick–Slip Instability

#### 6.3.1.1. Effect of the *L*/*D* Ratio

The dependence of the critical shear stress for the onset of the stick–slip instability on the $L/D$ ratio varies with the polymer and the experimental conditions: $\sigma_{c2}$ may increase or decrease or remain constant as the $L/D$ ratio increases.

Experiments with HDPE melts have shown that as the capillary length increases, the stick–slip regime is shifted to lower shear rates while its size and the value of the upper critical wall shear stress, $\sigma_{c2}$, as well as the stress difference $(\sigma_{c2} - \sigma_{c3})$ are increased. This effect is illustrated in Fig. 6.8, where we see four flow curves obtained by Durand et al. (51) for different $L/D$ ratios. Similar observations on HDPEs have been made by Myerholtz (4), Sato and Toda (58), and by Hatzikiriakos and Dealy (24), who attributed the fact that $\sigma_{c2}$ increases with the $L/D$ ratio to the pressure dependence of wall slip. Increasing the die diameter eventually leads to a continuous flow curve. As shown in Fig. 6.8, in the case of orifice die ($L/D = 0$), the flow curve is smooth (with no slope change) and no pressure oscillations are observed, although the extrudate is distorted above a certain shear rate, which agrees with earlier observations (50). The slope invariance agrees with the observations of Li et al. (55) on HDPE and LLDPE, but is in contrast to those of Piau et al. on PDMS (15) and LLDPE (18) melts (see, e.g., Fig. 6.9).

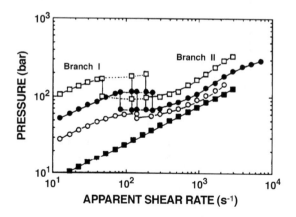

**FIGURE 6.8** Apparent flow curves for a HDPE and $L/D = 0$ (■), 4 (○), 8 (●), and 16 (□). The pressure oscillations for the last three $L/D$ ratios are shown in Fig. 6.4; $D = 1.3$ mm. (From Ref. 51.)

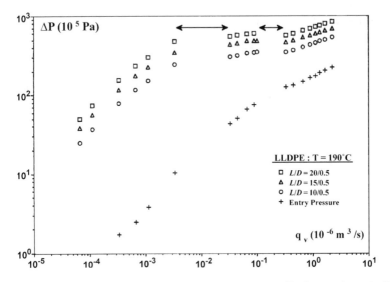

**FIGURE 6.9** Flow curves of a LLDPE with two oscillating regimes indicated by the arrows; $D = 0.5$ mm. (From Ref. 18.)

Observations similar to those of Durand et al. (51) have been made by Vergnes et al. (61) who carried out speed-controlled capillary rheometer experiments with EPDM compounds. They also observed the existence of stick–slip regimes at large $L/D$ values, in which the produced extrudates were smooth, although the pressure was oscillatory.

Experimental works with other materials show that $\sigma_{c2}$ is not always an increasing function of the $L/D$ ratio. The experiments of Vinogradov et al. (8) with PBs carried out at low $L/D$ ($\leq 25$) showed that $\sigma_{c2}$ slightly decreases with $L/D$, while those of Wang and Drda (25,26) with entangled linear polyethylenes showed that $\sigma_{c2}$ is virtually independent of $L/D$. Ramamurthy (12) found that for LLDPE resins, $\sigma_{c1}$ and $\sigma_{c2}$ are essentially independent of die diameter and $L/D$ ratio. The data of El Kissi and Piau (16) for PDMS, obtained with constant pressure experiments, show a slight decrease of $\sigma_{c2}$ when $L/D$ is increased from 20 to 40. In constant piston speed experiments with a LLDPE, Kalika and Denn (13) also observed shifting of the stick–slip regime to lower shear rates and reduction of both $\sigma_{c2}$ and ($\sigma_{c2} - \sigma_{c3}$), as they increased $L/D$ from 33.2 to 66.2 and 100.1. For $L/D = 100.1$, $\sigma_{c2}$ was found to be slightly less than $\sigma_{c3}$. Thus, in this case, the corresponding flow curve is not multivalued, and, therefore, no hysteresis loop is observed in pressure-controlled experiments.

## 6.3.1.2.  Effect of the Capillary Diameter–Indirect Slip Velocity Measurements

The variation of the flow curve with the capillary diameter (or the gap thickness in slit dies) is an indication of the occurrence of slip. As explained in Chapter 4 (present volume, by Archer), the slip velocity can be indirectly measured as a function of the wall shear stress using Mooney or similar techniques on flow curve data obtained with different capillary diameters, and, preferably, fixed $L/D$ ratio, so that the effects of pressure on viscosity and slip velocity are eliminated. In the case of no-slip, all flow curves coincide; otherwise, for a given apparent shear rate, the wall shear stress increases with the diameter. The onset of slip corresponds to the critical shear stress at which the flow curves determined with a series of capillaries having different diameters diverge.

Experiments with fluids exhibiting the stick–slip instability indicate that dependence of the flow curve on the capillary diameter becomes stronger as the apparent shear rate increases. The lower part of the left branch is insensitive to the capillary diameter, which suggests that the no-slip boundary condition applies in this region. However, the upper part may be slightly dependent on the capillary diameter, which implies that weak slip may occur. Finally, the right branch of the flow curve is strongly dependent on the capillary diameter, which indicates the occurrence of strong slip.

The occurrence of slip in the left branch of the flow curve has been inferred by measurements on HDPEs (3,23,24) and LLDPEs (12,13). However, other studies on LLDPEs showed that slip is negligible in the sharkskin regime (25,26). Ramamurthy (12) observed that slip occurs at about the same shear stress as the critical stress $\sigma_{c1}$ at which sharkskin appears. The data of Hatzikiriakos and Dealy (23) on a HDPE indicate that the critical shear stress for the onset of slip is in the range of 0.1–0.18 MPa, depending on the molecular weight and the polydispersity of the polymer. Noting the excellent superposition of the data obtained from capillaries of different $L/D$ ratios, Hatzikiriakos and Dealy (23) concluded that the pressure has a small effect on the viscosity, and then argued that the dependence of the flow curve (above the critical stress $\sigma_{c1}$) on both the diameter and the $L/D$ ratio implies that slip occurs and that the slip velocity depends on both the wall shear stress and normal stress. They developed a modified Mooney technique to obtain estimates of the slip velocity and proposed a slip equation for the left branch of the flow curve.

The strong dependence of the right branch of the flow curve on the capillary diameter has been shown for HDPE (3,23,51,62,76), LLDPE (12,13), PB (8), PDMS (16), and PIB (75) melts. Because of the strong slip occurring in the right branch of the flow curve, flow in this region is nearly plug. For a HDPE melt, Lupton and Regester (3) reported slip velocities more than 10 times those of the left branch. Hatzikiriakos and Dealy (24) found the slip velocities in the right branch to be a substantial fraction of the average velocity in the capillary. Observations of plug or nearly plug flow in the right branch of the flow curve have also been made by other researchers on HDPEs (51,76), LLDPE (13), and EPDM compounds (61).

Based on slip velocity estimates, Hatzikiriakos and Dealy (24) proposed a power-law slip equation for the right branch of their flow curve. El Kissi and Piau (17) also derived a nonmonotonic slip equation, which holds on both branches of the flow curve.

### 6.3.1.3. Effects of Temperature, Molecular Weight, and Molecular Weight Distribution

Experiments with different polymer melts have shown that an increase in temperature shifts the onset and the development of the sharkskin and the stick-slip instabilities to higher shear rates, but has less effect on the critical shear stresses, $\sigma_{c1}$ and $\sigma_{c2}$ (4,7,8,12,50). (The lower critical shear stress, $\sigma_{c3}$, has not been so much studied in the literature.) Vinogradov et al. (8) reported that temperature has no effect on the size of the stick–slip regime of narrow-distributed PB and PI melts. However, experiments with HDPE showed that the right branch of the flow curve is temperature-insensitive and that an increase of the temperature reduces the size of the stick–slip regime (63).

The experiments of Ramamurthy (12) with LLDPEs and those of El Kissi and Piau (16) with PDMS showed that the critical wall shear stress $\sigma_{c2}$ for the onset of the stick–slip instability is essentially independent of temperature. Other experiments with PBs (8,77), PIs (8), and Teflon fluoropolymer resins (1,68,69) showed that $\sigma_{c2}$ slightly increases with temperature, while more pronounced increases have been observed with HDPEs (24,53) and an ethylene–propylene copolymer (5). Yang et al. (77) pointed out that the increase of $\sigma_{c2}$ with temperature indicates that the stick–slip transition is most likely not produced by a chain desorption process.

For most tested polymers, $\sigma_{c2}$ varies usually between 0.01 and 1.0 MPa, but this variation is much less than a decade within any particular homologous series (8,43). Hatzikiriakos and Dealy (24) reported values in the range of 0.22–0.50 MPa for HDPE, which are consistent with the value of 0.43 MPa reported by both Ramamurthy (12) and Kalika and Denn (13) for LLDPEs. For PBs of various molecular weights and polydispersities, Vinogradov et al. (8) reported values of $\sigma_{c2}$ in the range 0.1–1 MPa. They also noted that the ratio of $\sigma_{c2}$ to the plateau modulus for narrow-distributed PBs and PIs is fairly constant, ranging from 0.4 to 0.5 (7). Similarly, in experiments with monodisperse PBs, Yang et al. (77) found that $\sigma_{c2} \approx 0.36$ MPa with the plateau modulus being equal to 1.0 MPa.

Obviously, $\sigma_{c1}$ is a fraction of $\sigma_{c2}$. Vinogradov et al. (7) reported that $\sigma_{c1} \approx 0.2\sigma_{c2}$ for polybutadienes, Kalika and Denn (13) obtained $\sigma_{c1} \approx 0.6\sigma_{c2}$ for LLDPE, and Hatzikiriakos and Dealy (22) found that $\sigma_{c1} \approx 0.45\sigma_{c2}$ for HDPE. Although the sketch in Fig. 6.3 implies that $\sigma_{c1}$ is smaller than $\sigma_{c3}$, it should be noted that this is not always true; $\sigma_{c1}$ may be between $\sigma_{c3}$ and $\sigma_{c2}$ or even coincide with $\sigma_{c2}$, in which case the sharkskin regime is not identifiable.

The effects of lowering the molecular weight and broadening the MWD are quite similar to that of increasing temperature, in the sense that they also shift the instabilities to higher shear rates and have no significant effects on $\sigma_{c1}$ and $\sigma_{c2}$ (4,62). Experiments with HDPEs (4,24,62), LLDPEs (12), PBs (7,59,77), PIs (7,11), and PSs (10,11) have shown that reducing the molecular weight and broadening the MWD tend to suppress the stick–slip instability regime, which is also true for the preceding sharkskin regime (12,47,50,52). Thus the stick–slip instability is not observed with low-molecular weight or broad-MWD polymers.

Experiments with nearly monodisperse or narrow-MWD PBs (7,8,59, 77) and PIs (7,8,10,11), and with systems of the same polydispersity ($I = M_w / M_n$), e.g., HDPEs (4,62,78) and PDMS (16), showed that increasing the molecular weight shifts the sharkskin and the stick–slip instability regimes to lower shear rates and increases their sizes, while the critical stresses $\sigma_{c1}$ and $\sigma_{c2}$ remain essentially the same. The measurements of Myerholtz (4) and Blyler and Hart (62) on HDPE melts and those of Ramamurthy (12) on a wide range of LLDPE resins have shown that $\sigma_{c2}$ is essentially independent of the molecular weight. On the other hand, the experiments of Utracki and Gendron (64) on a LLDPE showed that $\sigma_{c2}$ is reduced as the molecular weight is increased. The data of Durand et al. (78) on HDPEs also showed that the stress difference ($\sigma_{c2} - \sigma_{c3}$) increases considerably with the molecular weight.

On the other hand, broadening the MWD shifts the sharkskin and stick–slip instability regimes to higher shear rates, with $\sigma_{c1}$ and $\sigma_{c2}$ again being essentially unaffected (4,8,12,77). Myerholtz (4) demonstrated that the size of the hysteresis loop is reduced as the MWD of HDPEs becomes broader and collapses into a line resulting in a smooth flow curve when the MWD is very broad. Reduction of the stick–slip instability regime with MWD broadening has also been observed on PBs by Vinogradov et al. (8).

The experiments of Wang et al. (26,77) showed that for highly entangled HDPE melts of both and narrow MWD, $\sigma_{c2}$ scales with the weight-average molecular weight as $\sigma_{c2} \sim M_w^{-0.5}$, while for sufficiently entangled monodisperse PB melts, $\sigma_{c2}$ is independent of both $M_w$ and MWD, which is attributed to a different state of PB chain adsorption on steel surface.

### 6.3.2 Pressure and Flow Rate Oscillations

As already mentioned, the pressure and the mass flow rate in the stick–slip instability regime oscillate with the same period, $T$, following the hysteresis loop defined by the two-branched flow curve (3,9,24,51). The period of the oscillations steadily decreases, as the constant piston speed experiment proceeds, while the amplitude remains fairly constant and starts gradually decreasing only at the very end of the run (58,69). In capillary rheometer experiments with a linear PE, Sato and Toda (58) observed that the oscillations cease below a critical value of the piston height, which increases with the imposed piston velocity.

As discussed below, the value of the imposed flow rate has practically no effect on the amplitude, but affects the period and the waveform of the pressure oscillations. The die entry angle has a slight or no influence on the oscillations (4,63).

Figure 6.10 depicts the pressures and mass flow rates measured by Hatzikiriakos and Dealy (24) during three different cycles and plotted as functions of the normalized time $t/T$. During the compression phase, the flow rate is low and the fluid more or less sticks to the wall. When the pressure reaches its maximum value, the flow rate suddenly increases, which indicates the occurrence of strong slip. During the relaxation phase, the flow rate

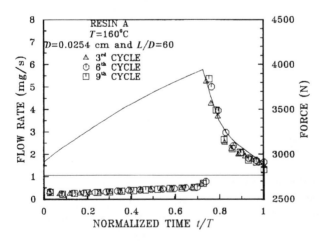

**Figure 6.10** Pressures (continuous curve) and mass flow rates (points), obtained in the stick–slip instability regime during three dierent cycles versus the normalized time $t/T$, where $T$ is the period of oscillation for the particular cycle. The horizontal line is the average mass flow rate calculated assuming incompressibility. HDPE, $\dot{\gamma}_A = 742$ sec$^{-1}$. (From Ref. 24.)

decreases slowly; when the pressure reaches its minimum value, the flow rate suddenly decreases, because of the reestablishment of adhesion at the wall, and the process is repeated. The horizontal line in Fig. 6.10 is the average mass flow rate.

### 6.3.2.1. Effect of the L/D Ratio

Because the variations of $\sigma_{c2}$ and $(\sigma_{c2} - \sigma_{c3})$ with the $L/D$ ratio depend on the material and the experimental conditions, this is also the case with the amplitude and the period of the pressure oscillations. Experiments with HDPEs have shown that both the amplitude and the period of the pressure oscillations increase with $L/D$ (51). This effect is illustrated in Fig. 6.4, where the pressure oscillations for the three nonzero $L/D$ ratios of Fig. 6.8 are shown. The increases in the amplitude and the period can also be directly deduced from the flow curves of Fig. 6.8, because the resulting hysteresis loop becomes wider as the $L/D$ ratio is increased. According to Den Doelder et al. (79), HDPEs do not exhibit pressure oscillations for short dies ($L/D < 5$), because these are overruled by the entry and exit pressure losses. The experiments of Rosenbaum et al. (69) with a TFE–HFP–PAVE resin showed that, for a fixed length of the barrel occupied by the polymer, the period of the oscillations increases from $L/D = 10$ to 20 but decreases from $L/D = 20$ to 40. In contrast to observations with HDPE (24), the difference between the two extreme shear stress values does not scale with the $L/D$ ratio.

The experiments of Hatzikiriakos and Dealy on a HDPE (24) showed that, for a given apparent shear rate in the unstable regime, the period of the oscillations is dramatically reduced if the diameter of the capillary is increased. Because the apparent shear rate, $\gamma_A = 32Q/\pi D^3$, is fixed, increasing the value of $D$ leads to a much higher piston speed. As a result, the polymer in the reservoir is compressed in a shorter period of time. Similar observations on HDPEs have been made by Durand et al. (78), who noted that the oscillations tend to disappear as the die diameter is increased.

### 6.3.2.2. Effect of the Reservoir Length

The important role of the polymer compressibility is evident from the fact that, for a given flow rate, the period of the oscillations scales roughly with the volume of the polymer melt in the reservoir (see Refs. 3,4,9,13,24,51,58, 63,69,80). This effect is shown in Fig. 6.11, where the periods measured by Hatzikiriakos and Dealy (24) for three different $L/D$ ratios and the same apparent shear rate are plotted vs. the reservoir length. It is also clear that the period of the oscillations increases with the $L/D$ ratio. This is expected, as longer dies result in higher pressures in the reservoir, which implies that more material is accumulated there that takes a longer time to flow out.

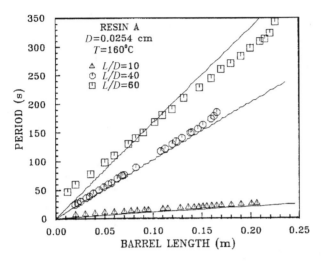

**FIGURE 6.11** Periods of oscillations for three different $L/D$ ratios vs. the reservoir length. The continuous curves are the predictions of the phenomenological model of Hatzikiriakos and Dealy (24), which is discussed in Sec. 6.4. HDPE, $\dot{\gamma}_A$ = 742 sec$^{-1}$. (From Ref. 24.)

It should be noted in Fig. 6.11 that the period of the oscillations is a linear function of the reservoir length, i.e., it decreases linearly with time. Moreover, the extrapolations of the experimental data do not pass through the origin. This was observed by many authors (9,50,51) and indicates that extrapolation at zero reservoir volume corresponds to a finite period of oscillations. The existence of an oscillatory (i.e., stick–slip) regime at zero absolute pressure cannot be confirmed for PE and other common melts in sliding plate rheometer experiments because of the fact that with such a rheometer, the high shear rates required cannot be achieved. At these shear rates, polymer melts assume a high elastic state and exhibit stick–slip transitions similar to those observed with elastomers when they are sliding on metallic surfaces. However, Archer et al. (81,82) reported shear stress oscillations of PB melts in planar Couette flow devices with aluminum and stainless steel substrates. At a critical shear stress value of about 0.26 MPa, a dramatic transition from simple shear to stick–slip flow has been observed with the period of the stress oscillations initially being close to the longest relaxation times of the polymers. Stress oscillations become more pronounced at higher shear rates and their frequency increases with shear rate until steady flow is reestablished at even higher shear rates with shear stresses substantially lower than those observed before the onset of the stick–slip flow regime.

Similar observations in sliding plate rheometer experiments on a PB have been made by Kazatchkov (83).

Weill (9) and Durand et al. (51) studied the effect of the reservoir length on the durations of compression and relaxation and found that both times increase linearly with the reservoir length, which indicates that the effect of the reservoir length on the waveform of the pressure oscillation is not significant. For larger values of the reservoir length, this effect is even weaker, which explains the fact that the normalized oscillations in Fig. 6.10 coincide (24).

### 6.3.2.3. Effect of the Imposed Shear Rate

Experiments with HDPEs (4,9,24,51), LLDPE (13), and PBs (7) have shown that the period and the shape of the pressure oscillations vary also with the imposed flow rate, whereas their amplitude remains unchanged. As illustrated in Fig. 6.12, where the periods for two apparent shear rates are plotted as functions of the reservoir length occupied by a HDPE, the period decreases as the apparent shear rate is increased within the stick–slip regime (24). This was attributed by Hatzikiriakos and Dealy (24) to the fact that the melt is compressed and relaxed in shorter periods of time. A period reduction has also been reported in experiments on other HDPEs (51,79), LLDPE (13), and

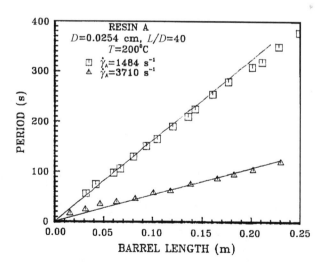

**FIGURE 6.12** Effect of the apparent shear rate on the period of oscillations. The continuous curves are the predictions of the phenomenological model of Hatzikiriakos and Dealy (24), which is discussed in Sec. 6.4. HDPE. (From Ref. 24.)

PB (59). In the case of PB, Lim and Schowalter (59) also observed reduction of the magnitude of the pressure oscillations with increasing flow rate.

Lupton and Regester (3) and Bergem (52) also observed that the period of the oscillations varies with the imposed flow rate but they did not give any information about this variation. Nevertheless, the experiments of Myerholtz (4) and Weill (9) on HDPEs showed that the period of the oscillations decreases with the imposed shear rate only up to the middle of the stick–slip domain, where it reaches a minimum, and then starts increasing. Myerholtz (4) noted that lowering the molecular weight shifts the minimum of the period to the right of the stick–slip regime. A minimum of the period near the end of the stick–slip regime has also been observed by Okubo and Hori (84) with HDPEs, and by Vinogradov et al. (7) with PB.

As the apparent shear rate increases, the duration of the descending portion of the pressure oscillation is not significantly affected, and thus the normalized duration of the ascending portion is relatively reduced (4,7,13,24,51,80,84). The effect of the imposed shear rate on the waveform of the pressure oscillation is illustrated in Fig. 6.13, which shows three pressure oscillations associated with flow rates ranging from the beginning to the end of the unstable zone (51).

### 6.3.3 Secondary Oscillation Regime

A second oscillation regime at higher shear rates has been observed in capillary rheometer experiments with different highly entangled polymers, such as LLDPE (17–20,24,55,85), HDPE (55,76,80), and PB (18). This second area

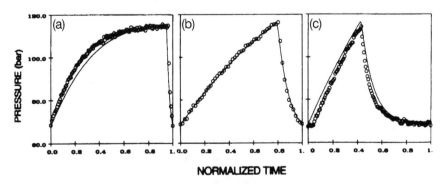

FIGURE 6.13 Influence of the imposed flow rate on normalized pressure oscillations. The three data sets correspond to (a) $\dot{\gamma}_A = 83$ sec$^{-1}$, (b) $\dot{\gamma}_A = 139$ sec$^{-1}$, and (c) $\dot{\gamma}_A = 232$ sec$^{-1}$. The solid lines are the predictions of the phenomenological model of Durand et al. (51), which is discussed in Sec. 6.4. HDPE, $D = 1.3$ mm, $L/D = 8$. (From Ref. 51.)

of oscillations occurs in slip conditions during both the compression and relaxation phases (54), and may be explained by means of a slip law exhibiting two maxima and two minima, similar to that proposed by Piau and El Kissi (18) for LLDPE and PB melts. The size and the position of the secondary stick–slip regime depend on the polymer and the experimental flow conditions.

In the experiments of Li et al. (55) on LLDPE and HDPE, the second oscillation region was observed at shear rates much higher than those of the primary stick–slip region. This was also the case with the experiments of Piau and El Kissi (18) and Hatzikiriakos and Dealy (24) with LLDPEs. The LLDPE flow curves shown in Fig. 6.9 have been constructed by Piau and El Kissi (18), who noted that the second area of instability is difficult to demonstrate, because of the small range of flow rates for which it occurs, and it disappears for larger capillary diameters or for lower temperatures. On the other hand, the experiments of Robert et al. (80) with a linear HDPE showed that the secondary stick–slip region may occur at flow rates just above, and may move within the primary stick–slip region by increasing the capillary diameter.

In studies with high-molecular weight PE (86) and LLDPE (65), islands of stable flow inside the stick–slip regime have been reported to exist in a narrow temperature range very close to the melting point. In both cases, a substantial pressure drop reduction has been observed, which is attributed to a mesophase transition. Windows of apparent shear rates where the pressure oscillations cease have also been reported for a TFE–HFP–PAVE resin by Rosenbaum et al. (69), who attributed the phenomenon to temperature fluctuations.

In general, the period and the amplitude of the pressure oscillations in the secondary oscillation region depend on the size and the position of the corresponding hysteresis cycle. In the experiments of El Kissi and Piau (20) on a LLDPE, the pressure oscillations in the secondary oscillation regime were larger in both amplitude and period than those in the primary oscillation regime. However, in the experiments of Robert et al. (80) on a HDPE, the observed pressure oscillations were relatively small.

Robert et al. (80) performed experiments with two different capillary diameters and the same $L/D$ ratio. With the smaller diameter, the second hysteresis cycle was very close to the primary one. The amplitude and the period of the pressure oscillations were respectively about 10 and 4 times smaller than those of the primary oscillations. Experiments at different shear rates within the secondary oscillation regime indicated that the period of the oscillations decreases and the relative duration of the compression phase is reduced as the imposed flow rate is increased, which is consistent with the experimental observations in the primary oscillation region. Results with the larger diameter revealed a much more complex situation, in which the secondary hysteresis cycle moves inside the primary one. The smaller secondary oscillations occur within the cycle of the primary oscillations, which results in a rather complex pressure waveform with two secondary maxima. Note that pressure oscillations with a reproducible secondary maximum have been recently observed by Merten et al. (35) in slit die experiments with a HDPE.

### 6.3.4 Extrudate Appearance

In general, the extrudate obtained during one cycle of oscillations consists of two segments of distinctly different appearance. These are associated with the ascending and descending parts of the pressure oscillation, i.e., with the compression phase along the left branch and the relaxation phase along the right branch of the flow curve. During the compression phase, the extrudate usually exhibits sharkskin. During the relaxation phase, the extrudate may be grossly distorted or rather smooth, depending on the scale of melt fracture along the right branch (i.e., on whether the lower part of the right branch corresponds to gross melt fracture or to superextrusion). Hence when compared to the segment associated with the relaxation phase, the shark-skinned segment of the compression phase appears to be relatively smooth in the former case and rougher in the latter. Moreover, with polymers exhibiting stick–slip but not sharkskin, there is also the possibility of both extrudate segments to be smooth. In this case, the extrudate shows a periodic diameter increase and decrease during a cycle (see, e.g., Refs. 4 and 80). Therefore, the association of the extrudate appearance with the pressure oscillation must be carried out very carefully to avoid confusion.

The relatively smooth (sharkskinned) and the distorted segments have been associated with the ascending and descending parts of the pressure oscillation, respectively, by Lupton and Regester (3) and Hatzikiriakos and Dealy (24) for the case of HDPE, and by Ramamurthy (12) for a LLDPE. The two extrudate parts shown in Fig. 6.5 show the changes that occur at the pressure maximum and minimum. It is clear that the transitions from a relatively smooth extrudate (ascending pressure) to a distorted extrudate (descending pressure) and vice versa abruptly occur. Hatzikiriakos and Dealy (24) noted that the distortions in the two segments are not uniform. In the relatively smooth part, as the pressure increases the extrudate becomes gradually rougher; in the distorted segment, the extrudate gradually becomes smoother as the pressure decreases.

In contrast to the above observations, Kalika and Denn (13) reported that for a LLDPE, the smooth segment of the extrudate corresponds to the relaxation and not to the compression phase of the pressure oscillation. Similar observations have been made with HDPE (57,84), PB (7,18,54), PDMS and LLDPE melts (15,18,54), a metallocene LLDPE (87), Teflon resins (69), and a styrene–butadiene rubber copolymer (70). In all these studies, the extrudate segment corresponding to the compression phase is covered with sharkskin-type cracks, while the segment corresponding to the relaxation phase is, more or less, smooth depending on the intensity of melt fracture. Let us note here that the disappearance of sharkskin during the relaxation phase, which corresponds to strong slip along the die walls, challenges theories considering slip as the cause of this instability.

Figure 6.14 depicts photos of PB extrudates obtained by El Kissi and Piau (54) at different flow rates. The smooth parts of the extrudate obtained in the stick–slip regime correspond to the relaxation phase. The succession of relatively smooth and cracked sections gives the extrudate the characteristic appearance of what Piau and El Kissi call *cork flow* (15).

Because they are determined by the relative durations of the compression and relaxation phases, the relative sizes of the two segments of the extrudate are not affected by the volume of the melt of the reservoir, but depend on the imposed flow rate. As the latter is increased in the unstable regime, the segment corresponding to the compression phase is relatively reduced. This is clear in Fig. 6.14, where the sharkskinned segments are reduced in size as the volumetric flow rate is increased, and disappear beyond the stick–slip regime (54).

### 6.3.5 Slip at the Wall

The role of slip in polymer extrusion instabilities, discussed in detail in Chapter 4 (present volume, by Archer), has been recognized as early as in

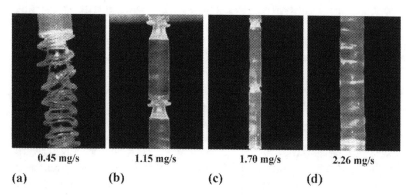

|  0.45 mg/s | 1.15 mg/s | 1.70 mg/s | 2.26 mg/s |

(a)  (b)  (c)  (d)

FIGURE 6.14 Instabilities in PB flow at the exit from an axisymmetric capillary in (a–c) and above (d) the stick–slip regime. $D = 2$ mm, $L = 20$ mm. (From Ref. 54.)

the mid-1960s (3,4), because then, both indirect and direct measurements have indicated that slip may occur in the upper part of the left branch and certainly occurs in the right branch of the flow curve following different molecular mechanisms (88).

When it occurs in the left branch, slip is weak (microscopic) and is associated with adhesive failure, that is stress-induced chain detachment/ desorption of polymer molecules from a weakly absorbing wall (12,13, 23,28,89). The number of the desorbed chains increases with shear stress (or shear rate) and, as a result, the slip velocity increases accordingly. Adhesive failure is inferred by experiments showing that there is a strong correlation between the work of adhesion of the polymer/wall interface and the onset of slip (52,90). Conversely, the findings of Wang and Drda (88) assert that it is not the work of adhesion, but stress that determines when polymer desorption may take place.

In the right branch, slip is strong (macroscopic) and is associated with cohesive failure on a highly absorbing wall; that is, the polymer molecules at the wall are disentangled from those in the bulk (21,52,53). As a result, the polymer slips freely, and essentially plug flow is obtained.

Obviously, the two molecular slip mechanisms strongly depend on the interfacial conditions, e.g., the material of die construction, the surface roughness, and the surface energy of the capillary die. Slip on high-energy surfaces that are highly absorbing, such as steel and aluminum, occurs through the second mechanism. The experiments of Wang and Drda (28) with highly entangled PEs have shown that the wall roughness leads to an increase of $\sigma_{c2}$ because of the increased resistance to interfacial disentanglement.

The critical wall shear stress, $\sigma_{c2}$, for the onset of the stick–slip instability corresponds to a transition from weak slip (chain detachment) to a strong one, as a result of sudden disentanglement of the polymer chains attached to the wall from the rest of the chains in the bulk. A theoretical basis for such a transition has been provided by de Gennes (91), who introduced the notion of the slip extrapolation length. This is defined as the distance of the polymer/ wall interface where the velocity of the bulk melt would be extrapolated to zero, and depends only on molecular properties of the polymer and the polymer/wall interface. According to the theoretical framework proposed by Wang et al. (25–28), the entire die wall takes part in the disentanglement of chains. After the disentanglement, the shear stress starts decreasing (relaxation) and when it assumes the value of the lower critical wall shear stress, $\sigma_{c3}$, another transition from strong to weak slip occurs, which is due to the reestablishment of entanglements at the polymer/wall interface. As a result of the two transitions from a weak to a strong slip and vice versa, a hysteresis loop is obtained, when the imposed shear rate is in the unstable regime.

The magnitude of the oscillations as well as the rate of transition from weak to strong slip strongly depend on the molecular weight of the polymers and should both scale with the reptation time (91). The experimental data of Wang and Drda (26) on HDPEs show that the slip extrapolation length varies with the molecular weight as $M_w^{3.4}$. This is in good agreement with de Gennes' (91) reptation theory for the rheology of entangled molecules, which states that the slip extrapolation length is proportional to $M_w^3$. Wang and Drda (26) also noted that the extrapolation length strongly scales with the degree of entanglement and is independent of temperature.

### 6.3.5.1. Die Materials and Polymer Processing Aids

The onset of extrudate distortion can be appreciably delayed by changing the die material or the surface roughness of the die wall and by using processing aids, such as surface coatings and additives. Most of these techniques target at slip promotion that leads to stresses lower than the critical stresses $\sigma_{c1}$ and $\sigma_{c2}$, and provide further confirmation that the stick–slip transition is an interfacial phenomenon.

In a series of experiments with a blown film die and LLDPE, Ramamurthy (12) found that $\sigma_{c1}$ shows some dependence but $\sigma_{c2}$ is less dependent on the materials of construction, which is consistent with the observations of Tordella on HDPEs (68). In pressure-controlled experiments with PBs, Vinogradov et al. (8) found that the flow curves obtained with identical steel and glass dies were the same, while the flow curve obtained with a Teflon die coincided with them only up to a shear stress roughly equal to $\sigma_{c2}/5$, beyond which the slope was sharply reduced so that the spurt instability was observed at a higher shear rate and a slightly lower critical shear stress. They attributed the slope change and the shifting of the instability regime to higher shear rates to the poor adhesion of PB to Teflon. In contrast to these observations, Ramamurthy (12), having noted that sharkskin and stick–slip flow can be eliminated by using α-brass instead of a stainless steel die, suggested that the elimination of the instabilities was a consequence of improved adhesion. The experiments of Hill et al. (92) on peeling of LLDPE from different metallic surfaces supported Ramamurthy's hypothesis. However, recent experiments have confirmed Vinogradov's observation that what suppresses the instabilities is wall slip, and not improved adhesion (30,31). This effect is illustrated in Fig. 6.15, where the flow curves obtained by Ghanta et al. (30) using identical brass and stainless steel dies are shown. The slope of the flow curve obtained with the brass die is quickly reduced, suggesting early occurrence of slip. Sharkskin and stick–slip flow are not observed, because the corresponding critical wall shear stresses are much higher, and the data smoothly reach the right branch of the flow curve obtained with the stainless steel die.

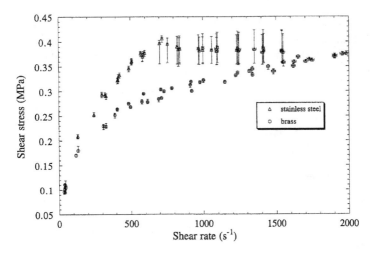

**FIGURE 6.15** Flow curves obtained with extrusion of LLDPE through identical stainless-steel and brass capillaries. The vertical lines show the amplitudes of the pressure oscillations at a given shear rate. (From Ref. 30.)

Fluoropolymer-coated dies have been reported to promote interfacial slip, by reducing adsorption of polymer chains to the die surface, and to significantly delay or completely eliminate sharkskin and, in certain cases, the stick–slip instability (8,19,27,28,93). Fluorinated surfaces are characterized by low surface energy and are often referred to as slippery surfaces (19). Additives used as dispersions at low concentrations in the base polymer migrate toward the die wall and act as lubricants inducing slippage between themselves and the main polymer (94). They promote slip through reduction of the apparent viscosity, and, as a result, the required shear stresses are reduced and the flow curve is shifted to higher shear rates (32).

An example of the shift of the flow curve and instability onset to higher flow rates is depicted in Fig. 6.16, where the flow curve of a HDPE containing a Teflon resin is compared to that of the virgin polymer. The sharkskin and stick–slip instabilities are eliminated in the presence of the processing additive, and distortions are only obtained at the point where the two flow curves merge, which is indicated as the onset of gross melt fracture. Comparison of Figs. 6.16 and 6.15 shows that the effect of the additive is quite similar to that of a slippery die material. The experiments of Rosenbaum et al. (69) showed that polyethylene and Teflon resins may interchange roles; that is, polyethylene can also be used as a processing aid in the extrusion of Teflon resins, eliminating sharkskin and stick–slip melt fracture. Conventional fluoropolymer and new polymer processing additives are discussed respec-

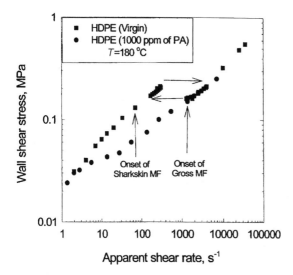

**FIGURE 6.16** Flow curve data of a HDPE extruded in a pressure-driven capillary rheometer as virgin and in the presence of a fluoroelastomer. The sharkskin and stick–slip instabilities are eliminated when the processing additive is used. (From Ref. 32.)

tively in Chapters 8 (present volume, by Lavallee) and 9 (present volume, by Hatzikiriakos).

## 6.3.5.2. Direct Wall Slip Measurements and the Stick–Slip Instability

The first direct wall slip measurements in the stick–slip instability regime were reported by Paskhin (75) who used Laser-Doppler velocimetry (LDV) and studied the extrusion of PIB melts. Much later, Lim and Schowalter (59) used hot-film anemometry to study the extrusion of nearly monodisperse PB melts in rectangular dies. They detected wall slip in the stick–slip regime, and found that the slip velocity and the pressure oscillate with the same period and a 180° phase difference, which means that the slip velocity is at a maximum when the pressure is at a minimum. They also observed that the oscillations occur globally throughout the slit die.

Measurements of polymer wall slippage at a submicrometric scale were reported by Migler et al. (94), who studied the slip of a PDMS melt on a polished silica surface in a Couette flow cell. Using fluorescent-labeled chains, they measured a transition from weak to strong slip, which was explained in terms of a coil-stretch transition of the chains bound to the surface. A similar fluorescence technique was used by Legrand and Piau (95) to study wall slip of a high-molecular weight PDMS in a rough stainless steel die. Their measurements showed that slip occurs in a plane localized at a distance from the ridges of the surface roughness and not at the polymer/wall interface, which agrees with earlier experimental observations (28,52).

Münstedt et al. (33) investigated the flow behavior of LDPE and HDPE melts in a slit die using high spatial and temporal resolution LDV. They did not detect any indications of measurable wall slip for the LDPE, but for the HDPE they reported pronounced slip velocities even at low apparent shear rates before the appearance of pressure oscillations. In the stick–slip regime, they found that the velocity oscillates between two profiles with the same frequency as the pressure. These low and high flow rate profiles are observed at the pressure minimum and maximum, respectively, and clearly indicate the occurrence of wall slip. Beyond the stick–slip regime, the observed velocity profiles are nearly plug, which verifies the occurrence of strong slip at the wall. More recently, Merten et al. (35) carried out similar experiments with a HDPE measuring pressure and velocity oscillations in the stick–slip regime. Their results showed that the pressure oscillates in phase at all positions in the slit and that the minimum of the pressure oscillation occurs slightly after the maximum of the velocity. This phase shift was attributed to the effect of the melt compressibility (35).

Robert et al. (34,76) investigated the stick–slip instability on HDPEs in a transparent slit die using LDV and flow birefringence. Their measurements confirm that the velocity oscillates between two steady state profiles: A classical profile with no-slip on the left branch and a profile showing strong macroscopic wall slip on the right branch of the flow curve. Such velocity profiles are shown in Fig. 6.17. Robert et al. (34) also noted that wall slip seems to be initiated at the entry of the die land and occurs all along, but it is not homogeneous across the die width. In a recent article (96), they reported a

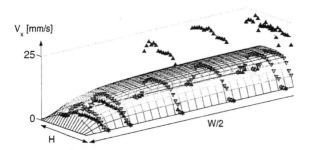

FIGURE 6.17   Maxima (▲) and minima (▼) profiles of the oscillating velocity in the stick–slip regime measured by LDV; HDPE, slit die. (From Ref. 34.)

phase shift between the pressure and velocity oscillations which is much larger than that observed by Münstedt et al. (33), i.e., the maximum of the velocity was found to be approximately a quarter out of phase with the maximum of the pressure. This effect was attributed to the viscoelasticity of the HDPE, which tends to slow the transients during the jumps between the two branches of the flow curve. On the other hand, the oscillations of stresses were found to be in phase with velocity oscillations (96).

## 6.4 ONE-DIMENSIONAL PHENOMENOLOGICAL MODELS

One-dimensional phenomenological models have been developed by various groups to quantitatively describe the pressure changes during the capillary flow of a polymer melt. These are based on the principle of mass conservation and take into account the compressibility effects in the reservoir. In addition to the melt compressibility, a necessary condition for obtaining pressure oscillations is the multivaluedness of the flow curve. A generic multivalued flow curve consisting of two branches is shown in Fig. 6.18. This is, of course, similar to the experimental flow curves and can be theoretically justified in terms of a multivalued slip equation or a nonmonotonic constitutive equation, as explained in previous sections. Hence when nonlinear slip is employed, fluid elasticity is not an essential factor for explaining the oscillations (24,97).

What is different in the various one-dimensional models proposed in the literature is the way the multivalued flow curve is incorporated. A rough approach is to use the experimental flow curve as input, which makes a model purely heuristic (9,84). A more refined approach is to "construct" the flow curve using the conservation of momentum in the capillary. However, even in

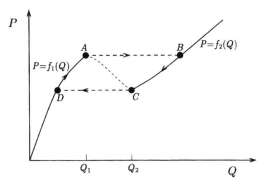

**FIGURE 6.18** A generic multivalued flow curve and the resulting hysteresis loop.

this case, the proposed models still require fitting to the experimental data of certain parameters concerning the limits of the hysteresis loop ABCD, the occurrence of slip, and the constitutive equation.

The geometry of the capillary rheometer is shown in Fig. 6.19. A number of simplifying assumptions are made: (1) the entrance effects in the capillary and the pressure drop in the barrel are negligible; (2) the density, $\rho$, of the melt is related to pressure, $P$, in the barrel by $\rho(P) = \rho_0(1 + \beta P)$, where $\rho_0$ is the reference density (at atmospheric pressure), and $\beta \equiv (d\rho/dP)/\rho$ is the isothermal melt compressibility, assumed to be uniform and constant; (3) the volume of the melt in the capillary is negligible compared to that in the barrel, and, thus, the total volume of the melt is $V(t) = Ah(t)$, where $A$ is the cross-sectional area of the barrel and the plunger, and $h(t)$ is the distance from the plunger to the capillary entrance; and (4) the plunger moves at a constant speed, $v_p$, i.e., the volumetric flow rate at the inlet, $Q_i = A\,v_p = -A\,(dh/dt)$, is constant. With the above assumptions, the conservation of mass yields the following ordinary differential equation:

$$Ah\beta\,\rho_0\,\frac{dP}{dt} = \rho Q_i - \rho_0 Q_e \qquad (6.8)$$

where $Q_e$ is the volumetric flow rate at the exit of the die (i.e., of the extrudate). Given that $\rho \approx \rho_0$, it is clear that if $Q_i - Q_e > 0\ (<0)$, the total mass in the reservoir will increase (decrease), causing a positive (negative) change in the pressure.

It is instructive at this point to refer to the explanation provided by Pearson (40) for the pressure oscillations and the time scale of the periodicity, which are dictated by Eq. (6.8) and the shape of the flow curve. Assuming that the flow curve is described by the discontinuous function shown in Fig. 6.18 and denoting $\bar{\rho}$ as the density which effectively makes Eq. (6.8) hold over the whole length of the capillary, one can replace the mass flow rate $\rho_0 Q_e$ by $\bar{\rho}\,Q$. If $Q_i$ lies in the unstable regime, i.e., if $Q_1 < Q_i < Q_2$, one expects $P$ to

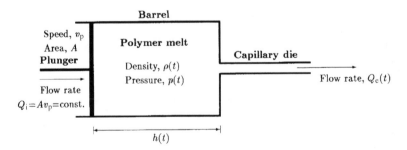

FIGURE 6.19  Geometry of the capillary rheometer.

oscillate around the hysteresis cycle ABCD and the period of the oscillation to be of the order

$$T = O \left[ \frac{Ah\beta P}{Q_i(1 - Q_1/Q_2)} \right] \tag{6.9}$$

This result explains the experimental observations for the period of the oscillations, which linearly increases with the volume of the polymer melt in the reservoir and decreases (at least initially) with the value of the imposed flow rate.

Hatzikiriakos and Dealy (24) developed a phenomenological model based on the assumption that slip occurs at the capillary wall according to the slip equations they obtained for the two branches of the flow curve. By integrating along the capillary, the velocity profile obtained by the lubrication approximation for a power-law fluid, they derived an expression relating the flow rate $Q_e$ as a function of the wall shear stress, $\sigma_w(t,z)$ and the slip velocity, $u_s(t,z)$. They then integrated the momentum equation and made certain assumptions for the stresses, to obtain equations describing the variation of the wall shear stress, $\sigma_w$, and the wall normal stress, across the capillary.

By solving numerically Eq. (6.8) together with the above equations, Hatzikiriakos and Dealy (24) obtained pressure and flow rate oscillations in the unstable regime. The trajectories of their solutions closely follow the hysteresis loop of the flow curve. Because of the jumps between the two branches of the flow curve, the flow rate oscillations are characterized by abrupt changes, as in the experiments. The period of the oscillations linearly decreases with the volume of the reservoir, as expected. The comparison of the model predictions with the experimental data for three $L/D$ ratios in Fig. 6.11 reveals good agreement between theory and experiment. However, the period is slightly underestimated and, in contrast to the experimental data, the predictions pass through the origin. The model also captures the reduction of the period and the changes in the waveform of the pressure oscillations that are observed experimentally as the imposed flow rate is increased in the unstable regime. As shown in Fig. 6.12, the calculated periods for two different shear rates compare well with the corresponding HDPE data.

Piau and El Kissi (18) proposed a one-dimensional model that requires, as input, the experimental flow curve and employs an empirical slip equation with one or two minima. Assuming that density variations are not large ($\rho/\rho_0 \approx 1$), Eq. (6.8) can be integrated to

$$t = \frac{h_0}{v_p} \left[ 1 - \exp\left( -\beta \int_{P_0}^{P(t)} \frac{dP}{1 - Q_e/Q_i} \right) \right] \tag{6.10}$$

where $h_0$ is the initial position of the piston, and $P_0$ is the initial pressure. Piau and El Kissi (18) assumed that the pressure and the flow rate follow the experimental hysteresis loop ABCD. Along the left branch, they assumed that no slip occurs and gradually incremented the pressure, calculating $Q_e$ from the flow curve. Along the right branch, they gradually decreased the slip velocity, calculating the wall shear stress from the slip equation, and then $Q_e$ from the analytical solution for the steady state Poiseuille flow of a power-law fluid, and $P$ from the flow curve. The time required at each step of both phases was calculated by Eq. (6.10). Piau and El Kissi (18) presented only some pressure oscillations obtained for different geometries, which compared well with their experimental observations with a PDMS.

Adewale and Leonov (43) proposed a more sophisticated model consisting of a viscoelastic constitutive equation and a nonmonotonic slip equation, which is a modification of a stochastic slip law for cross-linked elastomers. The constitutive equation contains a *hardening* parameter, which is solely determined by the polymer molecular characteristics. The critical conditions for the transition from no-slip to slip are obtained under the assumption that hardening causes loss of adherence of the polymer melt to the die wall and initiation of slip. Adewale and Leonov (43) estimated the parameters of the model from the experimental data of Vinogradov et al. (7) for a high-molecular weight, narrowly distributed PI. They noted that although a thin loop represents the data, a deeper hysteresis loop is more appropriate for demonstration purposes. In qualitative agreement with experimental observations on HDPE (4,9,84) and PB melts (7), the model predicts a minimum of the period near the end of the stick–slip regime. It also predicts shortening of the normalized duration of the compression phase, as the imposed flow rate is increased in the unstable regime, which also agrees with experiments (4,13,24,51).

Sato and Toda (58) developed a statistical stick–slip model of microscopic springs describing the behavior of polymer chains at the wall, which results in a nonmonotonic flow curve. They carried out a linear stability analysis for a Newtonian fluid, which indicated that steady state solutions in the negative-slope regime of the flow curve are unstable when the piston height is greater than a critical value. This result was verified with numerical calculations as well as with experiments with a linear PE melt. The model predicts a linear relation between the period of oscillations and the piston height (above the critical value). However, the oscillations follow smooth limit cycles and not the hysteresis loop defined by the flow curve.

The relaxation–oscillation model developed by Molenaar and Koopmans (98) has been the basis for the development of a number of one-dimensional models. Following Weill (9), Molenaar and Koopmans (98) assumed that density variations in the reservoir are not large ($\rho/\rho_0 \approx 1$), and simplified Eq. (6.8) to

$$\frac{dP}{dt} = \frac{1}{C}(Q_i - Q_e) \tag{6.11}$$

where $C = A\beta h$. Assuming that $h$ is slowly varying, they took $C$ as a constant and noted that there must exist a functional relationship between the pressure and the flow rate, such that

$$\frac{dQ_e}{dt} = \frac{1}{K}F(Q_i, P) = \frac{1}{K}[P - f(Q_i)] \tag{6.12}$$

where $K$ is an unknown factor that depends on the geometry and the material but is constant in time (98). This factor expresses the ability of the melt to adjust the pressure via adjustment of the flow rate. The coupled Eqs. (6.11) and (6.12) are integrated in time for given $F$, $Q_i$, $C$, and $K$ and initial conditions.

Durand et al. (51) have subsequently applied experimental results for HDPE to the model of Molenaar and Koopmans, assuming that $f$ coincides with the two stable branches of the experimental steady state flow curve and linearly varies between points A and C in the unstable regime. Two parameters of the model together with the flow rate limits of the oscillating regime are identified from the experimental data. Fig. 6.20 shows a set of computed

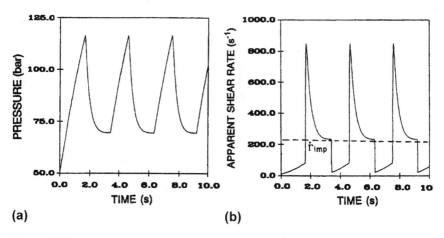

**FIGURE 6.20** Pressure (a) and shear rate oscillations (b) during hysteresis cycle, calculated with the phenomenological model of Durand et al. (51); $D = 1.3$ mm, $L/D = 8$, $\dot{\gamma} = 232$ sec$^{-1}$. (From Ref. 51.)

pressure and shear rate oscillations corresponding to the end of the unstable zone. The predictions of the model agree with experimental observations with a HDPE, and the period of the oscillations linearly decreases with the volume of the reservoir. However, if the slope of the period variation is adjusted to that of the experiments, then the period is underestimated and the ordinate of the period at zero reservoir volume is zero, which is not the case with experiments. As already mentioned, this drawback is also exhibited by the model of Hatzikiriakos and Dealy (24). As shown in Fig. 6.13, the model of Durand et al. (51) describes very well the decrease in the period and the reduction of the normalized duration of the ascending portion of the pressure oscillation that are observed as the imposed flow rate is increased in the unstable regime.

Den Doelder et al. (79) refined the model of Molenaar and Koopmans by introducing the momentum equation, including a modified slip law based on the ideas of Greenberg and Demay (99), taking into account pressure corrections, and using the Carreau–Yasuda model to account for the shear-thinning behavior of a HDPE melt. Their model reproduces very well the experimental pressure oscillations shown in Fig. 6.4, which were obtained with a HDPE and different capillary lengths. It also describes well data showing the effect of the reservoir length on the period of the oscillations as well as on the durations of the compression and relaxation phases. As with the period predictions of other phenomenological models, the computed straight lines pass through the origin, in contrast to the experimental data. The model initially predicts a decrease in the period of the oscillations as the imposed volumetric flow rate is increased in the unstable regime. However, for higher values of the volumetric flow rate near the critical value $Q_2$, the model predicts a significant increase in the period. As already mentioned, this variation of the period, also predicted by the model of Adewale and Leonov (43), is consistent with some experimental data for HDPE and PB melts.

More recently, Dubbeldam and Molenaar (100) presented a phenomenological model incorporating a nonmonotonic slip equation taken from Ref. 101. They obtained a two-dimensional dynamical system for the wall slip velocity and the pressure in the barrel, which is amenable to analytical evaluation in the case of Newtonian and power-law fluids. Their explicit expressions for the oscillation frequency and the size of the instability regime show that shear thinning moves the instability region to higher flow rates, reducing its size and slightly enhances the period of the pressure oscillations.

Despite the progress made with one-dimensional models in describing the experimental stick–slip instabilities, they are still descriptive and not predictive, because they require, as input, certain experimental parameters. These models cannot predict the onset of instability, the wall slip, and the geometry effects. Moreover, they do not include the extrudate region, and, therefore, they cannot relate the distortions of the extrudate to the pressure oscillations. A necessary step toward achieving the goal of predicting stick–slip and flow curve characteristics from basic polymer–polymer bulk and polymer–wall interfacial properties is the development and use of dynamic molecular slip models.

## 6.5 MECHANISMS FOR THE STICK–SLIP INSTABILITY

Macroscopic mechanisms for the stick–slip instability had been the subject of debate for quite a long time (36–39), until the physics of stick–slip transition was revealed and experimentally confirmed (27–29). The theoretical explanations suggested in the literature for the stick–slip instability are based on the nonmonotonicity of the flow curve, which exhibits a maximum and a minimum, and the fact that steady state solutions corresponding to the negative-slope regime of the flow curve are unstable. Nonmonotonicity of the flow curve can be obtained by a nonmonotonic slip law or by a nonmonotone constitutive equation. The transitions from the maximum of the flow curve to the right positive-slope branch and from the minimum to the left positive-slope branch lead to a hysteresis cycle, which is used to describe the observed sudden flow rate changes in pressure-controlled experiments and the pressure and flow rate oscillations in flow rate-controlled ones.

The compressibility/slip mechanism has been the most popular explanation for the stick–slip instability, and is the only one that is consistent with experimental observations. According to this mechanism, the periodic transitions between weak and strong slip at the capillary wall (i.e., the jumps between the two branches of the flow curve), which lead to the pressure and flow rate oscillations and generate waves on the extrudate surface, are sustained by the compressibility of the melt in the reservoir. This is the mechanism employed in the one-dimensional phenomenological models discussed in the previous section. The compressibility/slip mechanism has also been used in two-dimensional finite-element simulations of the extrusion process (97), in which the reservoir region, where most of the fluid is compressed and decompressed, has been omitted.

Viscoelasticity may replace compressibility and, when combined with nonmonotonic slip, can act as a storage of elastic energy generating self-sustained pressure oscillations and waves on the extrudate surface in the stick–slip regime. However, the elasticity/slip mechanism cannot generate oscillations that follow the hysteresis loop of the flow curve. It leads only to small-amplitude, high-frequency distortions of the extrudate surface that are consistent with sharkskin rather than with the stick–slip instability (41,43). These oscillations may be superimposed to the much larger oscillations caused by the compressibility/slip mechanism.

The constitutive instability mechanism has initially raised high expectations for explaining the stick–slip instability (37,44,102). Huseby (102) and, much later, Lin (10,11) conjectured that the maximum in the constitutive equation, which is predicted by some reptation models, is a manifestation of a physical instability that promotes the stick–slip phenomenon. A supporting argument for this mechanism came from the fact that it can be theoretically shown that the stress maximum can be eliminated by lowering the molecular weight and/or broadening the MWD, which agrees with the experimental observations on linear and flexible polymers (see Sec. 6.3.1). Hence the nonmonotonicity of the resulting flow curve allowed an explanation of the spurt effect and the fitting of certain flow curve data for nearly monodisperse systems. An analogous hypothesis was proposed by Vinogradov et al. (5–8). Based on their finding that the shear rate at the stick–slip transition correlates with the oscillation frequency at which the marked change of the $G'$ and $G''$ spectra occurs, they conjectured that the stick–slip transition under controlled pressure is due to a bulk transition from a liquidlike to a solidlike state (7). However, this theory lacks a clear theoretical basis, and is not supported by experimental observations showing that the stick–slip transition results from a breakdown of chain entanglements between absorbed and next-layer unbound chains; that is, it is not a manifestation of a constitutive instability (see, e.g., Ref. 77).

The drawbacks of the constitutive instability mechanism can be summarized as follows:

1. Because this mechanism employs no-slip boundary conditions, the resulting flow curves are not diameter-dependent, which is contrary to experimental observations (37,43).

2. Most of the polymer melts and elastomers exhibiting the stick–slip instability do not obey a nonmonotonic constitutive equation (43). Nonmonotonic constitutive equations may be used only for polymers having a very high molecular weight and a very narrow MWD, and not for most commercial polymers used in studies of oscillatory flow and flow curve hysteresis, which have moderate molecular weight and a rather broad MWD (24).

3. The apparent slip predicted by nonmonotonic constitutive models is solely dependent on the molecular characteristics of the polymer, and, consequently, independent of the roughness and the surface energy of the capillary die, in contrast to experimental observations (19,28,77).

4. This mechanism cannot lead to apparent slip in the sharkskin regime. Thus it can be used only for molten polymers exhibiting no-slip in this regime, such as nearly monodisperse PBs (59,103,104).

5. Self-sustained pressure oscillations in Poiseuille flow of fluids obeying nonmonotonic constitutive equations have been reported only for extreme values of certain material parameters, at which, however, the accuracy of the numerical schemes is questionable (45).

## 6.6 CONCLUDING REMARKS

In the present chapter, the stick–slip extrusion instability has been reviewed. Despite the accumulation of experimental evidence for almost 50 years, significant progress in understanding the interfacial nature of the stick–slip phenomenon was made only during the past decade. It is now well understood that the onset of the instability corresponds to a transition from weak to strong slip, as a result of sudden disentanglement of the polymer chains attached to the die wall from chains in the bulk. Progress also has been made in determining the effect of molecular weight on the magnitude of the oscillations as well as on the critical shear stress $\sigma_{c2}$, in the case of PEs. Nevertheless, to verify the molecular-scale slip mechanisms leading to the stick–slip instability, further research is required. Slip mechanisms should be tested for other highly entangled linear polymer melts with well-characterized polymer/wall interfaces and correlated to the chain entanglement density and other molecular structure parameters. Although direct velocity measurements at micrometric or submicrometric scale are useful, the development of experimental tools for studying polymer chain behavior within one radius of gyration is a necessary step for elucidating the nature of wall slip and stick–slip transition.

## ACKNOWLEDGMENTS

I would like to thank the referees for their useful comments and constructive criticism. Financial assistance from the University of Cyprus is also gratefully acknowledged.

## REFERENCES

1. Tordella, J.P. Fracture in the extrusion of amorphous polymers through capillaries. J. Appl. Phys. 1956, *27*, 454–458.
2. Bagley, E.B.; Cabott, I.M.; West, D.C. Discontinuity in the flow curve of polyethylene. J. Appl. Phys. 1958, *29*, 109–110.
3. Lupton, J.M.; Regester, J.W. Melt flow of polyethylene at high rates. Polym. Eng. Sci. 1965, *5*, 235–245.

4. Myerholtz, R.W. Oscillating flow behavior of high-density polyethylene melts. J. Appl. Polym. Sci. 1967, *11*, 687–698.

5. Vinogradov, G.V.; Ivanova, L.I. Viscous properties of polymer melts and elastomers exemplified by ethylene–propylene copolymer. Rheol. Acta 1967, *6*, 209–222.

6. Vinogradov, G.V.; Driedman, M.L.; Yarlykov, N.V.; Malkin, A.Y. Unsteady flow of polymer melts: polypropylene. Rheol. Acta 1970, *9*, 323–329.

7. Vinogradov, G.V.; Malkin, A.Y.; Yanovskii, Y.G.; Borisenkova, E.K.; Yarlykov, B.V.; Berezhnaya, G.V. Viscoelastic properties and flow of narrow distribution polybutadienes and polyisoprenes. J. Polym. Sci., A-2 1972, *10*, 1061–1084.

8. Vinogradov, G.V.; Protasov, V.P.; Dreval, V.E. The rheological behavior of flexible-chain polymers in the region of high shear rates and stresses, the critical process of spurting, and supercritical conditions of their movement at $T > T_g$. Rheol. Acta 1984, *23*, 46–61.

9. Weill, A. About the origin of sharkskin. Rheol. Acta 1980, *19*, 623–632.

10. Lin, Y.H. Explanation for stick–slip melt fracture in terms of molecular dynamics in polymer melts. J. Rheol. 1985, *29*, 605–637.

11. Lin, Y.H. Unified molecular theories of linear and non-linear viscoelasticity of flexible linear polymers—explaining the 3.4 power law of the zero-shear viscosity and the slip–stick melt fracture phenomenon. J. non-Newton. Fluid Mech. 1987, *23*, 163–187.

12. Ramamurthy, A.V. Wall slip in viscous fluids and influence of materials of construction. J. Rheol. 1986, *30*, 337–357.

13. Kalika, D.S.; Denn, M.M. Wall slip and extrudate distortion in linear low-density polyethylene. J. Rheol. 1987, *31*, 815–834.

14. El Kissi, N.; Piau, J.M. Écoulement de fluides polymères enchevêtrés dans un capillaire. Modélisation du glissement macroscopique à la paroi. C. R. Acad. Sci. Paris, Sér. II, 1989, *309*, 7–9.

15. Piau, J.M.; El Kissi, N.; Tremblay, B. Influence of upstream instabilities and wall slip on melt fracture and sharkskin phenomena during silicones extrusion through orifice dies. J. non-Newton. Fluid Mech. 1990, *34*, 145–180.

16. El Kissi, N.; Piau, J.M. The different capillary flow regimes of entangled polydimethylsiloxane polymers: macroscopic slip at the wall, hysteresis and cork flow. J. non-Newton. Fluid Mech. 1990, *37*, 55–94.

17. El Kissi, N.; Piau, J.M. Adhesion of LLDPE on the wall for flow regimes with sharkskin. J. Rheol. 1994, *38*, 1447–1463.

18. Piau, J.M.; El Kissi, N. Measurement and modelling of friction in polymer melts during macroscopic slip at the wall. J. non-Newton. Fluid Mech. 1994, *54*, 121–142.

19. Piau, J.M.; El Kissi, N.; Toussaint, F.; Mezghani, A. Distortions of polymer melt extrudates and their elimination using slippery surfaces. Rheol. Acta 1995, *34*, 40–57.

20. El Kissi, N.; Piau, J.M. Stability phenomena during polymer melt extrusion. In *Rheology for Polymer Melt Processing*; Piau, J.M., Agassant, J.F., Eds.; Elsevier: New York, 1996; 389–420.

21.  El Kissi, N.; Piau, J.M.; Toussaint, F. Sharkskin and cracking of polymer melt extrudates. J. non-Newton. Fluid Mech. 1997, *68*, 271–290.
22.  Hatzikiriakos, S.G.; Dealy, J.M. Wall slip of molten high density polyethylenes. I. Sliding plate rheometer studies. J. Rheol. 1991, *35*, 495–523.
23.  Hatzikiriakos, S.G.; Dealy, J.M. Wall slip of molten high density polyethylenes. II. Capillary rheometer studies. J. Rheol. 1992, *36*, 703–741.
24.  Hatzikiriakos, S.G.; Dealy, J.M. Role of slip and fracture in the oscillating flow of HDPE in a capillary. J. Rheol. 1992, *36*, 845–884.
25.  Wang, S.Q.; Drda, P.A. Superfluid-like stick–slip transition in capillary flow of linear polyethylene. 1. General features. Macromolecules 1996, *29*, 2627–2632.
26.  Wang, S.Q.; Drda, P.A. Stick–slip transition in capillary flow of linear polyethylene. 2. Molecular weight and low-temperature anomaly. Macromolecules 1996, *29*, 4115–4119.
27.  Wang, S.Q.; Drda, P.A.; Inn, Y.W. Exploring molecular origins of sharkskin, partial slip, and slope change in flow curves of linear low density polyethylene. J. Rheol. 1996, *40*, 875–897.
28.  Wang, S.Q.; Drda, P.A. Stick–slip transition in capillary flow of linear polyethylene. 3. Surface conditions. Rheol. Acta 1997, *36*, 128–134.
29.  Wang, S.Q. Molecular transitions and dynamics at polymer/wall interfaces: origins of flow instabilities and wall slip. Adv. Polym. Sci. 1999, *138*, 227–275.
30.  Ghanta, V.G.; Riise, B.L.; Denn, M.M. Disappearance of extrusion instabilities in brass capillary dies. J. Rheol. 1999, *43*, 435–442.
31.  Pérez-Gonzàlez, J. Slip phenomenon via electrical measurements. J. Rheol. 2001, *45*, 845–853.
32.  Achilleos, E.; Georgiou, G.; Hatzikiriakos, S.G. The role of processing aids in the extrusion of polymer melts. J. Vinyl Addit. Technol. 2002, *8*, 7–24.
33.  Münstedt, H.; Schmidt, M.; Wassner, E. Stick and slip phenomena during extrusion of polyethylene melts as investigated by laser-Doppler velocimetry. J. Rheol. 2000, *44*, 413–427.
34.  Robert, L.; Vergnes, B.; Demay, Y. Experimental investigation during the stick–slip flow of a HDPE with Laser-Doppler velocimetry and flow birefringence. Procs 6th European Conf Rheol, Erlangen, 2002; 145–146.
35.  Merten, A.; Schwets, M.; Münstedt, H. Simultaneous measurements of pressure and velocity oscillations during spurt flow of a high-density polyethylene. Procs 6th European Conf Rheol, Erlangen, 2002; 147–148.
36.  Denn, M.M. Issues in viscoelastic fluid mechanics. Annu. Rev. Fluid Mech. 1990, *22*, 13–34.
37.  Larson, R.G. Instabilities in viscoelastic flows. Rheol. Acta 1992, *31*, 213–263.
38.  Leonov, A.I.; Prokunin, A.N. *Nonlinear Phenomena in Flows of Viscoelastic Polymer Fluids*; Chapman and Hall: London, 1994.
39.  Denn, M.M. Extrusion instabilities and wall slip. Annu. Rev. Fluid Mech. 2001, *33*, 265–287.
40.  Pearson, J.R.A. *Mechanics of Polymer Processing*; Elsevier: London, 1985.
41.  Brasseur, E.; Fyrillas, M.M.; Georgiou, G.C.; Crochet, M.J. The time-

dependent extrudate-swell problem of an Oldroyd-B fluid with slip along the wall. J. Rheol. 1998, *42*, 549–566.

42. Leonov, A.I. On the dependence of friction force on sliding velocity in the theory of adhesive friction of elastomers. Wear 1990, *141*, 137–145.

43. Adewale, K.E.P.; Leonov, A.I. Modeling spurt and stress oscillations in flows of molten polymers. Rheol. Acta 1997, *36*, 110–127.

44. McLeish, T.C.B.; Ball, R.C. A molecular approach to the spurt effect in polymer melt flow. J. Polym. Sci., B 1986, *24*, 1735–1745.

45. Fyrillas, M.; Georgiou, G.C.; Vlassopoulos, D. Time-dependent plane Poiseuille flow of a Johnson-Segalman fluid. J. non-Newton. Fluid Mech. 1999, *82*, 105–123.

46. Mair, R.W.; Callaghan, P.T. Shear flow of wormlike micelles in pipe and cylindrical Couette geometries as studied by nuclear magnetic resonance microscopy. J. Rheol. 1997, *41*, 901–924.

47. Venet, C.; Vergnes, B. Experimental characterization of sharkskin in polyethylenes. J. Rheol. 1997, *41*, 873–892.

48. Wise, G.M.; Denn, M.M.; Bell, A.T.; Mays, J.W.; Hong, K.; Iatrou, H. Surface mobility and slip of polybutadiene melts in shear flow. J. Rheol. 2000, *44*, 549–567.

49. Inn, Y.W.; Fischer, R.J.; Shaw, M.T. Visual observation of development of sharkskin melt fracture in polybutadiene extrusion. Rheol. Acta 1998, *37*, 573–582.

50. Cogswell, F.N. Stretching flow instabilities at the exits of extrusion dies. J. non-Newton. Fluid Mech. 1977, *2*, 37–57.

51. Durand, V.; Vergnes, B.; Agassant, J.F.; Benoit, E.; Koopmans, R.J. Experimental study and modeling of oscillating flow of high density polyethylenes. J. Rheol. 1996, *40*, 383–394.

52. Bergem, N. Visualization studies of polymer melt flow anomalies in extrusion. Procs 7th Int Congr Rheol, Gothenberg, 1976; 50–54.

53. Drda, P.A.; Wang, S.Q. Stick–slip transition at polymer melt/solid interfaces. Phys. Rev. Lett. 1995, *75*, 2698–2701.

54. El Kissi, N.; Piau, J.M. Enhancing processability of polymer melts during extrusion. Procs 6th European Conf Rheol, Erlangen, 2002; 3–8.

55. Li, H.; Hürlimann, H.P.; Meissner, J. Two separate ranges for shear flow instabilities with pressure oscillations in capillary extrusion of HDPE and LLDPE. Polym. Bull. 1986, *15*, 83–88.

56. Ovaici, H.; Mackley, M.R.; McKinley, G.H.; Crook, S.J. The experimental observation and modeling of an Ovaici necklace and stick–spurt instability arising during the cold extrusion of chocolate. J. Rheol. 1998, *42*, 125–157.

57. Weill, A. Capillary flow of linear polyethylene melt: sudden increase of flow rate. J. non-Newton. Fluid Mech. 1980, *7*, 303–314.

58. Sato, K.; Toda, A. Physical mechanism of stick–slip behavior in polymer melt extrusion: temperature dependence of flow curve. J. Phys. Soc. Jpn. 2001, *70*, 3268–3273.

59. Lim, F.J.; Schowalter, W.R. Wall slip of narrow molecular weight distribution polybutadienes. J. Rheol. 1989, *33*, 1359–1382.

60. Durand, V. Écoulement et instabilité oscillante des polyéthylènes haute densité. Ph.D. thesis, Ecole des Mines de Paris, France, 1993.

61. Vergnes, B.; d'Halewyn, S.; Boube, M.F. Wall slip and instabilities in the flow of EPDM compounds. In *Theoretical and Applied Rheology*; Moldenaers, P., Keunings, R., Eds.; Elsevier: Amsterdam, 1992; Vol. 1, 399–401.

62. Blyler, L.L.; Hart, A.C. Capillary flow instability of ethylene polymer melts. Polym. Eng. Sci. 1970, *10*, 93–203.

63. Uhland, E. Das anormale fleissverhalten von polyäthylen hoher dichte. Rheol. Acta 1979, *18*, 1–24.

64. Utracki, L.A.; Gendron, R. Pressure oscillation during extrusion of polyethylene II. J. Rheol. 1984, *28*, 601–623.

65. Pudjijanto, S.; Denn, M.M. A stable "island" in the stick–slip region of linear low-density polyethylene. J. Rheol. 1994, *38*, 1735–1744.

66. Hatzikiriakos, S.G.; Kazatchkov, I.B.; Vlassopoulos, D. Interfacial phenomena in the capillary extrusion of metallocene polyethylenes. J. Rheol. 1997, *41*, 1299–1316.

67. Tordella, J.P. An unusual mechanism of extrusion of polytetrafluoroethylene at high temperature and pressure. Trans. Soc. Rheol. 1963, *7*, 231–239.

68. Tordella, J.P. Unstable flow in molten polymers. In *Rheology*; Eirich, F.R., Ed.; Academic Press: New York, 1969; Vol. 5, 57–91.

69. Rosenbaum, E.E.; Hatzikiriakos, S.G.; Stewart, C.W. Flow implications in the processing of tetrafluoroethylene/hexafluoropropylene copolymers. Int. Polym. Process. 1995, *X3*, 204–212.

70. Goutille, Y.; Guillet, J. Influence of filters in the die entrance region on gross melt fracture: extrudate and flow visualization. J. non-Newton. Fluid Mech. 2002, *102*, 19–36.

71. Plucktaveesak, N.; Wang, S.Q.; Halasa, A. Interfacial flow behavior of highly entangled polybutadiene solutions. Macromolecules 1999, *32*, 3045–3050.

72. Kraynik, A.M.; Schowalter, W.R. Slip at the wall and extrudate roughness with aqueous solutions of polyvinyl alcohol and sodium borate. J. Rheol. 1981, *25*, 95–114.

73. Britton, M.M.; Mair, R.W.; Lambert, R.K.; Callaghan, P.T. Transition to shear banding in pipe and Couette flow of wormlike micellar solutions. J. Rheol. 1999, *43*, 897–909.

74. Kulikov, O.L.; Hornung, K. Wall detachment and high rate surface defects during extrusion of clay. J. non-Newton. Fluid Mech. 2002, *107*, 133–144.

75. Paskhin, E.D. Motion of polymer liquids under unstable conditions and in channel terminals. Rheol. Acta 1978, *17*, 663–675.

76. Robert, L.; Vergnes, B.; Demay, Y. Experimental investigations of the spurt instability in the flow of molten high density polyethylene. Procs XIIIth Congr Rheol, Cambridge, 2000; Vol. 3, 158–160.

77. Yang, Y.; Wang, S.Q.; Halasa, A.; Ishida, H. Fast flow behavior of highly entangled monodisperse polymers. 1. Interfacial stick–slip transition of polybutadiene melts. Rheol. Acta 1998, *37*, 415–423.

78. Durand, V.; Vergnes, B.; Agassant, J.F.; Koopmans, R.J. Influence of the

molecular weight on the oscillating flow of HDPE melts. In *Theoretical and Applied Rheology*; Moldenaers, P., Keunings, R., Eds.; Elsevier: Amsterdam, 1992; Vol. 1, 416.

79. Den Doelder, C.F.J.; Koopmans, R.J.; Molenaar, J. Quantitative modelling of HDPE spurt experiments using wall slip and generalised Newtonian flow. J. non-Newton. Fluid Mech. 1998, *79*, 503–514.

80. Robert, L.; Vergnes, B.; Demay, Y. Complex transients in the capillary flow of linear polyethylene. J. Rheol. 2001, *44*, 1183–1187.

81. Mhetar, V.; Archer, L.A. Slip in entangled polymer melts. 2. Effect of surface treatment. Macromolecules 1998, *31*, 8617–8622.

82. Dao, T.T.; Archer, L.A. Stick–slip dynamics of entangled polymer liquids. Langmuir 2002, *18*, 2616–2624.

83. Kazatchkov, I.B. Influence of molecular structure on rheological and processing behaviour of molten polymers. Ph.D. thesis, University of British Columbia, Canada, 1998.

84. Okubo, S.; Hori, Y. Model analysis of oscillating flow of high-density polyethylene melt. J. Rheol. 1980, *24*, 253–257.

85. Becker, J.; Bengtson, P.; Klason, C.; Kubát, J.; Sáha, P. Pressure oscillations during capillary extrusion of high density polyethylene. Int. Polym. Proc. 1991, *4*, 318–325.

86. Waddon, A.J.; Keller, A. The temperature window of minimum flow resistance in melt flow of polyethylene: further studies on the effect of strain rate and branching. J. Polym. Sci., Part B, Polym. Phys. 1992, *30*, 923–929.

87. Pérez-Gonzàlez, J.; Pérez-Trejo, L.; de Vargas, L.; Manero, O. Inlet instabilities in the capillary flow of polyethylene melts. Rheol. Acta 1997, *36*, 677–685.

88. Wang, S.Q.; Drda, P.A. Molecular instabilities in capillary flow of polymer melts: interfacial stick–slip transition, wall slip and extrudate distortion. Macromol. Chem. Phys. 1997, *198*, 673–701.

89. Brochard, F.; de Gennes, P.G. Shear-dependent slippage at a polymer/solid interface. Langmuir 1992, *8*, 3033–3037.

90. Anastasiadis, S.H.; Hatzikiriakos, S.G. The work of adhesion of polymer/wall interfaces and its association with the onset of wall slip. J. Rheol. 1998, *42*, 795–812.

91. de Gennes, P.G. Mécaniques des fluides: écoulements viscométriques de polymers enchevêtrés. C. R. Acad. Paris 1993, *288*, 219–220.

92. Hill, D.A.; Hasegawa, T.; Denn, M.M. On the apparent relation between adhesive failure and melt fracture. J. Rheol. 1990, *34*, 891–918.

93. Hatzikiriakos, S.G.; Hong, P.; Ho, W.; Stewart, C.W. The effect of Teflon coatings in polyethylene capillary extrusion. J. Appl. Polym. Sci. 1995, *55*, 596–603.

94. Migler, K.B.; Hervet, H.; Leger, L. Slip transition of a polymer melt under shear stress. Phys. Rev. Lett. 1993, *70*, 287–290.

95. Legrand, F.; Piau, J.M. Wall slip of a polydimethylsiloxane extruded through a slit die with rough steel surfaces: micrometric measurement at the wall with fluorescent-labeled chains. J. Rheol. 2000, *42*, 1389–1402.

96. Robert, L.; Vergnes, B.; Demay, Y. Flow birefringence study of the stick–slip instability during extrusion of high-density polyethylenes. J. non-Newton. Fluid Mech. 2003, *112*, 27–42.

97. Georgiou, G. The time-dependent compressible Poiseuille and extrudate-swell flows of a Carreau fluid with slip at the wall. J. non-Newton. Fluid Mech. 2003, *109*, 93–114.

98. Molenaar, J.; Koopmans, R.J. Modelling polymer melt-flow instabilities. J. Rheol. 1994, *38*, 99–109.

99. Greenberg, J.M.; Demay, Y. A simple model of the melt fracture instability. Eur. J. Appl. Math. 1994, *5*, 337–357.

100. Dubbeldam, J.L.A.; Molenaar, J. Dynamics of the spurt instability in polymer extrusion. J. non-Newton. Fluid Mech. 2003, *112*, 217–235.

101. Georgiou, G.C.; Crochet, M.J. Compressible viscous flow in slits, with slip at the wall. J. Rheol. 1994, *38*, 639–654.

102. Huseby, T.W. Hypothesis on a certain flow instability in polymer melts. Trans. Soc. Rheol. 1966, *10*, 181–190.

103. Deiber, J.A.; Schowalter, W.R. On the comparison of simple non-monotonic constitutive equations with data showing slip of well-characterized polybutadienes. J. non-Newton. Fluid Mech. 1991, *40*, 141–150.

104. Vlassopoulos, D.; Hatzikiriakos, S.G. A generalized Giesekus constitutive model with retardation time and its association to the spurt effect. J. non-Newton. Fluid Mech. 1995, *57*, 119–136.

# 7

## Gross Melt Fracture in Extrusion

**John M. Dealy**
McGill University, Montreal, Quebec, Canada

**Seungoh Kim**
Verdun, Quebec, Canada

### 7.1  WHAT IS GROSS MELT FRACTURE?

Because there are no universally accepted terms describing the various types of extrudate distortion, each author must define the terms he intends to use. In this chapter, gross melt fracture (GMF) will be used to describe a pronounced, chaotic distortion of extrudate that arises from a severe instability of the flow at the entrance to a converging die. A distorted extrudate of this type is shown in Fig. 7.1. This phenomenon affects the entire cross section of the extrudate and is therefore sometimes called *volume melt fracture* to distinguish it from sharkskin, which affects only the surface of the extrudate and is often called *surface melt fracture*. However, volume melt fracture also includes the helical extrudate that is caused by a swirling motion of the melt as it enters the converging section of the die. Although this phenomenon does involve the entire cross section of the melt, there is no actual fracture, but it was decided to include it in this chapter anyway.

It can be quite misleading to try to define various types of die flow instability solely on the basis of the appearance of the extrudate. For example,

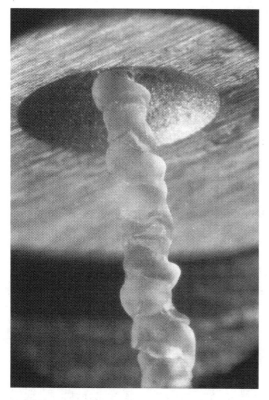

**FIGURE 7.1**  Photo of typical gross melt fracture of silicone at the exit of an orifice. (Photo courtesy of J.M. Piau.)

sharkskin can cause a helical extrudate distortion that is a surface effect resulting from an instability very near the exit of a die and not at the entrance. Likewise, it is not correct to say that sharkskin is always a much less severe distortion than that resulting from GMF. In fact, sharkskin, if defined as a defect arising from the stress field at the exit of a die, can be quite severe.

It is important to note that much of the literature on extrudate distortion is based on laboratory studies of the flow from a capillary die, where the flow is generated by a piston moving through a reservoir or by gas pressure imposed on the reservoir. In addition, the phenomena observed do not always have counterparts in industrial processing. Sharkskin occurs in industrial processing, particularly film blowing, and in capillary and slit flow. In addition, since sharkskin occurs at lower flow rates than GMF, it is more often the factor that limits production rates than GMF. Thus, there are very few industrial processes in which gross melt fracture has been reported as a production problem. One important exception is wire coating, which is discussed in some detail at the end of this chapter.

## 7.2  SPURT AND OSCILLATORY FLOW

Spurt and oscillatory flow are curious phenomena that occur in the capillary flow of most high-molecular weight, linear, molten polymers whose molecular weight distributions are not too broad. These phenomena are sometimes associated with extrudate distortions, but they are quite independent of entrance and exit flow instabilities and are in fact manifestations of wall slip. Linear polypropylene and polystyrene do not exhibit oscillatory flow/spurt, whereas high-molecular weight, linear polyethylene exhibits these phenomena even at very low flow rates. These observations suggest that side groups suppress wall slip and that slip occurs when the wall shear stress is sufficiently high. Low-density polyethylene (LDPE), which has significant levels of long-chain branching, also does not exhibit spurt. This may be because at the entrance to a capillary this polymer develops a large corner vortex in the reservoir (see Fig. 2.4b in Chapter 2 of the present volume), which streamlines the flow of the melt through the contraction. This reduces the peak tensile stress, possibly keeping it below the critical stress for GMF. On the other hand, isotactic polypropylene (iPP) with low levels of long-chain branching does exhibit spurt/oscillating flow (1). Chapter 6 of the present volume is devoted entirely to these phenomena, and a brief discussion is presented here only to clarify issues involving gross melt fracture.

As mentioned in Chapter 2, there are two types of capillary rheometer; in one the flow is generated by the application of gas pressure to the reservoir, and in the other it is generated by a piston that moves down in the reservoir at a constant velocity. Spurt occurs in pressure-driven instruments, while oscillatory flow occurs in piston-driven instruments. In pressure-driven capillary flow of some polymers, at a critical pressure the flow rate suddenly increases dramatically, often by a factor of 10 or more. This phenomenon was first reported by Vinogradov for a series of elastomers (2), and for the highest molecular weight elastomer, the flow rate increased by a factor of about 100! This is the spurt effect. On a plot of pressure drop vs. flow rate or apparent wall shear rate (i.e., a flow curve), spurt appears as a sudden shift to a high-flow branch of the curve at the critical pressure for spurt.

A typical flow curve for a molten linear polyethylene is shown in Fig. 7.2. There are two branches of the curve, a no-slip (or weak slip)/low-flow branch, and a slip/high-flow branch. At the critical pressure, identified as point A, the system suddenly jumps to point B on the high-flow branch. This

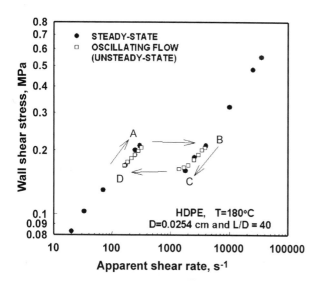

FIGURE 7.2  Wall shear stress vs. apparent wall shear rate for flow of HDPE in a constant-piston-speed, capillary rheometer. As the flow rate (proportional to $\dot{\gamma}_A$) increases, at point A, oscillatory flow begins and continues until point C is reached, whereupon the system proceeds along the right-hand branch. If $\dot{\gamma}_A$ is now reduced, the system proceeds to point C and then oscillates until point A is reached. In the oscillatory flow region, $\dot{\gamma}_A$ and $\sigma_w$ oscillate between points in segments D–A and C–B. (Data from Ref. 3.)

results in the dramatic increase in flow rate that is called spurt. If, after reaching the high-flow branch, the pressure is steadily reduced, the system will descend along the high-flow branch but will not jump to the low-flow branch until the pressure reaches another, lower critical pressure, at point C, where it jumps to point D. The result is a hysteresis loop. There is at present no theory that can predict the upper and lower critical pressures.

In piston-driven capillary flow, there is a range of piston speeds over which stable flow is not possible, and the driving pressure and the flow rate both oscillate periodically with time. A typical flow curve obtained in this type of operation can be described by reference to Fig. 7.2. If the piston speed is increased after reaching point A, the rate of extrusion from the capillary and the driving pressure start to oscillate periodically. The oscillations continue at all piston speeds between values corresponding to the apparent shear rates at points A and B. At higher piston speeds, the system moves steadily up the high-flow branch. How can the extrusion rate oscillate, while the piston speed remains constant? As the driving pressure increases as the system moves up the low-flow branch, the mass of fluid in the reservoir increases as a result of melt compressibility. As the piston speed starts to exceed that corresponding the point A, slip starts, the driving pressure drops, and the excess mass in the reservoir is rapidly depleted. Now the flow rate decreases, slip stops, and the pressure starts to rise again. At piston speeds in the unstable region, the pressure oscillates between values corresponding to the wall shear stresses at points A and B. When the apparent wall shear rate reaches the level of point B in Fig. 7.2, the flow becomes steady again, and data follow the high-flow branch. If the piston speed is now decreased, oscillatory flow will start at point C and continue until the apparent wall shear rate reaches the level of point D, when the flow becomes stable again and the system moves down the low-flow branch.

If the viscosity is sufficiently high, and the capillary is small and long, the oscillations occur sufficiently slowly that it is possible to track the flow rate during a cycle. Hatzikiriakos and Dealy (3) used this technique and found that the system path is a continuous loop from A to B to C to D. Kay et al. (4) used an extruder to feed melt to a capillary and monitored both the flow rate and pressure continuously. Their flow curves show in detail the flow behavior throughout a cycle. In another study (5), an extruder was used to study the effect of the material of construction of the die on oscillating flow. The results revealed that the use of brass instead of stainless steel eliminated the unstable flow regime and produced a smooth transition from the lower to the upper branch of the flow curve.

The dynamics of this oscillation can be modeled by taking into account the compressibility of the melt in the reservoir and its non-Newtonian behavior in the capillary (3,6,7), but the critical pressures cannot be predicted and must be determined experimentally.

Spurt (and oscillatory flow) do not occur in the flow of LDPE, very low density polyethylene, linear polypropylene, poly(vinyl chloride), and polymers with large side groups such as polystyrene and polymethylmethacrylate. This implies that side groups may suppress wall slip. Whereas very small oscillations have been reported for some LDPEs and radiation-cross-linked polypropylene, their flow curves show no sign of a discontinuity.

Although it has been suggested (8) that spurt and oscillatory flow are manifestations of a physical property of the polymer, it seems more likely that they are system instabilities that result from the interaction between compression in the reservoir and the flow through the die. We note that capillary flow is in fact a complex process in which the kinematics of the flow varies greatly between the reservoir and the point in the capillary where the flow is fully developed. Furthermore, the pressure varies substantially along the capillary, and the viscosity, and probably the slip velocity, vary significantly with pressure. Another complexity of the spurt effect is that while it occurs at constant driving pressure, the wall shear stress in the capillary is actually lower after than before the onset of spurt. This is because the entrance pressure drop increases sharply when the flow rate suddenly increases as spurt occurs. In addition, if the entrance pressure drop increases while the total pressure drop is constant, the pressure drop, and thus the wall shear stress, in the capillary must decrease. Finally, spurt is not observed in sliding plate rheometers, which can generate high shear rates while keeping the pressure and shear rate uniform throughout the sample.

Although gross melt fracture may occur simultaneously with spurt, the two phenomena are not related. This is clear from the fact that GMF occurs in flow from a converging die, even when no capillary is attached to it and no spurt occurs; spurt and oscillatory flow only occur in the flow through a capillary. Confusion on this point arises because the very great increase in flow rate that occurs at the onset of spurt (from A to B in Fig. 7.2) often takes the entrance flow past its threshold for gross melt fracture. However, this is not always the case, and sometimes the onset of gross melt fracture occurs at a pressure above that at which spurt occurs (to the right of point B). This results in a *superextrusion* range of flow rates in which there is slip but no GMF. Finally, the recent careful work of Meller et al. (9) has confirmed that oscillatory flow is not related to GMF.

When the critical entrance pressure drop for the onset of GMF does correspond to a flow rate within the range where steady flow is unstable and oscillatory flow occurs, the appearance of the extrudate oscillates along with the flow rate. Depending on the polymer, in the course of one cycle, the extrudate can have segments exhibiting sharkskin, moderate GMF, severe GMF, and even a smooth surface (3). Figure 7.3 is a photograph showing the variation in extrudate appearance during one cycle.

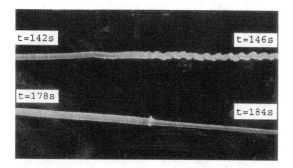

**FIGURE 7.3** Appearance of extrudate during oscillatory flow of same HDPE as in Fig. 7.2. $\dot{\gamma}_A$ (nominal) = 1480 sec$^{-1}$. At $t$ = 142 sec, system is just below point A in Fig. 7.2; at $t$ = 146 sec, it is just below point B; at $t$ = 178 sec, it is just above point C, and at $t$ = 184 sec, it is just above point D. (From Ref. 3.)

## 7.3 HELICAL EXTRUDATE FROM SWIRLING ENTRANCE FLOW

Smooth, helical extrudates from a capillary were first reported by Tordella (10), who associated them with a swirling flow at the entrance. It has been clearly demonstrated, for example, by Oyanagi (11) and by Bergem (12) that such extrudates do indeed result from an entrance flow instability that causes the spiraling of the flow as it enters a capillary from a larger reservoir. Piau et al. (13) found that this occurs even when there is no capillary, i.e., for an orifice die. When a given polymer exhibits both helical extrudate and gross melt fracture, the spiral flow occurs first as the flow rate is increased. Figure 7.4a shows a helical extrudate, and Fig. 7.4b shows a photo of the same melt with carbon black tracer, frozen in the capillary during helical flow and microtomed to reveal the flow pattern.

Oyanagi (11) used birefringence to visualize swirling entrance flow and hypothesized that it involved slow, localized slip along a conical surface that rotated along with the streamlines. On the other hand, McKinley et al. (14) did not see any evidence of slip in their thorough experimental study of hydroelastic entrance flow instabilities in abrupt contractions, and McKinley et al. (15) later reported a theoretical framework for analyzing the swirling flow as a hydroelastic instability with no slip surface. A recent attempt to provide a theoretical explanation for this phenomenon (16) is based on an analysis of the stability of Poiseuille flow, i.e., fully developed flow in a tube. However, this approach is not appropriate, as it does not address the flow in the entrance region.

Oyanagi (11) also observed two additional types of unstable flow in his studies of high-density polyethylene (HDPE). At flow rates above those where

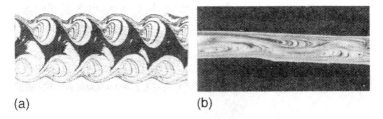

(a)                                              (b)

FIGURE 7.4   (a) Microtome of HDPE extrudate at a flow rate slightly above the oscillatory flow regime. Carbon-black tracer shows the perfectly periodic, nonchaotic nature of the flow. (b) Microtome of HDPE frozen in the capillary during the flow whose extrudate is shown in (a). Shows that the flow pattern that gives rise to the helical extrudate arises at the entrance and is transmitted downstream with the flow to the exit where is produces a helical swell. (From Ref. 12.)

swirling flow occurred, he observed a "switching" motion in which the angle of the center streamline oscillated periodically between two limiting values, producing a "zigzag" or "wavy" extrudate. This is the type of flow observed in slit flow under conditions that would produce swirling flow in a capillary, i.e., the extrudate is smooth and there is a periodic oscillation of the angle of the center surface of the flow about the center plane of the die (17). Oyanagi (11) reported yet another type of flow when the capillary was short. Instead of the wavy or zigzag extrudate, the extrudate consisted of a series of connected "beads" and was axially symmetric. His birefringence photos revealed that these resulted from a periodic pulse flow at the inlet in which the flow rate oscillated periodically between relatively high and low values. It is likely that the extraordinary phenomena observed by Oyanagi arose from the unusual structure of the polymer he studied. This was a very high molecular weight film resin with a broad, bimodal MW distribution manufactured using the Mitsui slurry process.

Yesilata et al. (18) measured the pressure fluctuations in a small tube downstream of a four-to-one contraction and also visualized the flow by means of streakline photography. The fluid was a polyisobutylene solution. They interpreted their observations in terms of three unstable regimes. At the lowest flow rates, the pressure was quite constant with time. At a fairly modest flow rate, this gave way to a regime in which there were quite small periodic variations in the pressure, and this was called a *type A* instability. This was observed only over a small range of flow rates and was replaced suddenly by a *type B* instability characterized by pressure oscillations of a lower frequency and higher amplitude. Finally, at substantially higher flow rates, the pressure fluctuations became nonperiodic, and this was described as a *type C* instability.

In the remainder of this chapter, the term gross melt fracture will be used to imply a situation in which there is a chaotic (aperiodic) entrance instability involving true rupture and leading to a very rough extrudate. However, GMF can sometimes be superposed on a swirling motion (19).

## 7.4  HELICAL SHARKSKIN AND THE DOUBLE HELIX

The helical extrudate that results from an entrance flow instability is not to be confused with the helical *screw-thread* distortion that is a type of sharkskin surface melt fracture. Figure 7.5A is a photo of such an extrudate taken by Bergem (12). Bergem microtomed a slice of frozen polymer taken from the exit region of the capillary, and his photograph of this slice (Fig. 7.5B) shows clearly that this is an exit effect. As is explained in detail in Chapter 5 of the present volume, sharkskin deformation usually has a small amplitude and can be either of the screw-thread or quasiperiodic type. However, sharkskin

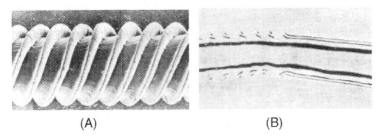

(A)                                                    (B)

FIGURE 7.5  (A) Micrograph of capillary extrudate of *trans*-1,5-polypentenamer showing quadruple screw-thread pattern. (B) Microtome slice of capillary exit region for same conditions as in (A). Carbon-black tracer shows that the screwthread pattern is caused by a surface fracture at the exit. (From Ref. 12.)

(defined as an exit effect) can sometimes have an amplitude that is comparable to the diameter of the extrudate. For example, there has been a report (20) of an extrudate, said to be exhibiting severe screw-thread sharkskin, which appeared as two identical but separate, intertwined, helical strands, as shown in Fig. 7.6. At a critical flow rate the two strands separated to form two independent helixes. Although it was identified as an exit effect, and a

FIGURE 7.6  Micrograph of extrudate of ethylene–propylene copolymer as flow rate was being reduced from the 2-branch to the double-helix zones. The critical wall shear stress for this transition was 0.28 MPa. (From Ref. 20.)

conceptual mechanism was sketched, the detailed origin of this instability is unknown at present.

## 7.5 MANIFESTATIONS OF GROSS MELT FRACTURE

In this section, we consider only phenomena that involve chaotic entrance flow and a very rough extrudate. Early studies of GMF involved only inspection of the outside appearance of capillary extrudate, and Fig. 7.1 shows a typical observation. It is clear that the irregularity is of major proportions and quite irregular. The tracer studies of HDPE flow by Bagley and Birks (21) revealed that the rough extrudate resulted from a chaotic flow regime at the entrance. As the flow rate was steadily increased from zero, a stable torroidal corner vortex slowly formed. At a critical flow rate, the flow became highly unsteady. The centerline streamline broke and sprang back at erratic, very brief intervals of time, allowing melt from the quiescent corner vortex to enter the capillary. After a very brief period, the centerline flow reestablished itself only to break again.

More recently, tracer studies have revealed the interior of the extrudate to show clearly that the melt was truly ruptured (22). Figure 7.7 shows several photos of extrudate from a converging die, without a capillary, that has been sliced to reveal its interior. A thin layer of carbon black-loaded polymer was loaded into the capillary with unloaded polymer above and below it. As flow developed, the marked polymer dipped down in the center and eventually formed an annular ring in the capillary and extrudate. Figure 7.7A shows a smooth extrudate with no visible sharkskin. As the flow rate increases, the sharkskin becomes increasingly severe, as shown in Fig. 7.7B. In Fig. 7.7C we see a more severe sharkskin superposed on gross melt fracture, with the polymer at the center actually broken into distinct segments. Figure 7.7D shows a more severe GMF. At even higher flow rates, in place of a continuous strand leaving the die there were small, completely detached pieces of polymer. Comparing Fig. 7.7C and D, we can see that it is not always possible to identify the types of distortion occurring simply by inspecting the extrudate.

## 7.6 ORIGINS OF GROSS MELT FRACTURE

As explained above, there is ample evidence that GMF arises from a flow instability in a converging die. Furthermore, it is clear that there is a genuine rupture of the melt, with clearly defined rupture surfaces. It is therefore tempting to relate this phenomenon to the behavior of melts in experiments designed to determine the conditions for their rupture. In metals and rubbers, fracture is generally assumed to result when the largest principal stress within

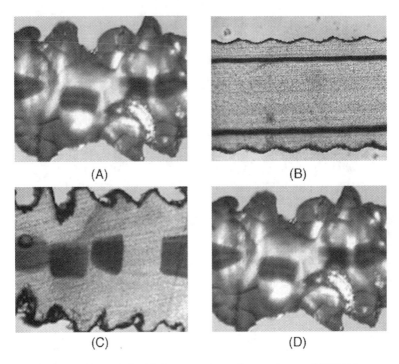

(A)                  (B)

(C)                  (D)

**FIGURE 7.7**  Photos of midplane section of extrudates of HDPE from an orifice die at four flow rates. Carbon-black tracer shows streamlines. (A) Undistorted flow, 0.18 g/min; (B) mild sharkskin, 0.52 g/min; (C) sharkskin plus GMF, 1.6 g/min; (D) severe sharkskin plus severe GMF, 2.1 g/min. The extrudates were sliced using a razor blade, and there may be some carbon black on the outer surface in some cases. (From Ref. 22.)

the material reaches a critical level. In a melt, there may be substantial viscous flow before rupture, and the rate of deformation and the duration of the application of the deforming stress may also be factors. When stretched very rapidly, however, a melt behaves like rubber, and critical stress would be the appropriate criterion. The rupture of elastomers and molten plastics is discussed in more detail in a later section of this chapter.

## 7.7  DEPENDENCE ON DIE DESIGN

While the occurrence of GMF is independent of whether or not a capillary is part of the die, the presence of a capillary can alter the appearance of the extrudate. In particular, as the length of a capillary is increased, the severity of the extrudate distortion decreases somewhat. It seems clear that this is a result of partial "healing" of the broken polymer as it flows along the capillary. The presence of a capillary also has the effect of increasing the pressure in the entrance region, and this will affect the rheological properties of the melt there. Carreras et al. (23) have measured this effect, and they found that increasing the pressure at constant flow rate can shift the flow into the oscillatory flow region.

Reducing the angle of convergence of the entrance channel delays the onset of GMF and also reduces its severity (24). This results from the reduction in tensile stress associated with the more gradual convergence of streamlines. Kim and Dealy (22) confirmed this finding but reported that the angle of convergence affected the onset of GMF only when it was less than 90°. The effect of die angle on the occurrence of GMF in the wire-coating process is discussed in a later section of this chapter. Piau et al. (13) found that GMF depends very little on the material of construction of the die, adding support to the idea that GMF depends primarily on the nature of the polymer.

There have been reports that the insertion of a filter at the bottom of the reservoir significantly reduces the severity of GMF. The filters consisted of closely woven wire grids, which markedly increased the entrance pressure drop. Piau et al. (25) found that such filters suppressed oscillatory flows/spurt and concluded that the transition from stick to slip occurs gradually instead of suddenly. Goutille and Guillet (26) carried out flow visualization studies in a slit die to show that the entrance flow instability that is associated with GMF is substantially stabilized by the presence of a filter.


## 7.8  CRITERIA FOR THE ONSET OF GMF

Although it has sometimes been proposed that there is a critical flow rate or critical apparent wall shear rate (calculated using the equation for capillary flow) for GMF, this is inconsistent with observations. For one thing, we know that GMF occurs even when there is no capillary attached to a converging die, so we should not expect any relationship with quantities calculated on the basis of capillary flow. In addition, in fracture mechanics it is the principal tensile stress that is assumed to provide the driving force for rupture. In entrance flow, there is a high tensile stress on the axis of symmetry because of the strongly converging streamlines. At low rates, there will be a substantial viscous component of the deformation of a melt, and the extension rate and/ or recoverable strain will be contributing factors for rupture. Denn and Marrucci (27) presented a model and some data to support their hypothesis that there is a maximum extension ratio to which a polymeric liquid can be subjected. However, rupture is much more likely to occur at high rates, and under these conditions, the melt will be rubbery in behavior, and a critical tensile stress is likely to be the appropriate criterion.

Paul and Southern (28) reported that the critical condition for GMF scaled with the plateau modulus of solutions of varying concentration that they studied. Hürlimann and Knappe (29) proposed that there is a critical extensional stress for the onset of gross melt fracture, and they used the entrance pressure drop as a measure of this stress. Cogswell (30,31) and Shaw (32) also concluded that there is a critical tensile stress for the onset of GMF.

Kim and Dealy (22,33) correlated their observations of GMF in a series of polyethylenes having various structures by using the maximum tensile stress that occurred in the die. They used a converging die, with no capillary attached, and estimated the maximum tensile stress from the pressure drop using techniques proposed by Cogswell (31) and Binding (34). They found that the stresses calculated using the two methods were nearly proportional to each other, and they adopted Cogswell's method, as it is the easiest to use. Kim and Dealy (22) found that the critical tensile stress for the onset of GMF was independent of temperature and contraction ratio. As shown in Fig. 7.8, they also found that while the critical stress decreased as the die angle increased, it was independent of this angle when it was greater than 90°. As mentioned in the next section, they found that the maximum tensile stress

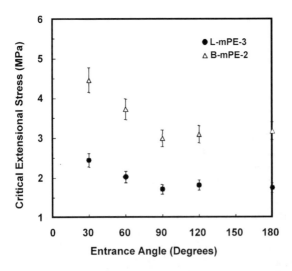

**FIGURE 7.8** Critical tensile stress for onset of GMF of HDPE extruded from orifice die as function of entrance angle. (From Ref. 22.)

correlated with molecular structure and concluded that the critical tensile stress at the onset of GMF is a well-defined property of a polymer.

## 7.9 EFFECT OF MOLECULAR STRUCTURE

Because the effect of the polymer on GMF has usually been reported in terms of quantities calculated using the capillary flow equations, the numbers have no absolute significance. However, they do show trends among a series of polymers being compared using the same apparatus. Oyanagi (11) reported critical wall shear stresses in the range of 0.3 to 0.5 MPa for a number HDPE resins. Vlachopoulos et al. (35,36) concluded from their studies of HDPE, polypropylene, and polystyrene that the critical wall shear stress was independent of molecular weight distribution (MWD), inversely proportional to $M_w$, and increased somewhat with temperature. Baik and Tzoganakis (37) reported that the critical wall shear stress for controlled-rheology propylenes decreases with $M_w$ and decreases as the MWD is broadened.

Some polymer comparisons have been based on the identification of a critical tensile stress. Cogswell (38) reported values of 1.0 MPa for LDPE, 3.0 MPa for HDPE, and 2.0 MPa for polypropylene, but these figures do not take molecular weight into account. Meller et al. (9) found that for HDPE, the critical stress for rupture decreased from 7.5 to 1.0 MPa as the molecular weight was increased.

Kim and Yang (39) studied the effect of the peroxide modification of linear low-density polyethylene (LLDPE) on the onset of GMF. They found that the dilution of dicumyl peroxide (DCP) in PP before adding it to the LLDPE enhanced its effectiveness in delaying GMF to significantly higher flow rates. They concluded that the dilution step enhanced the dispersion of the DCP and suppressed cross-linking.

A serious barrier to progress in our understanding of the detailed effects of molecular structure on GMF has been the unavailability of commercial samples having well-defined structures. The presence of very small amounts of long-chain branching or high-molecular weight material can have an important effect on flow behavior, but it was often not possible to be certain about these structural features. Yamane and White (40) observed capillary extrudates of a variety of polyethylenes and concluded that some of their samples contained small but unknown levels of long-chain branching, which were thought to have an important effect on the onset of GMF.

The advent of single-site catalysts has made it possible to produce polymers on a commercial scale that have well-defined molecular weight distributions and branching structures (41,42). Kim and Dealy (33) observed GMF in a series of single-site polyethylenes having various structures and correlated their results with the estimated maximum tensile stress that occurred in the die. They found that the critical stress was independent of the average molecular weight but varied linearly with polydispersity index ($M_w/M_n$) and level of long-chain branching. Figure 7.9 is their plot of critical tensile stress vs. degree of long-chain branching. The structure of the materials used in this study was later elucidated by modeling the polymerization reaction (42), and we now know that at the lowest levels of branching nearly all the branch points involve free arms so that the molecules are stars. Inner backbones only arise in significant numbers as the overall branching level increases. An inner backbone is a segment of the polymer chain that has no free end. Segments of this type are known to be essential for strain hardening, and this implies that they would also have a much stronger effect on the onset of GMF than free arms.

Meller et al. (9) carried out a numerical simulation of entrance flow to determine the stress profile and found that the onset of GMF is directly related to the calculated value of the tensile stress. They concluded that rupture occurs in uniaxial extension when the Weissenberg number (Wi) is greater than 0.5 and the Deborah number (De) is greater than 1. In Chapter 2 of the present volume it is explained that Wi > 0.5 implies a significant degree of orientation, while De > 1 implies a strong elastic component of the stress. By using a similar analysis, it has also been found that the effectiveness of

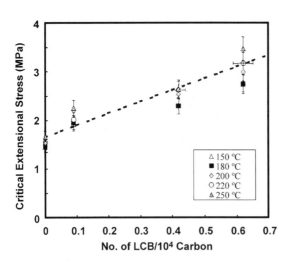

**FIGURE 7.9** Critical tensile stress for onset of GMF of a metallocene polyethylene containing low levels of long-chain branching at several temperatures. (From Ref. 22.)

converging flow for the dispersive mixing of immiscible blends is related to the elastic fracture of the dispersed phase (43).

## 7.10   TENSILE STRENGTH OF MOLTEN POLYMERS

Because it is widely believed that chaotic GMF is caused by the rupture of melt in the entrance region, it is of interest to examine what is known about the rupture of molten polymers. All liquids can withstand significant tensile stress (negative pressure) without rupturing or vaporizing, if care is taken to eliminate nucleation sites for cavitation. The traditional criterion for the onset of cavitation is a critical state of "pressure." However, Joseph (44) has pointed out that in a deforming fluid there is no well-defined "pressure" that has any physical meaning. In a Newtonian fluid, the trace of the viscous stress tensor is zero, and it is convenient to define pressure as the negative of the mean normal stress, as this is equal to the deviatoric stress. However, the liquid has no way of averaging stresses, and the mean normal stress thus has no physicochemical significance in a deforming liquid.

Molten, highly entangled polymers differ in several important ways from classical fluids, and this has important implications for their rupture behavior. First, the force attracting an individual molecule to its neighbors is very large because of the enormous length of the molecules. In addition, when deformed, polymer molecules are oriented and often stretched in a manner that depends on the deformation field. The induced anisotropy gives rise to large differences in normal stresses, and the mean normal stress is not equal to the deviatoric stress. It is common practice to define a viscous stress tensor as the difference between the total stress and some isotropic stress that is often referred to as the "pressure." However, this isotropic contribution to the total stress only takes on a quantitative meaning in the context of a particular constitutive equation.

Joseph (44) has proposed a criterion for cavitation that involves the three principal stresses, and Tschoegl (45) has worked out geometrical representations of failure surfaces in principal stress space. The principal stresses are strongly influenced by the normal stress differences, which thus play a central role in cavitation and rupture. Cavitation has been observed in the drawing of copper wire (46), the cold-drawing of polypropylene (47), and the shearing of polymer melts (48). Son and Migler (49) observed the transient formation of cavities near the wall, just before the end of a capillary through which polyethylene was flowing. The appearance of these cavities appeared to be associated with unstable flow at the entrance of the capillary. There have been no reported observations of cavitation in the uniaxial extension of melts, with the possible exception of that of Torres (50).

The role of "flaws" or material inhomogeneities as nucleation sites for cavities and cracks in melts is not clear at the present time. In the fracture of metals, nucleation sites are an essential component of fracture mechanics. Chen et al. (51) studied the possible role of nucleation sites in melt rupture. They found that a substantial induction time was required for the onset of fracture, which supports the hypothesis that the growth of a cavity is essential for rupture. They also observed that bubbles formed near the edge of the sample and that fractures started at the edge and then migrated inward. However, after steps were taken to eliminate tiny bubbles or dissolved gases from samples, there was no change in the fracture behavior. Also, microscopic examination revealed no dust particles or voids larger than 0.1 μm, which might avoid collapse long enough to start a crack, and this dimension is much too large to be generated by thermal fluctuations at the molecular level. Furthermore, similar values of the critical stress for rupture (about 0.1 MPa) have been reported by various groups working on different polymers. Finally, Chen et al. (51) noted that the addition of solvent reduced the critical stress for rupture. They concluded that flaw size is an intrinsic property of a polymer melt or solution at a given temperature. An intrinsic flaw size has also been put forward by Kinlock and Young (52) to explain the failure of rubbers and glassy polymers.

It is obviously of interest to measure the rupture strength of molten polymers in uniaxial extension. In fact, the only data in the literature are for uncured elastomers. While these are technically molten polymers, they differ greatly from thermoplastics in that the recoverable strain is a much larger fraction of the imposed strain. This makes it relatively easy to study the response of elastomers to large rapid stretching deformations up to the rupture point. The review by Malkin and Petrie (53) summarizes what is known about the rupture of elastomers. The most important work was done by Vinogradov (2), who studied the rupture of a series of monodisperse elastomers and concluded that for stretching at a constant strain rate it is the recoverable strain that governs rupture. He organized his observations in terms of four regimes of rupture behavior based on the strain rate, as shown in Fig. 7.10. In zone I, at the lowest rates of deformation, the recoverable strain is small, and the polymer is liquidlike and able to deform without limit. In zone II, there is a transition from viscous to elastic behavior, and rupture occurs in response to the increasing level of recoverable strain. In zone III, the deformation is entirely elastic, there is no flow, and the recoverable strain is equal to the total strain. Finally, in zone IV, the deformation occurs so rapidly that molecular motions are too slow to respond to the stress, and the behavior becomes glassy, leading to a brittle fracture at very small strains. In zones III and IV, the behavior of the melts and uncured elastomers approaches that of cured rubber. As a result of the high strain rate involved, it is likely that the

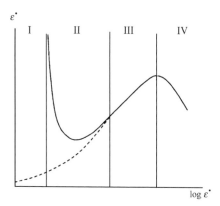

$\varepsilon^{\cdot}$

I    II    III    IV

$\log \varepsilon^{\cdot}$

FIGURE 7.10   Sketch showing Vinogradov's representation of rupture behavior of polymeric liquids. Critical strain for rupture vs. log of strain rate. Dashed line shows recoverable (elastic) strain. (From Ref. 53.)

rupture that occurs in the GMF of melts involves behavior at the high end of zone II and the low end of zone III.

In the right half of zone II and in zone III, where most of the strain is recoverable, rupture is directly related to stored elastic energy. Vinogradov was able to obtain reproducible data in these zones and found that the tensile stress at break was linearly related to the recoverable strain at break. For a sample stretched at a constant stress rather than a constant strain rate, he found that there was a measurable time to break or "durability," which was related to the applied stress by a power law. He also proposed an equation for calculating the durability for an arbitrary loading history. More recently, Joshi and Denn (54) have proposed a molecular model that predicts when rupture will occur and obtained reasonable agreement with some elastomer data.

However, it has not been possible to determine the tensile strength of molten thermoplastics in uniform, uniaxial extension. Unless the melt is strain hardening, it always fails in a ductile manner (9), which is not a true (brittle) rupture. Ductile failure is in fact a flow instability that can be modeled using a viscoelastic constitutive equation (55). In any event, ductile failure cannot occur in an extrusion die, as there is no free surface. Even with strain hardening materials, it is not possible to stretch melts fast enough to achieve a true rupture in an extensional rheometer capable of generating uniform strain at a controlled rate. Figure 7.11 shows Meissner's data (56) for the recoverable strain of a LDPE following stress release during tensile start-up flow. At the highest strain rate (1 $\sec^{-1}$) and after very small amounts of

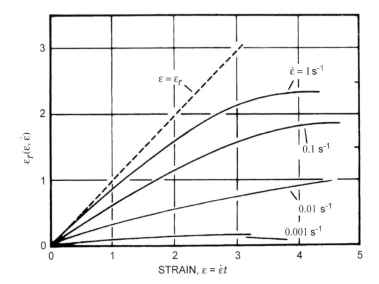

**FIGURE 7.11** Recoverable strain vs. total strain for uniaxial extension of LDPE at 150°C. The dashed line corresponds to 100% strain recovery. (From Ref. 56.)

stretch, all the strain is recoverable, but it levels off to a constant value at large deformations, i.e., long stretching times. This constant value is the maximum recoverable strain at each strain rate. We note that the maximum recoverable strain increases rather slowly with strain rate. Since the maximum strain rate in extensional rheometers is about 1.0 sec$^{-1}$, we see that it is impossible to generate rupture in this polymer using this apparatus.

Referring to Vinogradov's diagram (Fig. 7.10), we see that the behavior of molten plastics in extensional rheometers falls in zone I and the left half of zone II. Thus, little of the deformation is recoverable, and a very large strain rate is required to achieve failure. Limitations of the apparatus and/or flow instabilities make it impossible to reach such high rates. Torres (50) made a determined effort to observe rupture in a series of polyethylenes of various types using a sophisticated extensional rheometer. Although he did occasionally see something that looked like cavitation or rupture, his results were too irreproducible to allow drawing any quantitative conclusions.

Melt rupture has been observed in more complex flows involving draw down of a filament (57–59). However, since the variables are not well controlled in these experiments, it is not possible to draw quantitative conclusions from the results. Thus, it has not proven possible, to date, to measure the tensile strength of molten thermoplastics in a well-controlled experiment, and as a result, one cannot compare the rupture stresses determined in GMF experiments with those measured in uniform stretching.

## 7.11 GROSS MELT FRACTURE IN WIRE AND CABLE COATING

Extrudate distortions described as surface roughness have been reported in the wire and cable coating industry since LDPE was first used as an insulation material in the early 1950s. Polymer structure and die design have been varied on an ad hoc basis during this period in order to increase the production rate at which this defect occurs, and it would be useful to have a reliable criterion for its onset. However, no detailed discussion of the cause of this defect has appeared. We here present a brief description of the wire-coating process and discuss the possibility that gross melt fracture is caused by extensional flow in the coating die.

The coating processes used in the wire and cable industry are of two types, depending on the production speed and the size of substrate being coated. The substrate can be either a single conductor or a bundle of coated wires forming a cable. In general, the larger the substrate, the slower the coating process. A "wire" typically consists of a single copper conductor such as that used for telephone lines. On the other hand, a "cable" consists of wound conductors and an insulation layer or sometimes a bundle of wires rather than a single conductor. Thus, "wire coating" is the term usually used in the telecommunications industry, while "cable coating" is the term used in power cable manufacturing. However, one sometimes hears these two terms used interchangeably.

Surface roughness in the form of small, random bumps on the surface of final products has been observed in both coating processes, but the present discussion will be limited to wire coating, as the highest speeds are used in this process, and GMF is therefore most likely to occur. The meaning of the term "high-speed" has changed a great deal over the years. In the early 1950s, a line speed of 300 m/min was considered high, but by the 1960s this had increased to 600 m/min, and line speeds reached 1200–1500 m/min in the 1970s (60). For our present purpose, we consider high-speed coating to involve line speeds between 1800 and 3000 m/min, which are now common in the industry. Thus, we will not discuss jacket coating, where the line speed is in the range of 200–300 m/min (61–63). Although surface roughness is also observed during jacketing and is one of limiting factors in production rate, it is believed that the origin of this defect is at the die exit and that it is therefore a surface melt fracture.

In the wire-coating process, a preheated copper wire having a diameter from 0.4 to 0.75 mm passes through a guide tip at a speed usually higher than 2000 m/min. The wire then passes through the center of a specially designed extrusion die called a "pressure-tooling" die such as those shown in Fig. 7.12A and B. Molten polymer is extruded through an annular passage surrounding the wire, called the gum space, and adheres to the moving wire. The molten polymer first makes contact with wire in the "impact region" within the die after passing through several converging zones. Because of the wire's high speed, the melt is stretched at a high rate as it exits the die.

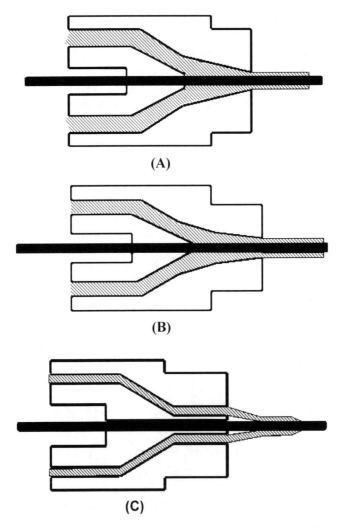

**(A)**

**(B)**

**(C)**

FIGURE 7.12 Sketch of three types of wire coating die: (A) Pressure tooling die with two-stage taper, (B) pressure tooling die with three-stage taper, (C) tube-tooling die.

Figure 7.12C shows a "tube-tooling" die of the type used for jacket coating. The annular cross section gradually decreases, keeping the same gap, leading to a straight, annular section. Thus, slip-promoting processing aids such as fluoropolymers are effective in eliminating surface roughness in jacket (or cable) coating. In a tube-tooling die, the melt contacts the moving cable some distance away from the die exit.

To evaluate the role of GMF in wire coating, we must know something about the flow pattern and stress field in a pressure-tooling die. There are two distinct flow regions: a shear flow within the annular die and a converging flow from the impact zone to the die exit, where the melt encounters the wire moving at high speed. Experimental study of the flow in the die is difficult because of the very small volume of the flow channel, and very few observations have been reported in the literature (59,63–68).

Because of the problems of direct observation, there have been a number of attempts at numerical simulation, usually using a finite element method (FEM), along with the lubrication approximation. Early attempts used a purely viscous fluid model (70–78). Caswell and Tanner (70) used this approach and predicted recirculation in the converging region and the shape of free surface at the die exit. The effect of compressibility was considered by Chung (73) who concluded that compressibility cannot be ignored if the wire speed is greater than 30 m/min. Mitsoulis (74) modeled the flow in a two-stage tapered die, which is now standard for HDPE (64). He found that shear thinning reduced the small extrudate swell and the amount of recirculation in the die. This methodology was extended to line speeds closer to industrial practice by Mitsoulis et al. (76), Lafleur et al. (77) and Wapperom and Hassager (78). Mitsoulis et al. (76) studied the flow in a die having a three-stage taper with a maximum line speed of 1200 m/min. They found that it was necessary to account for nonisothermality in order to predict the pressure distribution and that the experimental results of Haas and Skewis (60) were between those predicted assuming isothermal and adiabatic walls.

As line speed increases, both the overall pressure drop and the temperature increase dramatically (77,78). The steep temperature rise due to viscous dissipation would be expected to cause thermal degradation of the polymer, but the residence time in the die is very short ($10^{-2}$ to $10^{-3}$ sec), and degradation is actually not a problem, except after a long period of operation in any dead zone where molten polymer does not circulate. Most viscous dissipation occurs near the wall, so this is where the highest temperature is reached. This results in a low viscosity near the wall that lubricates the flow of the cooler bulk flow in the center, and it is this effect, rather than slip, that limits the pressure drop.

The use of a purely viscous model for melt flow has been useful for some purposes. However, the very short residence time in the die, together with the strong extensional component of the deformation, makes this a flow in which melt elasticity is very important. Binding and Blythe (79) demonstrated the inadequacy of inelastic models for the prediction of stresses and pressure drop. They pointed out that for high-speed coating using a pressure-tooling die the residence time is of the order of $10^{-2}$ sec, whereas typical wire-coating materials such as HDPE, MDPE, and LDPE have relaxation times that are many times larger than this. They recommended the use of the exponential version of the Phan-Thien–Tanner (PTT) constitutive equation, and this model has been used to simulate the flow in a wire-coating die (80–83). However, the highest line speed in these simulations, 20 m/min, is quite low compared to those used in wire coating, and the results cannot be compared directly with industrial observations. A comparison of the simulation results of Mitsoulis et al. (76) with the experimental observations of Haas and Skewis (60) indicate that conditions likely to produce GMF do indeed occur in high-speed coating. Haas and Skewis found that the overall pressure drop increased with wire speed and that there was a sudden increase in surface roughness at a speed of about 600 m/min.

As we have seen, it is now fairly certain that the surface roughness that occurs in high-speed wire coating is the result of gross melt fracture caused by the high extensional stress in the die. A brief review of the development of resins for this application will add support to this conclusion.

Low-density polyethylene was the first material used for wire coating in the 1950s because of its excellent dielectric properties, environmental stress crack resistance, and resistance to oxidative degradation. This polymer was the standard insulation material for some time but was eventually replaced by HDPE because of its better cut-through and abrasion resistances and better capacitance stability (84–86). In addition, the development of special grades of HDPE having broad MWD made it possible to increase line speeds (84). Later, linear, low-density polyethylene made possible a further increase in production rates to the range of 1500 to 2000 m/min before surface distortions appeared. Medium-density PE was also used, mainly in Europe, although it was less cost-effective. Strong pressure to reduce wire production costs then led to the development of a special grade of HDPE produced using the UNIPOL gas-phase polymerization process (86). This made it possible to increase production speeds to above 2000 m/min.

Finally, there has been a report of a new wire-coating resin made in Japan using a slurry process with a Ziegler-type, titanium catalyst that is said to be processable at rates up to 3000 m/min (87). This HDPE has a very broad, skewed molecular weight distribution, with a polydispersity index (PI $\equiv M_w/M_n$) in the range of 32 to 35. This contrasts with the gas-phase HDPE that is the current standard material, which has a PI in the range of 8 to 10. However, the gas-phase polymer is made with a chromium catalyst and may contain some long-chain branching, which would be expected to have a significant effect on processability. Most recently, iPP has been proposed as an alternative wire-coating material (68,69). Borke (68) reported that when conventional iPP was used as a primary insulation material, eccentricity occurred at line speeds as low as 150 m/min and that the maximum line speed was less than 1700 m/min. Lee and Schloemer (69) developed a new iPP to overcome this limitation; it is a two-phase, in situ, reactor-blended material and consists of a PP matrix with 23% rubber. He concluded that the eccentricity of the coating was the result of inlet melt fracture.

In summary, the evolution of materials for wire-coating suggests that GMF is the limiting factor in the process. Furthermore, the most reliable criterion for the onset of GMF is a critical tensile stress. This criterion predicts the following order of processability: HDPE (7.9 MPa)>LLDPE (3.72 MPa)>LDPE (0.7–0.8 MPa), although these materials have quite different MW and MWD (33). Furthermore, the presence of low levels of LCB in metallocene polyethylenes increases the critical tensile stress up to 3.2 MPa, which may explain why certain HDPEs can have good performance, despite the fact that their MWDs are narrow as compared to HDPE produced using Z-N catalyst. Furthermore, Jhang (65), Kim and Hah (66), and Rokunohe et al. (87) have shown that parameters such as the polydispersity index and the flow rate ratio, i.e., the ratio of the high flow to the standard melt indexes, $MI_{21.6 kg}/MI_{2.16 kg}$ (88), are not good indicators of the performance of polymers in wire coating. Therefore, they stressed the importance of the shape of the MWD.

At the same time that there were major developments in resin manufacture, there were also significant advances in die design, and these also helped to increase production rates. Tapered dies have always been used to streamline the flow of melt as it approaches the die exit. It was discovered early that it was advantageous to stage the tapering by using two successive angles of convergence. Kim and Hah (66) reported that they were able to reduce the tensile stress and resulting melt fracture by use of a die having successively decreasing entrance angles, with an angle at the tip of less than 7°. The detailed design of wire-coating dies varies from one machine manufacturer to another, each claiming an advantage of its design.

In conclusion, the polymer structural features and die designs that have been found to yield the highest line rates provide strong evidence that extrudate distortion in wire coating is caused by gross melt fracture.

## 7.12  AN ADDITIVE THAT SUPPRESSES GMF

The use of specially designed resins, filters, and tapered dies to delay the onset of GMF is discussed above. Another interesting development is the observation that a certain type of solid additive can suppress GMF in certain types of die. The additive is boron nitride (BN), and a 1997 patent (89) describes its use as a processing aid. The properties and performance of this additive are described in detail in Chapter 9 of this book.

The precise mechanism by which BN works is not yet clear, but one possibility is that the particles provide many nucleation sites for very small cracks, thus relieving the stresses that cause large-scale rupture before they reach the critical level. Results of a recent study (90) indicate that the polar component of the surface energy is the crucial factor for the effectiveness of BN both for suppressing sharkskin and delaying the onset of GMF. The original patent refers specifically to the use of BN for use in a wire-coating die. However, we note that the studies reported to date involve the relatively low-speed flow through capillaries and tube-tooling dies used for jacket coating rather than high-speed flow through pressure-tooling dies.

## 7.13 SUMMARY

Gross melt fracture is an extrusion defect that is known to result from a chaotic interruption of streamlines in a sharply converging flow channel. The ultimate cause appears to be the erratic rupture of the melt in response to the large tensile stresses that occur in such a flow. Many observations of this phenomenon have been reported, and it is thought to be responsible for extrusion defects that occur in high-speed wire coating. There are currently no useful models that describe gross melt fracture. Future progress on this front will require a theory of melt rupture and the development of a technique to study it under well-controlled conditions.

## REFERENCES

1. Sammler, R.L.; Koopmans, R.J.; Mangnus, M.A.; Bosnyak, C.P On the melt fracture of polypropylene. Soc. Plast. Eng. Annu. Tech. Conf. (ANTEC) Tech. Papers 1998, *44*, 957–961.
2. Vinogradov, G.V. Viscoelasticity and fracture phenomena in uniaxial extension of high-molecular linear polymers. Rheol. Acta 1975, *14*, 942–954.
3. Hatzikiriakos, S.G.; Dealy, J.M. Role of slip and fracture in the oscillating flow of HDPE in a capillary. J. Rheol. 1992, *36*, 845–884.
4. Kay, D.; Carrreau, P.J.; Lafleur, P.G.; Robert, L.; Vergnes, B. A study of the stick–slip phenomenon in single-screw extrusion of linear polyethylene. Polym. Eng. Sci. 2003, *43*, 78–90.
5. Ghanta, V.G.; Riise, B.L.; Denn, M.M. Disappearance of extrusion instabilities in brass capillary dies. J. Rheol. 1999, *43*, 435–442.
6. Uhland, E. Das anomale Fleissverhalten von Polyäthylen hoher Dichte. Rheol. Acta 1979, *18*, 1–24.

7.  Becker, J.P.; Bengtsson, P.; Klason, C.; Kubát, J.; Sáha, P. Pressure oscillations during capillary extrusion of high-density polyethylene. Int. Polym. Process. 1991, 6, 318–325.

8.  Wang, S.Q. Molecular transitions and dynamics at polymer/wall interfaces: Origins of flow instabilities and wall slip. Adv. Polym. Sci. 1999, 138, 227–275.

9.  Meller, M.; Luciani, A.; Sarioglu, A.; Manson, J. Flow through a convergence. Part I: Critical conditions for unstable flow. Polym. Eng. Sci. 2002, 42, 611–633.

10. Tordella, J.P. Fracture in the extrusion of amorphous polymers through capillaries. J. Appl. Phys. 1956, 27, 454–458.

11. Oyanagi, Y. A study of irregular flow behavior of high-density polyethylene. Polym. Symp. 1973, 20, 123–136.

12. Bergem, N. Visualization studies of polymer melt flow anomalies in extrusion. Proc. 7th Int. Congr. Rheol. Chalmers University: Gothenburg, Sweden, 1976; 50–54.

13. Piau, J.; El Kissi, N.; Tremblay, B. Influence of upstream instabilities land wall slip on melt fracture and sharkskin phenomena during silicone extrusion through orifice dies. J. Non-Newton. Fluid Mech. 1990, 34, 145–180.

14. McKinley, G.H.; Raiford, W.P.; Brown, R.A.; Armstrong, R.C. Nonlinear dynamics of viscoelastic flow in axisymmetric abrupt contractions. J. Fluid Mech. 1991, 23, 411–456.

15. McKinley, G.H.; Pakdel, P.; Öztekin, A. Rheological and geometric scaling of purely elastic flow instabilities. J. Non-Newton. Fluid Mech. 1996, 67, 19–47.

16. Meulenbroek, B.; Storm, C.; Bertola, V.; Wagner, C.; Bonn, D.; van Saarloos, W. Intrinsic route to melt fracture in polymer extrusion: A weakly nonlinear subcritical instability of viscoelastic Poiseuille flow. Phys. Rev. Lett. 2003, 90, 24502.

17. Nakamura, K.; Ituaki, S.; Nishimura, T.; Horikawa, A. Instability of polymeric flow through an abrupt contraction. J. Text. Eng. 1967, 36, 49–57. (Translation of J. Text. Mach. Soc. Jpn. 1987, 40 (1), 57–65).

18. Yesilata, B.; Ostekin, A.; Neti, S. Instabilities in viscoelastic flow through an axisymmetric sudden contraction. J. Non-Newton. Fluid Mech. 1999, 85, 35–62.

19. Koopmans, R. Extrusion defects. Unpublished manuscript, 2001.

20. Fernández, M.; Santamaría, A. A striking hydrodynamic phenomenon: Split of a polymer melt in capillary flow. J. Rheol. 2001, 45, 595–602.

21. Bagley, E.B.; Birks, A.M. Flow of polyethylene into a capillary. J. Appl. Phys. 1960, 31, 556–561.

22. Kim, S.; Dealy, J.M. Gross melt fracture of polyethylene. I: A criterion based on tensile stress. Polym. Eng. Sci. 2002, 42, 482–494.

23. Carreras, E.S.; El Kissi, N.; Piau, J.-M.; Tousaint, F. Pressure and temperature effects on flow stability of PE melts during extrusion. Unpublished manuscript, 1998.

24. Bagley, E.B.; Schreiber, H.P. Effect of die entry geometry on polymer melt fracture and extrudate distortion. Trans. Soc. Rheol. 1961, 5, 341–353.

25. Piau, J.-M.; Nigen, S.; El Kissi, N. Effect of die entrance filtering on mitigation of upstream instability during extrusion of polymer melts. J. Non-Newton. Fluid Mech. 2000, 91, 37–57.

26. Goutille, Y.; Guillet, J. Influence of filters in the die entrance region on gross melt

fracture: Extrudate and flow visualization. J. Non-Newton. Fluid Mech. 2002, *102*, 19–36.

27. Denn, M.M.; Marrucci, G. Stretching of viscoelastic liquids. AIChE J. 1971, *17*, 101–103.

28. Paul, D.R.; Southern, J.H. The role of entanglements in the elastic fracture of polymer solutions. J. Appl. Polym. Sci. 1975, *19*, 3375–3381.

29. Hürlimann, H.P.; Knappe, W. Der Zusammenhang zwischen der Dehnspannung von Kunststoffschmelzen im Düseneinlauf und im Schmelzbruch. Rheol. Acta 1972, *11*, 292–301.

30. Cogswell, F.N. Converging flow of polymer melt in extrusion dies. Polym. Eng. Sci. 1972, *12*, 64–73.

31. Cogswell, F.N. Converging flow and stretching flow: A compilation. J. Non-Newton. Fluid Mech. 1978, *4*, 23–38.

32. Shaw, M.T. Flow of polymer melts through a well-lubricated, conical die. J. Appl. Polym. Sci. 1975, *19*, 2811–2816.

33. Kim, S.; Dealy, J.M. Gross melt fracture of polyethylene. II: Effects of molecular structure. Polym. Eng. Sci 2002, *42*, 485–503.

34. Binding, D.M. An approximate analysis for contraction and converging flows. J. Non-Newton. Fluid. Mech. 1988, *27*, 173–189.

35. Vlachopoulos, J.; Lidorikis, S. Melt fracture of polystyrene. Polym. Eng. Sci. 1971, *11*, 1–5.

36. Vlachopoulos, J.; Alam, M. Critical stress and recoverable shear for polymer melt fracture. Polym. Eng. Sci. 1972, *12*, 184–192.

37. Baik, J.J.; Tzoganakis, C. A study of extrudate distortion in controlled-rheology polypropylene. Polym. Eng. Sci. 1998, *38*, 274–281.

38. Cogswell, F.N. Polymer melt rheology during elongational flow. Appl. Polym. Symp. 1975, *27*, 1–18.

39. Kim, Y.C.; Yang, K.S. Effect of peroxide modification on melt fracture of linear low density polyethylene during extrusion. Polym. J. 1999, *31*, 579–584.

40. Yamane, H.; White, J.L. A comparative study of flow instabilities in extrusion, melt spinning, and tubular blown film extrusion of rheologically characterized high density, low density and linear low density polyethylene melts. J. Rheol. Jpn. 1987, *15*, 131–140.

41. Wood-Adams, P.M.; Dealy, J.M. Using rheological data to determine the branching level in metallocene polyethylenes. Macromolecules 2000, *33*, 7481–7488.

42. Costeux, S.; Wood-Adams, P.M. Thermorheological behavior of polyethylene: Effects of microstructure and long chain branching. Macromolecules 2001, *34*, 6281–6290.

43. Meller, M.; Luciani, A.; Manson, J. Flow through a convergence. Part II: Mixing of high viscosity ratio polymer blends. Polym. Eng. Sci. 2002, *42*, 634–653.

44. Joseph, D.D. Cavitation and the state of stress in a flowing liquid. J. Fluid. Mech. 1988, *366*, 367–378.

45. Tschoegl, N. Failure surfaces in principal stress space. Polym. Sci. Symp. 1971, *32*, 239–267.

46. Wright, R.N. Mechanisms of wire breaks. Wire J. Int. May 1982, 86–90.

47. Zhang, X.C.; Butler, M.F.; Cameron, R.E. The ductile–brittle transition of irradiated isotactic polypropylene studied using simultaneous SAXS and tensile deformation. Polymer 2000, *41*, 3797–3807.
48. Archer, L.A.; Ternet, A.; Larson, R.G. Fracture phenomena in shearing flow of viscous liquids. Rheol. Acta 1997, *36*, 579–584.
49. Son, Y.; Migler, K.B. Cavitation of polyethylene during extrusion processing instabilities. J. Polym. Sci.: Part B: Polym. Phys. 2002, *40*, 2791–2799.
50. Torres, W. Rupture of Molten Polymers. M.Eng. dissertation, McGill University, Montreal, Canada, 2002.
51. Chen, Y.L.; Larson, R.G.; Patel, S.S. Shear fracture of polystyrene melts and solutions. Rheol. Acta 1994, *33*, 243–256.
52. Kinlock, J.; Young, R.J. Fracture Behavior of Polymers; New York: Elsevier, 1983.
53. Malkin, A.Ya.; Petrie, C.J.S. Some conditions for rupture of polymer liquids in extension. J. Rheol. 1997, *41*, 1–25.
54. Joshi, Y.; Denn, M.M. Rupture of entangled polymeric liquids in elongational flow. J. Rheol. 2003, *47*, 291–298.
55. McKinley, G.H.; Hassager, O. The Considère condition and rapid stretching of linear and branched polymer melts. J. Rheol. 1999, *43*, 1195–1212.
56. Meissner, J. Dehnungsverhalten von polyethylene-schmelzen. Rheol. Acta 1971, *10*, 230–242.
57. Kanei, E.; Onogi, S. Extensional and fractural properties of monodisperse polystyrenes at elevated temperatures. Appl. Polym. Symp. 1975, *27*, 19–46.
58. Takaki, T.; Bogue, D.C. The extensional and failure properties of polymer melts. J. Appl. Polym. Sci. 1975, *19*, 419–433.
59. Wagner, M.F.; Schulze, V.; Göttfert, A. Rheotens mastercurves and drawability of polymer melts. Polym. Eng. Sci. 1996, *36*, 925–935.
60. Haas, K.U.; Skewis, F.H. The wire coating process: Die design and polymer flow characteristic. Soc. Plast. Eng. Annu. Tech. Conf. (ANTEC) Tech. Papers 1974, *20*, 8–12.
61. Skewis, F.H. Processing cable jackets of polyethylene. Wire Wire Prod. April 1970, 65–69.
62. Viriyayuthakoru, M.; Deboo, R.V. A FEM model for cable jacketing simulation. Soc. Plast. Eng., Annu. Tech. Conf. (ANTEC) Tech. Papers 1983, *29*, 178–182.
63. Wasserman, S.H.; Adams, J.L. Rheology and crystallization in fiber optic cable jacket and conduit extrusion. Soc. Plast. Eng. Annu. Tech. Conf. (ANTEC) Tech. Papers 2001, *47*, 1144–1148.
64. Robinson, J.E. Rheology of polyethylene telephone singles insulation compounds. Proc. Int. Conf. Plastics in Telecommunications II, London; Plastics and Rubber Institute: London, 1982; 27.1–27.14.
65. Jhang, Y.X. The evaluation of high-speed extrusion of HDPE and PP insulating materials for telephone cable. Int. Wire & Cable Symp. Proc. IWCS Inc.: Eatontown, NJ, 1990; 59–63.
66. Kim, Y.S.; Hah, J.Y. A study on high-speed telephone wire extrusion line. Int. Wire & Cable Symp. Proc. IWCS Inc.: Eatontown, NJ, 1991; 492–498.

67. Tonyali, K.; Fernando, P.L. Characterization and development of foam/skin polyethylene insulation. Int. Wire & Cable Symp Proc. IWCS Inc.: Eatontown, NJ, 1992; 497–504.

68. Borke, B.H. Oscillatory flow of PP and its effect on conductor eccentricity. Int. Wire & Cable Symp. Proc. IWCS Inc.: Eatontown, NJ, 1998; 294–298.

69. Lee, C.D.; Schloemer, T.S. Alternative polyolefin insulation materials for communication cable applications. Int. Wire & Cable Symp. Proc. IWCS Inc.: Eatontown, NJ, 2002; 124–128.

70. Caswell, B.; Tanner, R.I. Wire coating die design using finite element methods. Polym. Eng. Sci. 1978, *18*, 416–421.

71. Carley, J.F.; Endo, T.; Krantz, W.B. Realistic analysis of flow in wire coating die. Polym. Eng. Sci. 1979, *19*, 1178–1187.

72. Riley, D.W. New flow measurements in wire coating extrusion. Soc. Plast. Eng. Annu. Tech. Conf. (ANTEC) Tech. Papers 1983, *29*, 65–67.

73. Chung, T.S. The effect of melt compressibility on a high-speed wire coating process. Polym. Eng. Sci. 1986, *26*, 410–414.

74. Mitsoulis, E. Finite element analysis of wire coating. Polym. Eng. Sci. 1986, *26*, 171–186.

75. Pittman, J.F.T.; Rashid, K. Numerical analysis of high-speed wire coating. Plast. Rubber Process. Appl. 1986, *6*, 153–159.

76. Mitsoulis, E.; Wagner, R.; Heng, F.L. Numerical simulation of wire coating low-density polyethylene: Theory and experiments. Polym. Eng. Sci. 1988, *28*, 291–310.

77. Lafleur, P.G.; Aprin, B.; Lenir, V. Computer aided design of wire coating dies. Adv. Polym. Technol. 1994, *13*, 297–304.

78. Wapperom, P.; Hassager, O. Numerical simulation of wire-coating: The influence of temperature boundary conditions. Polym. Eng. Sci. 1999, *39*, 2007–2018.

79. Binding, D.M.; Blythe, A.R.; Gunter, S.; Mosquera, A.A.; Townsend, P.; Webster, M.F. Modeling polymer melt flows in wire coating processes. J Non-Newton. Fluid Mech. 1996, *64*, 191–206.

80. Ngamaravaranggul, V.; Webster, M.F. Simulation of coating flows with slip effects. Int. J. Numer. Methods Fluids 2000, *33*, 961–992.

81. Ngamaravaranggul, V.; Webster, M.F. Simulation of pressure-tooling wire-coating flows with Phan-Thien/Tanner models. Int. J. Numer. Methods Fluids 2002, *38*, 677–710.

82. Baloch, A.; Matallah, H.; Ngamaravaranggul, V.; Webster, M.F. Simulation of pressure-tooling wire-coating flows through distributed computation. Int. J. Numer. Methods Heat Fluid Flow 2002, *12*, 458–493.

83. Matallah, H.; Townsend, P.; Webster, M.F. Viscoelastic computation of polymeric wire-coating flows. Int. J. Numer. Methods Heat Fluid Flow 2002, *12*, 404–433.

84. Scott, J.N.; Levett, C.T.; Leslie, C.F.; Hogan, J.P. Tailoring of high-density polyethylene for high speed wire coating. Wire J. March 1966, 409–463.

85. Hager, J.E. A guide to polyethylene wire & cable coating. Wire J. Oct 1970; 149–154.

86. Fisher, E.J. Trends in polyethylene materials for wire and cable. Wire J. Int. August 1983, 50–59 .

87. Rokunohe, M.; Okada, M.; Ueno, Y.; Konishi, J. Ultra-high speed extrusion of foamed polyethylene insulation. Int. Wire & Cable Symp. Proc. IWCS Inc.: Eatontown, NJ, 1976; 53–62.

88. Wang, H. A simple evaluation method for PE as telephone cable insulation material at high-speed extrusion and coating. Polym. Test. 2002, *21*, 333–335.

89. Buckmaster, M.D.; Henry, D.L.; Randa, S.K. High Speed Extrusion. US Patent 5,688,457; 1997.

90. Rathod, N.S.; Hatzikiriakos, S.G. The effect of surface energy of boron nitride on polymer processability. Soc. Plast. Eng. Annu. Tech. Conf. (ANTEC) Tech. Papers 2003, *49*, 17–21.

# 8

# Conventional Polymer Processing Additives

**Semen B. Kharchenko and Kalman B. Migler**
National Institute of Standards and Technology, Gaithersburg,
Maryland, U.S.A.

**Savvas G. Hatzikiriakos**
The University of British Columbia, Vancouver,
British Columbia, Canada

## 8.1 INTRODUCTION

Commercial polymer processing operations, such as fiber spinning, film blowing, blow molding, profile extrusion, and various coating flows, suffer by the onset of flow instabilities, which manifest themselves as distortions of the extruded product (Chapters 5–7). These phenomena, collectively known as *melt fracture*, pose a limitation in the rate production that threatens the economic feasibility of the processes (1–8).

In extrusion processes, the extrudate distortions are first observed when the volumetric flow rate exceeds a critical value. The severity of the distortions

---

Official contribution of the National Institute of Standards and Technology; not subject to copyright in the United States.

(amplitude and irregularity) increases with increasing flow rate (1,9–12). Thus the production rates of commercially acceptable products are limited to the ones below these critical values. To overcome these difficulties and to render the processes economically feasible, polymer processing aids (PPAs) are frequently used. Polymer processing aids can eliminate the flow instabilities known as sharkskin melt fracture and stick–slip or postpone them to higher flow rates. The end result is an increase of the productivity as well as an energy cost reduction, while high product quality is maintained.

The purpose of this chapter is to examine traditional processing aids that have been investigated over the past several decades and to discuss their performance in melt fracture elimination, so as to elucidate the role of these materials in instability elimination and polymer processability improvement. Newer processing aids such as boron nitride and its combinations with conventional processing aid are described in Chapter 10. The factors that determine the additive performance, such as concentration, dispersion quality, and interactions between additives, additive and polymer, and additive and die surface, are also examined. The various mechanisms by which different types of processing aids help eliminate the instabilities are reviewed as well.

The additives discussed in this chapter are primarily fluoropolymer based and their use is primarily for polyolefin extrusion processes. Additional examples of other conventional processing aids such as stearates, silicon-based polymers, and various polymer blend combinations are also briefly discussed. However, the most commercially relevant example is the use of fluoropolymer-based PPAs to eliminate sharkskin melt fracture in linear low-density polyethylene (LLDPE) industry and therefore this chapter is focused more on these PPAs.

## 8.2 FLUOROPOLYMER POLYMER PROCESSING AID TECHNOLOGY

The fluoropolymer PPA technology was discovered and developed in the early 1960s by DuPont Canada. It was introduced to the newly developed LLDPE to eliminate sharkskin melt fracture. The discovery was accidental as the scientists of DuPont tested the LLDPE on a blown film line whose extruder was used before for fluoropolymer extrusion. While sharkskin melt fracture was expected to be seen, the film was coming out clear and free of any distortions. Then, it was realized that the extruder was not properly purged, and the small amount of fluoropolymer still present was acting as a processing aid. It was only after a few hours that sharkskin came back again into the film appearance.

A widely used fluoropolymer additive for LLDPE is a copolymer of vinylidene fluoride ($VF_2$) and hexafluoropropylene and is often also referred to as fluoroelastomer, although it is not cross-linked. The general chemical structure of this is as follows:

-$(CF_2$-$CF)_n$-$(CF_2$-$CH_2)_m$-
  |
  $CF_3$

In $VF_2$, the hydrogens become acidic because of the electronegativity of the fluorine. The resulting dipole moment of this monomer gives it an attraction with the oxides in the steel. By itself, the $VF_2$ is crystalline; the incorporation of hexafluoropropylene breaks up the crystallinity and makes the material an un-cross-linked rubber. The hexafluoropropylene also has a poor interaction with the nonpolar polyethylene, reducing the adhesive between these two polymers. In typical use, the $VF_2$ copolymer is 60% by mass fraction. The relative molecular mass and viscosity is comparable to the polyethylenes that they typically are used with, and they are normally molten at processing temperatures. New generations of processing aids utilize combinations of this fluoropolymer with polyethylene glycol or other low relative molecular mass compounds.

In typical industrial usage, fluoropolymers are added to the polymer in small quantities (less than 0.1% of mass fraction of PPA in polymer); they are highly incompatible with polyolefins and the resulting material is a dilute two-phase blend. Some brand names of fluoropolymers used as PPAs include Dynamar, Viton Kynar, Teflon, and DFL (dry film type[*]).

A two-step procedure is often utilized to make the blend, in which a master batch is first prepared and pelletized, and then mixed with the process polymer; this is an effective route for preparing homogeneous dispersions (13,14). In the case of Viton, diluting the fluoropolymer in acetone before mixing was found to be beneficial (15). In addition to optimizing the additive level, there are other factors that critically affect the additive performance including interactions with other additives, dispersion quality, droplet size, and the rheology (some of these are discussed later). Many industrial patents have been issued for the incorporation of various processing aids in extrusion processes (16–18), which justifies the ongoing intense academic and industrial research in this area.

Beyond elimination of some extrusion instabilities, other benefits from the use of fluoropolymers include gel reduction (reduction of cross-linked,

---

[*]Certain equipment, instruments, or materials are identified in this paper to adequately specify experimental details. Such identification does not imply recommendation by the National Institute of Standards and Technology nor does it imply the materials are necessarily the best available for the purpose.

oxidized, and unmelted/unmixed gels that form in the extruder), reduced die buildup or die drool, decrease of the induction time for color change, and, to some extent, elimination of draw resonance in film casting, fiber spinning, and film blowing operations (19).

## 8.3 INSTABILITIES, SLIPPAGE, AND POLYMER PROCESSING AIDS

### 8.3.1 Sharkskin

The subject of sharkskin is covered in depth in Chapter 5, thus we mention here only a few salient features of it, particularly those relevant to the fluoropolymers. During the extrusion of *linear* polymers, in particular polyethylene, of sufficiently high and narrow molecular weight, a distortion of the surface occurs above a critical apparent shear rate. The amplitude and wavelength is an increasing function of shear rate. It has been shown to initiate at the air–polymer–wall interface as the polymer exits the die. The boundary condition between the wall and the polymer melt is crucial to the nature of flow instability occurring during the extrusion, but it has been somewhat controversial. Is there a stick condition, a partial slip condition, or perhaps stick–slip? As discussed in the previous chapter, while wall slippage may occur, it is not a necessary condition for sharkskin, as it has been observed in the absence of slip. Some scenarios for the boundary condition are shown in Fig. 8.1. When directly introduced to the die wall, fluoropolymer additives clearly cause (or enhance) wall slippage because their interaction with the mainstream polymer is weaker than that with the wall (Fig. 8.1B).

### 8.3.2 Slippage

To be effective, the fluoropolymer additives dispersed into the primary polymer must do two things. First, they must coat the die wall, in particular at the exit. Second, they must induce slippage between themselves and the primary polymer. We start with a discussion of slippage. Figure 8.2[*] shows the flow curve relating wall shear stress to apparent wall shear rate (defined in Chapter 7) in the presence and absence of a fluoropolymer additive. In the absence of the additive, the sharkskin, stick–slip, and gross melt fracture regimes are all present. The shift of the flow curve to higher flow rates in the presence of the PPA is depicted in Fig. 8.2 for a typical high-density polyethylene (HDPE) at a temperature of 180°C. The data were obtained

---

[*]The policy of the National Institute of Standards and Technology is to use the International System of Units (metric units) in all its publications. In this document, however, works of authors outside NIST are cited which describe measurements in certain non-SI units.

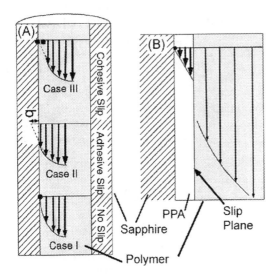

**FIGURE 8.1** (A) Possible boundary conditions at the polymer–wall interface: Case I—no slip; case II—polymer–wall slippage; case III—internal slip. (B) In the case of a fluoropolymer preferentially wetting the wall, slippage may occur at the polymer–polymer interface. (From Ref. 26.)

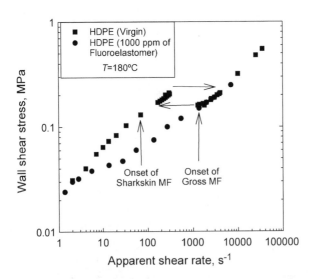

**FIGURE 8.2** The flow curve of a HDPE extruded in a pressure-driven capillary rheometer as virgin and in the presence of a fluoroelastomer. (From Ref. 35.)

by using a capillary rheometer and several dies of various lengths and diameters to determine and apply the *Bagley* correction (20) as described in Chapter 4. In the presence of the PPA, both the sharkskin and the stick–slip instability disappear, while the gross melt fracture regime still occurs. Similar results for a variety of polyolefins have been published (13,21–24). The reduced shear stress at a given throughput is a sign of slippage. However, in one study, it was found that DFL promoted adhesion and had no effect on the flow curve (21).

An additional benefit of the fluoropolymers can be deduced from Fig. 8.2. As a result of the slippage in the presence of the fluoropolymer, at a given apparent shear rate, the shear stress is reduced. Consequently, the power/torque requirement for a given production rate is lowered, and the production efficiency is increased (13,24) while the product maintains its high quality.

Recently, Migler et al. (25–28) employed stroboscopic optical microscopy to image the coating process of the PPA onto the capillary wall. They found that the PPAs migrate to the surface of the die, creating a thin coating that acted as a lubricant. In the case of extrusion of a 0.1% PPA/LLDPE blend, a new phenomenon, the formation of streaks on the die surface, which gradually propagate downstream, was observed (Fig. 8.3). As the onset of streaks was concomitant with pronounced decrease in shear stress and

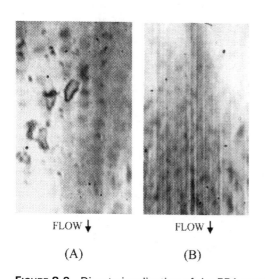

FLOW ↓                    FLOW ↓

(A)                        (B)

**FIGURE 8.3**  Direct visualization of the PPA coating of the die wall. While during the extrusion of LLPDE, the polymer–wall interface does not exhibit any special features except for a few gel-like particles moving slowly along it (A), a characteristic streaklike nature of the interface develops during the extrusion of the 0.1% PPA/LLDPE blend (B). (From Ref. 26.)

elimination of sharkskin, it was attributed to the formation of the fluoro-polymer layer. By monitoring the flow profiles during the extrusion of LLDPE with and without the PPA (Fig. 8.4), the authors demonstrated that the slip occurred at the interface between the two polymers, indicating that the polymers were fully disentangled.

### 8.3.3  Precoating Dies with Polymer Processing Aids

To separate the study of the slippage from that of the coating, the walls are often precoated with the fluoropolymer using a solution casting methods and subsequent curing. Hatzikiriakos and Dealy (29) used two different fluoro-polymer PPAs to coat the surface of the plates in a sliding plate rheometer for polyethylene and one fluoropolymer for polypropylene (PP) (30). They reported that the coatings resulted in a significant decrease of the critical shear stress for the onset of wall slip for both polyolefins and an increase of the slip velocity. The critical shear stress for the onset of slip, which is 0.1 MPa for the clean plate, was as low as 0.027 MPa.

In addition, Anastasiades and Hatzikiriakos (31) used a sessile drop method to measure the work of adhesion between HDPE/LLDPE and clean as well as fluoropolymer-coated stainless steel. They found that the fluoro-elastomers (FEs) reduce the work of adhesion between polyethylene and the surface. These findings support the hypothesis that slip is a result of adhesive failure, and that the slip velocity and critical shear stress for the onset of slip depend on the work of adhesion. Slip models taking into account these phenomena have been developed by Hatzikiriakos (32), Hill et al. (33), and Stewart (34).

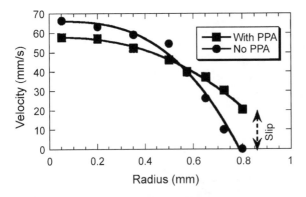

**FIGURE 8.4**  Flow velocimetry measurements of LLDPE with and without PPA. Measurements are conducted at 210°C and throughput of 3.9 g/min, using a sapphire die with $L/D = 15.9$ and $D = 1.6$ mm. (From Ref. 26.)

When an additive is present (either dispersed in the polymer or as a coating), proper die conditioning is needed before any equilibrium measurements. Kazatchkov et al. (30) found that when PP was extruded through Teflon-coated capillary dies, the shear stress was gradually reduced. A typical example is depicted in Fig. 8.5. Each run corresponds to the extrusion of a full capillary barrel. This indicates that the fluoropolymer coating was initially imperfect (perhaps porous as is typical for solution casting methods) and after a long induction time it was smoothened out, providing better lubrication and larger pressure reduction.

## 8.3.4  Coating Process of Fluoropolymers

For the PPAs to be effective, they must first coat the die wall; in fact, the major metric used to evaluate the efficacy of a given PPA is the time it takes to completely eliminate sharkskin. However, there are numerous factors that influence the coating and information about the *rate* at which the PPA coats the die wall, as well as characteristics of the coating that are less explored (15,26,27,35). For example, it was shown that an increase in the concentration of the PPA, its dispersion quality, and higher flow rates reduces coating times (19,23,35,36). Recently, Oriani and Chapman (37) showed that larger droplets eliminate sharkskin faster. Yet the overall mechanism whereby they coat the wall is only now being addressed in active research.

One factor that must influence the coating is the relative affinity of the polymers for the wall. It is reasonable to conclude that if the fluoropolymer has a greater affinity for the wall, then it should coat it. As described earlier, a

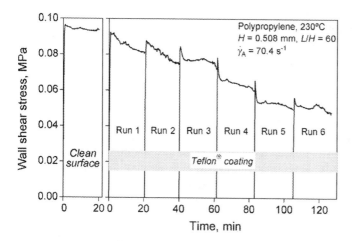

**FIGURE 8.5**  The effect of a Teflon® coating on the shear stress in the continuous extrusion of polypropylene (transient capillary experiments). (From Ref. 35.)

commonly used fluoropolymer contains a $VF_2$ comonomer, which does have a strong affinity to the oxides present in the steel. For example, fluoropolymers will displace nonpolar hydrocarbons with low work of adhesion, but will not displace polar polymers such as nylon, polyesters, or poly(methylmethacrylate) from metal surfaces because of the very high value for the work of adhesion of these polymers to metals.

However, the relative affinity is not the entire story, as will be described later, Shih (38) was able to create a coating where ethylene/propylene/1,4-hexadiene terpolymer (EPDM) is the minor phase in a fluoropolymer as well as the reverse, where the fluoropolymer is the minor phase in EPDM. Thus we are also led to consider the possibility that there is an important component from the flow kinetics that drives droplets (the minority phase) to the surface during extrusion. In a two-phase material, there are two mechanisms that may cause the droplets to migrate in a capillary flow. The first one is known as cross-stream migration studied experimentally (39,40) as well as via numerical simulations (41–45). However, what is known thus far (for Newtonian systems) is that it drives droplets *away* from the wall; thus it cannot cause the coating. The second mechanism is one whereby the lower viscosity component coats the wall. However, this mechanism is unlikely to cause the coating for several reasons: First, the concentrations used here are well below those where this low viscosity coating mechanism operates. Second, it also does not explain the results by Shih (38,46) whereby either phase can coat the wall when in the minority, although one of them is of higher viscosity. Third, in the case of fluoropolymer additives, lowering their viscosity generally results in poorer performance. Thus we consider other mechanisms to enhance coating; Kanu and Shaw (15) found an effect of entrance angles, whereby a sharp entrance angle enhanced the droplet migration to the wall, pointing to the importance of the entrance region.

The hypothesis that the coating may first form at the very *entrance* of the die was advanced in the work of Kharchenko et al. (47). First, it is known that the fluoropolymer need be present only at the very exit to eliminate sharkskin (see Chapter 5). Thus the following protocol was employed. Pure LLDPE was extruded through a capillary rheometer, followed by a 0.1% blend of fluoropolymer in LLDPE (Fig. 8.6). In this figure, the sharkskin fraction along with the reduced pressure values are plotted vs. the extrusion time for each load of the blend. During the middle of the extrusion of the 0.1% PPA/PE blend, the flow was stopped, and the die was "reversed" at a time when the pressure just started to decrease (some coating has occurred), but there was still 100% sharkskin in the extrudate. Therefore if the fluoropolymer first coats the die at the entrance region, then when the flow is turned back on, the coating would be present at the exit, exactly where sharkskin is initiated (Chapter 6). Indeed, as the extrusion was recommenced after the reversal, the sharkskin disappeared *instantaneously* and *completely*, while the reduced

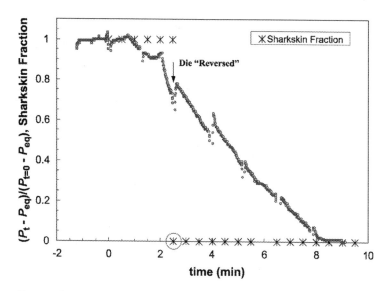

**FIGURE 8.6** Elucidation of the coating mechanism via die "reversal" experiment. Measurements are conducted with 0.1 % PPA/PE blend at $\dot{\gamma} = 225\ \text{sec}^{-1}$ and $T = 180°C$. $P_t$ is the pressure developed during extrusion of the PPA/LLDPE blend, $P_{eq}$ is the pressure at the steady state when PPA coating is fully developed, and $P_{t=0}$ is the pressure of pure PE. (From Ref. 47.)

pressure remained unchanged. This shows that in the time just before the flow stopped, the fluoropolymer had fully coated the die entrance, but the die exit contained no fluoropolymer. In contrast, when the experiment was conducted normally (without a reversal) there was a *gradual* and *slower* elimination of sharkskin, suggesting that the coating that started forming at the die entrance would require some *finite* time to migrate toward the die exit.

The idea of the "entrance-to-exit PPA migration" was further reinforced by performing the extrusion of LLDPE (certainly, under conditions of strong sharkskin) through the cylindrical sapphire die initially precoated with the fluoropolymer layer at its entrance (47). In that experiment, the first sharkskin-free streak appeared *after* the pressure dropped to its minimum ($\approx 80\%$ of the initial value). This tells us that the PPA material propagated along the die wall with a certain migration speed of its front, depending on the flow rate and the initial thickness of the PPA coating applied to the die entrance. Therefore a plausible mechanism of PPA coating can be envisaged as follows. During the extrusion of PPA/LLDPE blends, PPA droplets first adsorb at the die entrance (which explains the reduced pressure), and then they migrate as streaks toward the die exit under the influence of shear field.

## 8.3.5  Coating Thickness and a Simple Coating Model

The issue of coating thickness has recently been explored with the imaging technique based on the frustrated total internal reflection (47). Interestingly, it was found that the steady state PPA coatings developed during the extrusion of PPA/LLDPE at the die exit is only on the order of 150–450 nm, depending on the PPA concentration (Fig. 8.7). Moreover, even thinner coatings of about the size of a few PPA molecules (or about 25 nm) are found to be sufficient in elimination of the melt fracture. Additionally, the reflectivity data in that study conformed to streaklike nature of the PPA coating observed in earlier works through direct stroboscopic microscopy visualization (26).

The simple model developed in Ref. (47) is based on the balance between the mass flow rate of the fluoropolymer that coats the die ( $Q_{in}$ ) and the mass

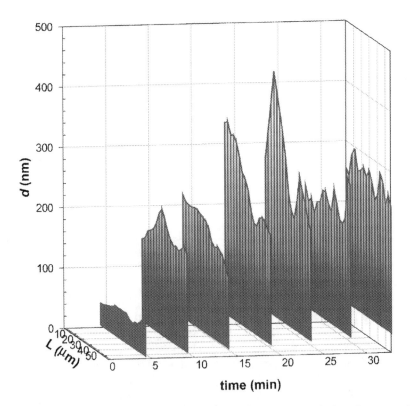

**FIGURE 8.7** "Visualization" of the PPA coating kinetics during the extrusion of 0.1% PPA/PE blend at $\dot{\gamma}$ = 215 sec$^{-1}$ and $T$ = 180°C; $d$ is the coating thickness and $L$ is the die circumferential length. (From Ref. 47.)

flow rate of the fluoropolymer leaving the die because of convection down-stream ($Q_{out}$). It assumes that a droplet (of radius $S$) will coat the wall if its streamline is within a distance $S$ from the wall; otherwise it will not coat it. This is schematically shown in Fig. 8.8 where upon contact with the wall the PPA droplet sticks and spreads downstream, thereby coating it. The mass flow rate of the coating material leaving the die is based on a linear velocity profile within the fluoropolymer layer. This allows us to write down the equations for the mass flow rates of the fluoropolymer in and out of the die:

$$Q_{in} = \int_0^{2\pi} \int_0^S C[v_{PE}(r')] R dr' d\varphi = 2RSC\pi \left( \frac{\dot{\gamma}_{PE}S}{2} + V_S \right) \tag{8.1}$$

$$Q_{out} = \int_0^{2\pi} \int_0^d \rho v_{PPA}(r') dr' R d\varphi = R\rho \dot{\gamma}_{PPA} d^2 \pi \tag{8.2}$$

where $r'$ is the distance from the wall, $R$ is the die radius, $C$ is the bulk concentration of the PPA in the PPA/PE blend, $\rho$ is the density of the PPA, and $v_{PE}(r')$ is the velocity of the polyethylene near the wall given by $v_{PE}(r') = \dot{\gamma}_{PE}(R, t)r' +$

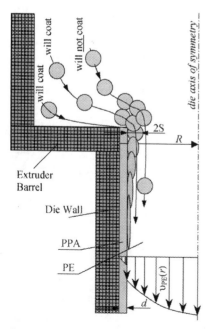

**FIGURE 8.8** Schematic representation of the PPA/PE flow in a circular die. (From Ref. 47.)

$V_S(\dot{\gamma},t)$. The slippage at the PE/PPA interface is taken into account through the slippage velocity $V_S(\dot{\gamma},t)$. $V_S$ is a function of $t$ because at the beginning of the process, the PE is in direct contact with the wall and we expect little/no slippage, whereas in steady state when the wall is fully coated, we anticipate strong slip. In steady state, we equate the right-hand sides of Eqs. (8.1) and (8.2):

$$2CS\left(\frac{\dot{\gamma}_{PE}S}{2} + V_S\right) = \rho\dot{\gamma}_{PPA}d^2 \tag{8.3}$$

At the beginning of the process, $(\dot{\gamma}_{PE}S/2)$ is an important factor for the coating kinetics. However, as the system reaches steady state and the die walls are fully coated, $V_S$ dominates the sum of Eq. (8.3) and $d$ can be expressed as:

$$d = \left(\frac{2CV_S S}{\rho\dot{\gamma}_{PPA}}\right)^{0.5} \tag{8.4}$$

The slippage velocity $V_S$ scales nearly linearly with apparent shear rate (26): $V_S$ = (0.1 mm) $\dot{\gamma}$. Predicted steady state coatings using the model are found to compare well to the experimentally measured ones (47). An important observation that is drawn from Eq. (8.4) is that the steady state coating thickness at the exit appears to be independent of die diameter and scales with the droplet size, which is consistent with the recent work by Oriani and Chapman (37) where it was found that larger droplets coat the walls significantly faster than smaller ones. As most of the droplets do not coat the wall and pass directly to the finished product, the upper limit on droplet size is presumably dictated by the largest size acceptable to the final product.

Note that Eq. (8.4) shows that the thickness (and coating time) is improved with increasing concentration. However, an excessive amount of lubricant should be avoided, first because of the cost of the PPA and second because it might lead to excessive lubrication of the extruder barrel, causing the polymer to slip in the extruder screw and which can have undesired effects (13).

### 8.3.6 Use of Fluoropolymer PPA in Combination with Interfacial Agents

Coadditives are compounds used to enhance the effectiveness of PPAs. Commercially available poly(ethylene)glycol (PEG) (37,48–57) and biodegradable aliphatic polyesters, such as polycaprolactone (PCL) when used as coadditives greatly enhance the performance of conventional PPAs (37) through a synergistic mechanism. They do not improve polyolefin processing when used *alone*, although they are often used to amplify mechanical properties of starch-based composites (58). They are believed to work by encapsulating the fluoropolymer droplets and are thus called interfacial agents (IA).

Thus for a given concentration of the PPA, they allow a reduction in the time for complete elimination of the melt fracture, and can also be used in lesser amounts if the PPA rheology is modified. If the PEG/fluoropolymer ratio becomes too high, a loss in the output capacity of an extruder (37,59) occurs because there is an excess of PEG that does not coat the PPA droplets, but instead contributes to the slippage in the feed zone of the extruder.

## 8.3.7 Interactions with Other Additives

There are cases where these PPAs negatively interact with other additives. These include antioxidants, antiblock agents, pigments, hindered amine light stabilizers (HALS), and other processing aids. Most of the studies in the literature refer to interactions of fluoropolymer-based processing aids with other additives and these are reviewed in the present section.

Antiblock and pigment additives are two of the types of polymer additive particles that have a deleterious effect on the fluoropolymer PPA efficiency. This is essentially because of the fact that fluoropolymer molecules easily adsorb on the surface of these particles and possibly hinder their migration to the wall. In addition, such fillers can abrade the PPA from the die surface, preventing the formation of the coating on the die surface and as a result higher extrusion pressures are obtained. In such cases, a higher amount of PPA is needed to maintain the PPA efficiency (19,54). Priester et al. (14) have also found that the presence of antiblock agents slow the conditioning curve (induction time or time needed for obtaining steady state) and that a higher additive concentration is required to have the same conditioning curve as in the case of no additive. Similarly to antiblocks, $TiO_2$ pigments have also been reported to interfere with fluoropolymer PPAs through the abrasion and adsorption mechanism (19).

Hindered amine light stabilizers are also known to interact with PPAs (19). Although the mechanism is not known, from the practical point of view, the use of HALS at high temperatures might be done with caution as the interaction increases. It has been reported that some HALS have a deleterious effect on the performance of PPAs at high temperature in terms of melt fracture elimination (19).

Apart from the negative effects that additives might have on the efficiency of fluoropolymer PPAs, positive effects have also been reported. For example, lack of antioxidants (AO) in the polymer reduces the ability of the PPA to perform (60). Essentially, the degradation products of the polymer (because of lack of AO) prevent the PPA coating formation. As discussed above in detail, the use of interfacial agents such as PEG also improves the PPA performance through a synergistic mechanism.

## 8.4 OTHER PROCESSING AIDS

### 8.4.1 Stearates

Stearates, such as calcium and zinc stearates, are present in several commercial resins of both linear and long-chain branched polyethylenes (61). There is strong evidence that stearates promote slip and aid in the reduction of extrudate instabilities in the case of metallocene low-density polyethylene copolymers containing a small degree of long chain branching (61). In typical transient capillary extrusion experiments of linear low-density polyethylenes (with no stearates), the load rises initially with time until it reaches a steady state. However, in the case of metallocene polyolefins that contain stearates, the die-conditioning phenomenon described before is observed. An example is depicted in Fig. 8.9 (61). It can be seen that the load (thus pressure and shear stress) to extrude the polymer through a capillary die initially rises and then starts dropping. Each load corresponds to the extrusion of a full barrel. After several consecutive capillary extrusions, i.e., by reloading the reservoir of the rheometer, the load attained its steady state value indicating that complete conditioning of the die surface has been achieved. Comparison of different resins has shown that the conditioning is longer as the amount of long chain branching is increased (61).

### 8.4.2 Silicon-Based Additives

Silicon-containing polymers and polydimethyl siloxane (PDMS) oils have been used as polymer processing aids (19). Because these fluids have a low

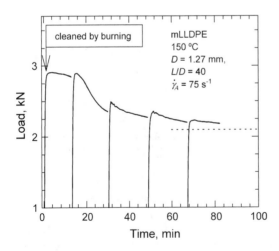

**FIGURE 8.9** The effect of stearates on the extrusion load in the continuous capillary extrusion of a metallocene LLDPE. (From Ref. 61.)

surface tension, they tend to reside on the polymeric surfaces and thus lower their frictional properties. However, their surface activity has limited the use of these materials in articles that are printed and/or painted. A high molecular weight of PDMS that is a solid at room temperature has been reported to be efficient in extrusion processes of LLDPE and PP (62,63). It has been essentially reported that use of such additives can significantly reduce surface roughness in extruded polymers as tapes.

### 8.4.3  Hyperbranched Polymers

In works of Hong et al. (64,65), it was shown that addition of small amounts of hyperbranched hexadecanote-terminated polyesters (HBP), synthesized based on Boltorn dendritic additives, significantly improved fiber extrusion and film blowing processing of LLDPE. Sharkskin elimination and substantial reduction in extruder power were achieved upon addition of only 0.5% of HBP. At the same time, the loss in the mechanical properties (tensile strength and elongation at break) was apparently minimal. By using the X-ray photoelectron spectroscopy and transmission electron microscopy, authors showed that HBP had a tendency to migrate to the wall surfaces, where it preferentially accumulated forming a phase-separated layer. The mechanism of fracture elimination by HBP was advocated to arise from either the reduced viscosity, or the interfacial slip.

### 8.4.4  Other Polymer Blends

Hydrocarbons blended with the process polymer can also act as PPAs. This technique might involve not only a small quantity of PPA dispersed into a process polymer, but also polymer blending to ratios of the same order of magnitude. Although the notion of processing aid has a meaning at small concentrations (typically less than 1% PPA mass fraction in resin), blending at higher concentrations might also improve processing. Examples from the literature are discussed below.

Blending of polypropylene (PP) with poly(ethyl vinyl acetate) (EVA) for flow instability reduction of PP was examined using capillary rheometry (66). It was shown that when blends of EVA mass fraction in PP ranging from 0% to 100% were tested, the critical shear rate for the onset of fracture increased with EVA concentration. Mechanical property measurements showed that while pure EVA had completely elastomeric behavior, blends of 50% PP and higher exhibited thermoplastic behavior with yield stress that increased with PP content. It was thus concluded that 15–20% EVA–PP blends provided good balance between the mechanical properties and melt fracture behavior.

Fujiyama and Kawasaki (67) studied the capillary flow properties of blends composed with various ratios of PP and HDPE with low (L), medium

(M), and high (H) melt flow index (MFI). They found that the flow curves of the different blends (PPM/PEH, PPM/PEM, PPM/PEL, PPH/PEM, PPL/PEM) were located between the flow curves of the two virgin components. In the blends that PP had a lower or equal MFI with the PE, the slope of the curves increased with decreasing PP content. The critical shear rate and the critical shear stresses for the onset of melt fracture of the blends were close to the ones determined by logarithmic additivity of the values that correspond to individual component.

Blending of two incompatible polymers, a fluoropolymer (Viton) and EPDM, in various proportions, to improve the performance of one or the other polymer was experimentally examined by Shih (38). Addition of a small amount of Viton (down to 0.4%) in EPDM significantly reduced the shear stress and improved the appearance of the extrudate. Ethylene/propylene/1,4-hexadiene terpolymer improved the processing of Viton in a similar fashion. This effect was more pronounced as the Viton concentration increased up to concentrations of 5%; no improvement was observed with further increase of Viton concentration. It was experimentally determined that, during the extrusion of blends, the polymer used in small concentration created a

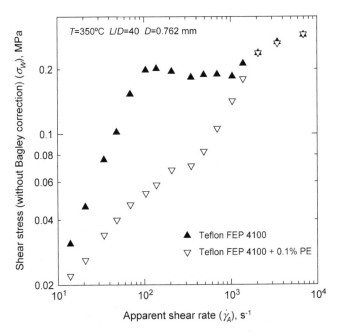

**FIGURE 8.10** The effect of a small amount of PE (GRSN-7047) on the flow curve of an FEP resin at 350°C. (From Ref. 35.)

buildup layer on the die wall, indicating that a phase separation occurred, in which the polymer used as additive accumulated on the wall.

Rosenbaum et al. (68) studied the use of PE as a processing aid in the extrusion of a fluoropolymer FEP (tetra-fluoro-ethylene-polypropylene) resin. Capillary extrusion experiments of LLDPE (GRSN/7047) dispersed in small quantity (0.1%) in FEP 4100 indicated that PE is an efficient PPA for eliminating melt fracture and stick–slip instability in the extrusion of fluoropolymers (Fig. 8.10). As a result of PE addition, the apparent flow curve is shifted to lower shear stresses allowing for smooth extrudates up to $1000 \text{ sec}^{-1}$. It is suggested that PE droplets, which have a lower viscosity than FEP, migrate to the wall. Furthermore, the low friction coefficient at the FEP–PE interface results in the slip of the extruded FEP over the thin PE layer. Consequently, this results in shear stress reduction and in instability suppression.

The gradual wall coating process discussed before for fluoropolymers and stearates (plotted in Figs. 8.5 and 8.9) was also reported in this case (polyethylene in the extrusion of fluoropolymers) (Fig. 8.11). The time required to obtain steady state response was found to depend on the process

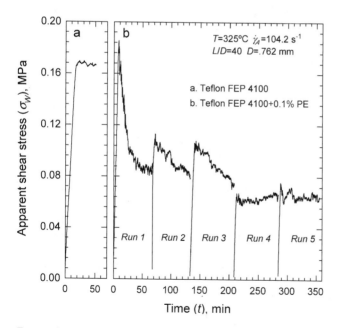

**FIGURE 8.11** Transient capillary extrusion: The effect of the addition of 0.1% of polyethylene on the transient response in the capillary extrusion of FEP 4100 at 325°C, $\dot{\gamma} = 104.2 \text{ sec}^{-1}$, $L/D = 40$ and $D = 0.762$ mm. (From Ref. 35.)

and geometry parameters. It decreased with increasing apparent shear rate, and with decreasing diameter or $L/D$ ratio. The latter suggests that this time is proportional to the time for complete surface coating. A decrease in the diameter or the $L/D$ ratio reduces the area to be coated, and an increase in the apparent shear rate increases the rate of the coating process. In both cases, the time required for complete coating and steady state operation is reduced.

## 8.5 CONCLUSION

Product quality and, in general, processing windows in extrusion of molten polymers are limited by the various flow instabilities that are observed at high production rates. These instabilities are attributed to the complex rheological and constitutive character of polymers in general, and have been scrutinized by researchers both experimentally and theoretically. While there are still unresolved issues in comprehending these phenomena, the industry demands for process optimization dictate the employment of processing aids for product quality improvement and energy requirement reduction.

Fluoropolymers have been traditionally used by the industry for elimination of the sharkskin and the stick–slip instabilities. Used as coatings, or more effectively in dispersion with the processing polymer, they typically increase the slip velocity of the extrudate, therefore reducing the energy requirements for a particular flow rate. Furthermore, they shift the critical shear rates for the onset of instabilities up to the point of the onset of gross melt fracture, therefore permitting higher production rates. Preferential coating of the walls of the die entrance by the fluoropolymer droplets, with consequent propagation of the fluoropolymer streaks to the die exit and lubrication at the polymer wall interface, was determined to be the dominant mechanism of instability suppression.

This was not only a direct conclusion of die-conditioning measurements, but also proven with optical microscopy. Fluoropolymer performance is highly dependent on the dispersion quality; this is achieved using larger droplets in the size range of 1–5 μm, thoroughly mixed with the polymer by a master batch. Optimization of the additive level, choice of the appropriate additive with the desired interactions with the polymer and/or the wall and other additives, and use of clean surfaces are also important.

Stearates, which are typically present in several commercial polymer resins, are believed to act as PPAs in the case of long-chain branched metallocene polyolefins and other polymers. Processing of these metallocene resins, which have no other additives but stearates, is very similar to the processing of fluoropolymer containing resins. Die conditioning associated with a steady state slip enhancement and instability elimination within a range of shear rates is observed.

Polymer blending is another traditional method for improving processing. Similarly to fluoropolymers, blending of hydrocarbons, or addition of a hydrocarbon in a fluoropolymer, can lead to significant reduction of the power requirement and elimination of the sharkskin and the stick–slip instability. Hydrocarbon blending results in processing behavior and mechanical properties that are intermediate of the polymers used. The mechanism by which a hydrocarbon enhances the processability of fluoropolymers was shown to be the same as in the case of polymer processing with fluoropolymers as processing aids.

Complete comprehension of these extrusion instabilities, as well as of the role of the additives in eliminating the instabilities, is important for process optimization. Elaborate techniques recently employed for flow characterization in extrusion processes, such as ATR/FTIR (69), fluorescence label chains for velocity measurements (70), NMR velocimetry (71), Laser Doppler velocimetry (72), detailed flow visualization through quartz/sapphire/glass capillary (26,28,73,74) and frustrated total internal reflectance (47) can prove valuable in studying the mechanisms of flow instability elimination when processing aids are present.

## ACKNOWLEDGMENTS

One of the authors (Savvas G. Hatzikiriakos) would like to dedicate this chapter to the memory of Dr. Charles W. Stewart, a researcher who made several important contributions to the subject of processing aids.

## REFERENCES

1. Ramamurthy, A.V. Wall slip in viscous fluids and influence of materials of construction. J. Rheol. 1986, *30*, 337–357.
2. Denn, M.M. Extrusion instabilities and wall slip. Annu. Rev. Fluid Mech. 2001, *33*, 265–297.
3. Larson, R.G. Instabilities in viscoelastic flows. Rheol. Acta 1992, *31*, 213–263.
4. Han, C.D. *Rheology in Polymer Processing*; Academic Press: New York, 1976.
5. Vinogradov, G.V.; Malkin, A.Y. *Rheology of Polymers*; Springer-Verlag: New York, 1980.
6. De Kee, D.; Wissbrun, K.F. Polymer rheology. Phys. Today 1998, *51*, 24–29.
7. Blyler, L.L.; Hart, A.C. Capillary flow instability of ethylene polymer melts. Polym. Eng. Sci. 1970, *10*, 193–203.
8. Galt, J.; Maxwell, B. Velocity profiles for polyethylene melts. Plast. Eng. 1964, 115–132.
9. Atwood, B.T.; Schowalter, W.R. Measurements of slip at the wall during flow of high-density polyethylene through a rectangular conduit. Rheol. Acta 1989, *28*, 134–146.

10. Hatzikiriakos, S.G.; Dealy, J.M. Wall slip of molten high-density polyethylenes. 2. Capillary rheometer studies. J. Rheol. 1992, 36, 703–741.
11. Hatzikiriakos, S.G.; Dealy, J.M. Role of slip and fracture in the oscillating flow of HDPE in a capillary. J. Rheol. 1992, 36, 845–884.
12. Lupton, J.M., Regester, R.W. Melt flow of polyethylene at high rates. Polym. Eng. Sci. 1965, 5, 235–241.
13. Rudin, A.; Blacklock, J.E.; Nam, S.; Worm, A.T. Improvements in polyolefin processing with a fluorocarbon elastomer processing aid. ANTEC Tech. Pap. 1986, 32, 1154–1158.
14. Priester, D.E.; Stika, K.M.; Chapman, G.R.; McMinn, R.S.; Ferrandez, P. Quality control techniques for processing additives. ANTEC Tech. Pap. 1993, 39, 2528–2532.
15. Kanu, R.C.; Shaw, M.T. Rheology of polymer blends—simultaneous slippage and entrance pressure loss in the ethylene-propylene-diene (EPDM) viton system. Polym. Eng. Sci. 1982, 22, 507–511.
16. Buckmaster, M.D.; Henry, D.L.; Randa, S.K. High Speed Extrusion. US Patent 5,688,4571997.
17. Kurtz, S.J.; Blakeslee, T.R.; Scarola, L.S. Methods for Reducing Sharkskin Melt Fracture During Extrusion of Ethylene Polymers. US Patent 4,282,1771981.
18. Rosenbaum, E.E.; Randa, S.K.; Hatzikiriakos, S.G.; Stewart, C.W. Extrusion Aid Combination. US Patent 00481792001.
19. Amos, S.E.; Giacoletto, G.M.; Horns, J.H.; Lavallée, C.; Woods, S.S. In Plastic Additives; Hanser: Munich, 2001.
20. Hatzikiriakos, S.G. Wall slip and melt fracture of HDPEs. Department of Chemical Engineering, McGill University, Montreal, Ph.D. Thesis, 1991.
21. Hatzikiriakos, S.G.; Stewart, C.W.; Dealy, J.M. Effect of surface-coatings on wall slip of LLDPE. Int. Polym. Process. 1993, 8, 30–35.
22. Hatzikiriakos, S.G.; Hong, P.; Ho, W.; Stewart, C.W. The effect of teflon (tm) coatings in polyethylene capillary extrusion. J. Appl. Polym. Sci. 1995, 55, 595–603.
23. Xing, K.C.; Schreiber, H.P. Fluoropolymers and their effect on processing linear low-density polyethylene. Polym. Eng. Sci. 1996, 36, 387–393.
24. Athey, R.J.; Thamm, R.C.; Souffle, R.D.; Chapman, G.R. The processing behavior of polyolefins containing a fluoroelastomer additive. ANTEC Tech. Pap. 1986, 32, 1149–1152.
25. Migler, K.B.; Gettinger, C.L.; Thalacker, V.P.; Conway, R. In Direct measurement of slippage induced by a polymer processing additive, Conference Proceedings of ANTEC; Society of Plastics Engineers: New York, NY, 1999; 3128–3131.
26. Migler, K.B.; Lavallée, C.; Dillon, M.P.; Woods, S.S.; Gettinger, C.L. Visualizing the elimination of sharkskin through fluoropolymer additives: coating and polymer–polymer slippage. J. Rheol. 2001, 45, 565–581.
27. Migler, K.B.; Lavallée, C.; Dillon, M.P.; Woods, S.S.; Gettinger, C.L. In Flow visualization of polymer processing additives effects, Conference Proceedings of ANTEC; Society of Plastics Engineers: Dallas, TX, 2001, 1132–1136.

28. Migler, K.B.; Son, Y.; Qiao, F.; Flynn, K. Extensional deformation, cohesive failure, and boundary conditions during sharkskin melt fracture. J. Rheol. 2002, *46*, 383–400.

29. Hatzikiriakos, S.G.; Dealy, J.M. Effects of interfacial conditions on wall slip and sharkskin melt fracture of HDPE. Int. Polym. Process. 1993, *8*, 36–44.

30. Kazatchkov, I.B.; Hatzikiriakos, S.G.; Stewart, C.W. Extrudate distortion in the capillary/slit extrusion of a molten polypropylene. Polym. Eng. Sci. 1995, *35*, 1864–1871.

31. Anastasiadis, S.H.; Hatzikiriakos, S.G. The work of adhesion of polymer/wall interfaces and its association with the onset of wall slip. J. Rheol. 1998, *42*, 795–812.

32. Hatzikiriakos, S.G. A slip model for linear-polymers based on adhesive failure. Int. Polym. Process. 1993, *8*, 135–142.

33. Hill, D.A.; Hasegawa, T.; Denn, M.M. On the apparent relation between adhesive failure and melt fracture. J. Rheol. 1990, *34*, 891–918.

34. Stewart, C.W. Wall slip in the extrusion of linear polyolefins. J. Rheol. 1993, *37*, 499–513.

35. Achilleos, E.C.; Georgiou, G.; Hatzikiriakos, S.G. Role of processing aids in the extrusion of molten polymers. J. Vinyl. Addit. Technol. 2002, *8*, 7–24.

36. Lavallée, C.; Woods, S.S. Die geometry and polymer processing additive (PPA) efficiency. ANTEC 2000.

37. Oriani, S.R.; Chapman, G.R. Fundamentals of melt fracture elimination using fluoropolymer process aids. SPE ANTEC Tech. Pap. 2003, *49*, 22–26.

38. Shih, C.K. Rheological properties of incompatible blends of 2 elastomers. Polym. Eng. Sci. 1976, *16*, 742–746.

39. Oliver, D.R. Influence of particle rotation on radial migration in the Poiseuille flow of suspensions. Nature 1962, *194*, 1269–1271.

40. Shizgal, B.; Goldsmith, H.L.; Mason, S.G. The flow of suspensions through tubes iv: oscillatory flow of rigid spheres. Can. J. Chem. Eng. 1965, *43*, 97–101.

41. Chan, P.C.H.; Leal, L.G. Motion of a deformable drop in a 2nd-order fluid. J. Fluid Mech. 1979, *92*, 131–170.

42. Zhou, H.; Pozrikidis, C. Pressure-driven flow of suspensions of liquid-drops. Phys. Fluids 1994, *6*, 80–94.

43. Coulliette, C.; Pozrikidis, C. Motion of an array of drops through a cylindrical tube. J. Fluid Mech. 1998, *358*, 1–28.

44. Li, X.F.; Pozrikidis, C. Wall-bounded shear flow and channel flow of suspensions of liquid drops. Int. J. Multiph. Flow 2000, *26*, 1247–1279.

45. Mortazavi, S.; Tryggvason, G. A numerical study of the motion of drops in Poiseuille flow. Part 1. Lateral migration of one drop. J. Fluid Mech. 2000, *411*, 325–350.

46. Shih, C.K. Capillary extrusion and mold flow characteristics of an incompatible blend of two elastomers. In *Science and Technology of Polymer Processing*; MIT Press: Cambridge, 1979.

47. Kharchenko, S.B.; McGuiggan, P.M.; Migler, K.B. Flow induced coating of fluoropolymer additives: development of frustrated total internal reflection imaging. J. Rheol. 2003, *47*, 1523–1545.

48. Leung, P.S.; Goddard, E.D.; Ancker, F.H. Process for Processing Thermoplastic Polymers. US Patent 4,857,5931989.
49. Duchesne, D.; Johnson, B.V. Extrudable Thermoplastic Hydrocarbon Polymer Composition. US Patent 4,855,3601989.
50. Duchesne, D.; Johnson, B.V. Extrudable Thermoplastic Hydrocarbon Polymer Composition. US Patent 5,015,6931989.
51. Blong, T.J., Greuel, M.P., Lavallée, C. Extrudable Thermoplastic Hydrocarbon Composition. US Patent 5,830,9471998.
52. Taylor, J.W.; Goyal, S.K.; Aubee, D.J.; Bonet, N.K.K. Polyethylene with reduced melt fracture. US Patent 5,459,187.
53. Blong, T.J.; Focquet, K.; Lavallée, C. Polymer processing additives and antioxidants . . . extruding under the influence. SPE ANTEC Tech. Pap. 1997, 43, 3011–3018.
54. Duchesne, D.; Blacklock, J.E.; Johnson, B.V.; Blong, T.J. Extrudable thermoplastic hydrocarbons and process aids. ANTEC Tech. Pap. 1989, 35, 1343–1347.
55. Leung, P.S.; Goddard, E.D.; Ancker, F.H. Process for Processing Thermoplastic Polymers. US Patent 4,925,8901990.
56. Focquet, K.; Dewitte, G.; Amos, S.E. Polymer Processing Additive Having Improved Stability. US Patent 6,294,6042001.
57. Oriani, S.R.; Chapman, G.R. Fundamentals of melt fracture elimination using fluoropolymer process aids. SPE ANTEC Tech. Pap. 2003, 49, 22–26.
58. Matzinos, P.; Tserki, V.; Gianikouris, C.; Pavlidou, E.; Panayiotou, C. Processing and characterization of LDPE/starch/PCL blends. Eur. Polym. J. 2002, 38, 1713–1720.
59. Slusarz, K.R.; Amos, S.E. The use of polymer processing additives to improve melt processing of M-LLDPE extrusion. SPE ANTEC Tech. Pap: San Francisco, CA, 2002.
60. Blong, T.J.; Focquet, K.; Lavallée, C. Polymer processing additives and antioxidants. ANTEC Tech. Pap. 1997, 43, 3011–3018.
61. Hatzikiriakos, S.G.; Kazatchkov, I.B.; Vlassopoulos, D. Interfacial phenomena in the capillary extrusion of metallocene polyethylenes. J. Rheol. 1997, 41, 1299–1316.
62. Lupton, K.E.; Pape, P.G.; John, V.B. RETEC Tech. Pap. 1988, 93–106.
63. Hauenstein, D.E.; Cimbalik, D.J.; Pape, P.G. ANTEC Tech. Pap. 1977, 21, 3002–3010.
64. Hong, Y.; Coombs, S.J.; Cooper-White, J.J.; Mackay, M.E.; Hawker, C.J.; Malmstrom, E.; Rehnberg, N. Film blowing of linear low-density polyethylene blended with a novel hyperbranched polymer processing aid. Polymer 2000, 41, 7705–7713.
65. Hong, Y.; Cooper-White, J.J.; Mackay, M.E.; Hawker, C.J.; Malmstrom, E.; Rehnberg, N. A novel processing aid for polymer extrusion: rheology and processing of polyethylene and hyperbranched polymer blends. J. Rheol. 1999, 43, 781–793.
66. Montoya, N.; Sierra, J.D.; del Pilar Noriega, M.; Osswald, T.A. Flow instability reduction of PP through blending of Eva. ANTEC Tech. Pap. 1999, 1270–1275.

67. Fujiyama, M.; Kawasaki, Y. Rheological properties of polypropylene high-density polyethylene blend melts. 1. Capillary-flow properties. J. Appl. Polym. Sci. 1991, *42*, 467–480.

68. Rosenbaum, E.E.; Hatzikiriakos, S.G.; Stewart, C.W. Flow implications in the processing of tetrafluoroethylene/hexafluoropropylene copolymers. Int. Polym. Process. 1995, *10*, 204–212.

69. Wise, G.M.; Denn, M.M.; Bell, A.T.; Mays J.W.; Hong, K.; Iatrou, H. Surface mobility and slip of polybutadiene melts in shear flow. J. Rheol. 2000, *44*, 549–567.

70. Legrand, F.; Piau, J.M.; Hervet, H. Wall slip of a polydimethylsiloxane extruded through a slit die with rough steel surfaces: micrometric measurement at the wall with fluorescent-labeled chains. J. Rheol. 1998, *42*, 1389–1402.

71. Britton, M.M.; Mair, R.W.; Lambert, R.K.; Callaghan, P.T. Transition to shear banding in pipe and Couette flow of wormlike micellar solutions. J. Rheol. 1999, *43*, 897–909.

72. Munstedt, H.; Schmidt, M.; Wassner, E. Stick and slip phenomena during extrusion of polyethylene melts as investigated by laser-Doppler velocimetry. J. Rheol. 2000, *44*, 413–428.

73. Inn, Y.W.; Fischer, R.J.; Shaw, M.T. Visual observation of development of sharkskin melt fracture in polybutadiene extrusion. Rheol. Acta 1998, *37*, 573–582.

74. Kazatchkov, I.B.; Yip, F.; Hatzikiriakos, S.G. The effect of boron nitride on the rheology and processing of polyolefins. Rheol. Acta 2000, *39*, 583–594.

# 9

## Boron Nitride Based Polymer Processing Aids

**Savvas G. Hatzikiriakos**
The University of British Columbia, Vancouver,
British Columbia, Canada

### 9.1 INTRODUCTION

The rate of production in many polymer processing operations including fiber spinning, film blowing, profile extrusion, and various coating flows is limited by the onset of flow instabilities (1,2). In particular, in extrusion processes when the throughput exceeds a critical value, small amplitude periodic distortions appear on the surface of extrudates (surface melt fracture or sharkskin). At higher throughput rates, more severe distortions of irregular form appear known as gross melt fracture (3). These two phenomena are independent and might occur together. In the gross melt fracture regime, the small amplitude periodic distortions may superpose on the irregular gross distortions. In other words, both phenomena (type of distortions) may be seen together depending on the scale under which they are examined. The surface melt fracture phenomena are believed to originate in the land of the die next to the die exit (Refs. 4–6 and this volume, Chapter 5 by Migler), while gross melt fracture is initiated at the die entry (Refs. 7–9 and this volume, Chapter 7 by Dealy).

To increase the rate of production by eliminating or postponing the melt fracture phenomena to higher shear rates, processing additives/aids may be used. These are mainly fluoropolymers and stearates that are widely used in the processing of polyolefins (HDPE, LLDPE). They are added to the base polymer at low concentrations (approximately 0.02–0.1 wt.% in the case of fluoropolymers and 0.1–0.3 wt.% in the case of stearates), and they essentially act as die lubricants, modifying the properties of the polymer–wall interface (increasing wall slip of the molten polymers). As a result of this lubrication effect, the onset of instabilities is postponed to much higher output rates, while the power requirement for extrusion is significantly reduced. It should be emphasized that these additives can eliminate only surface (sharkskin) and stick–slip (oscillating or cyclic) melt fracture. They do not appear to have an effect on the extrudate appearance in the gross melt fracture region (this volume, Chapter 8 by Kharchenco, Hatzikiriakos, and Migler).

It has been recently discovered that certain boron nitride (BN) powders can eliminate gross melt fracture (10). This significant discovery was made by Stu Randa and coworkers of DuPont, who were working for a number of years in the extrusion of foamed fluoroelastomers at the Experimental Station. Together with his coworkers, Stu Randa discovered that the use of boron nitride (BN) may eliminate gross melt fracture phenomena in the extrusion of fluoroelastomers and polyolefins (11). Details about the history of this discovery are given in the next section (Stu Randa, personal communication, 2002).

The present chapter discusses the use of boron nitride (BN) as a processing aid for the extrusion of fluoropolymers and polyolefins (10).

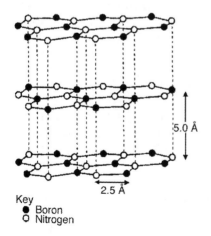

Key
● Boron
○ Nitrogen

FIGURE 9.1 Typical structure of boron nitride. (From Ref. 30.)

Boron nitride is a foam-nucleating agent, and when added to the polymer melt act as a very effective processing aid as will be seen in this chapter. Boron nitride (BN) is also used as a solid lubricant in many applications. Its structure resembles that of graphite (Fig. 9.1).

## 9.2 DISCOVERY OF BORON NITRIDE

This section is based on a personal communication (Stu Randa, 2002).

"Dr. John Tordella was a key associate with the Polychemicals Division of the E.I. duPont de Nemours Co. in the 1960s. Dr. Tordella studied the melt flow of various polymers using capillary rheometers. In the course of his work, Dr. Tordella used visualization techniques to observe land and gross melt fracture in the processing of Teflon® FEP resins (fluoropolymers). In such an experiment, as the rate was increased toward much higher levels, gross melt fracture occurred. To assist in tracing melt flow in the die inlet, Dr. Tordella placed very small black-pigmented particles at random in the neat resin extruded. The black particles assisted depicting the flow of the melt as viewed through the transparent die. During gross melt fracture, the flow within the inlet cone was unstable, producing vortices that moved from one side to the other. Just before such a flow upset occurs, one could note a black particle just entering the die land area on one side of the die. Then, at the very next moment, flow proceeded from the other side of the melt cone inlet. The black particle just observed moved back up into the melt cone several inches. This back and forth motion in the inlet created the kinky, back and forth, structure of the extrudate that exited the die. Along with this kinky nature, this gross melt fractured extrudate possessed an extremely coarse surface structure. Dr. Tordella studied the rheology of Teflon® FEP 100 and proposed that Teflon® FEP 100 could not be made into a smooth-surfaced extrudate above a shear rate of 20 s$^{-1}$ at 380°C (716°F). However, later, DuPont Polychemicals Researcher Mr. Reuben Fields proved otherwise. Mr. Fields showed that a superextrusion shear window exists for this FEP resin at rates in the range of about 400–800 s$^{-1}$. The trick in this hot melt extrusion is to use a die cooled to 315°C (600°F). Both FEP foam extrusions and some tubing extrusions make use of this patented technology in the industry today. Rheological studies by Dr. Carl Wolf showed that the same phenomenon exists for certain linear polyethylenes.

Boron nitride (BN) is extensively used as a foam cell nucleant in the extrusion-foaming of Teflon® CX 5010 and CX 5030 FEP fluoropolymer resins. A normal extrusion-foaming setup first involved the extrusion of neat Teflon® 100 FEP fluoropolymer resin at a melt temperature of 380°C (716 F). The next steps involved cooling the die to slide into the superextrusion shear rate window, where smooth surface extrudates of neat Teflon® FEP 100 resin can be made up to shear rates of 2000–3000 s$^{-1}$. Then the foam resin, CX 5010, would be added to the feed hopper and the nitrogen gas injected into the extruder. The FEP extrudate will begin to foam as the CX 5010 FEP resin exits the extrusion die at this same approximate rate that was initially stabilized. A slight rate increase was achieved as the injected gas plasticized the FEP to a certain extent. For years, this was the temperature profile (hot barrel temperatures of 380°C, 716°F and a cold die of 315°C, 600°F) used by the E.I. duPont de Nemours Co. Inc. and taught to their customers.

As the die temperature is raised from this low level to higher levels (ca. 340°C, 644°F), the shear rate can be substantially increased. At this temperature, the neat Teflon® 100 FEP resin also possess a supershear window, although much narrower compared to that at 315°C. Finally, the window of supershear completely disappears at die temperatures near 380°C (716°F). Thus, at such high die temperatures, a supershear window will not exist for neat Teflon® FEP resins in this particular die geometry.

Then, during one extrusion-foaming operation, the temperatures across the entire extruder, from feed hopper to die, were set at 382°C (720°F). The extrusion involved FEP foam resin CX 5010, which contains boron nitride. This molten resin exited the die at the shear rate of 4000 s$^{-1}$ having a smooth surface. Such a smooth surface should not be produced under these conditions. The occurrence is contradictory to known FEP resin extrusion technology. This prompted a study of this phenomenon over the next few years in FEP and other thermoplastic polymers." It is believed that the presence of this foam-nucleating agent initiates a large number of small, stable crazes/cracks instead of a large one when stressed, thus avoiding gross melt fracture (Dealy, private communication, 1997). This effect should be called the "Randa" effect, named after the person who discovered it.

## 9.3 TYPICAL FLOW CURVES

Various flow instabilities observed in the flow of polymeric liquids through capillary dies, generally known as melt fracture, are reflected in the apparent flow curve, determined by means of a capillary rheometer. This is essentially a log–log plot of the wall shear stress, $\sigma_w$, as a function of the apparent shear rate, $\dot{\gamma}_A$. These quantities are defined by Dealy (present volume, Chapter 2). A typical apparent flow curve for a linear polymer is shown in Fig. 5.2 (present volume, Chapter 5 by Migler). One may identify four different flow regions. Initially, there is a *stable* region where the extrudate appears smooth and glossy (region 1). Beyond some critical wall shear stress, $\sigma_{c1}$, the first visual manifestation of an extrusion instability appears as a high-frequency, small-amplitude distortion of the extrudate known as *sharkskin* (region 2). The onset of sharkskin appears to coincide with a change in the slope of the apparent flow curve, which in many cases marks the onset of wall slip (11–14). Whether or not slip plays a role in surface melt fracture phenomena is a conflicting issue in the literature. However, a polymer melt might slip over the die wall either by a local stick–slip mechanism similar to the Schallamach waves observed in the sliding of elastomers over metallic surfaces, or by a smooth continuous and constant slip velocity mechanism as in the case of a fluoropolymer-coated die wall. In the former case, surface melt fracture may be obtained whereas in the latter such distortions are often suppressed.

At a second critical value $\sigma_{c2}$ and within a certain range of apparent shear rates, the flow ceases to be stable (region 3). It is the region of *oscillating*, or *stick–slip* melt fracture, where the extrudate has the appearance of alternating smooth and distorted portions. In this region, the pressure drop oscillates between two extreme values when a flow rate-controlled capillary rheometer is used. The periodic variations of the pressure and apparent shear rate define a hysteresis loop that connects the two branches of the apparent flow curve. Finally, at still higher shear rates, there is a transition to a wavy chaotic distortion (*gross melt fracture*), which gradually becomes more severe with increase in $\dot{\gamma}_A$ (region 4). This typical behavior has been observed in the capillary extrusion of many linear polymers such as high-density and linear low-density polyethylene (13,14), polytetrafluoroethylene (3), polybutadiene (15–17), tetrafluoroethylene-hexafluoropropylene (18), polysiloxanes (19,20), and others.

## 9.4 TRADITIONAL PROCESSING AIDS

To increase the rate of production by eliminating or postponing the surface melt fracture phenomena to higher shear rates, processing additives/aids may be used. These are mainly fluoropolymers and stearates that are widely used in the processing of polyolefins and other commodity polymers (present volume, Chapter 8 by Kharchenco, Hatzikiriakos and Migler). These are added to the base polymer at low levels. They essentially act as die lubricants, and as a result of this lubrication effect, postpone instabilities to higher output rates as discussed before.

Figure 6.16 (present volume, Chapter 6 by Georgiou) depicts an example of the use of a fluoroelastomeric processing aid in the capillary extrusion of a HDPE. The addition of about 1000 ppm of a fluoroelastomer into a base HDPE resin significantly reduces the shear stress and, furthermore, eliminates surface melt fracture up to the point where gross melt fracture appears. This is the point in Fig. 6.15 where the two flow curves merge. This point also sets the onset of gross melt fracture. It should be emphasized that these additives, such as fluoroelastomers, can eliminate only surface melt fracture/sharkskin but not gross melt fracture.

## 9.5 BORON NITRIDE

As will be explained in this chapter, the use of boron nitride (BN) can extend the operability window for the processing of polymers within the gross melt fracture flow regime. We start with the examination of the effects of the addition of small amounts of BN (typically 200–2000 ppm) on the rheology of polymers. Consequently, the effects of BN on the processing of polyethylenes and fluoroelastomers are thoroughly discussed. Processes where BN can be used to enhance the processability of polymers include profile extrusion, wire coating, film blowing, blow molding, and fiber spinning. Moreover, the effects of physical properties of boron nitride on the processability of polymers are also examined, including particle size and its distribution, dispersion, and surface energy. Finally, the effects of combining BN with other traditional processing aids are discussed to a certain extent.

### 9.5.1 Effects on Rheology

#### 9.5.1.1. Linear Viscoelastic Measurements

It has been reported in the literature that the addition of a small amount of BN has almost no effect on the linear viscoelastic properties of polymers. Figure 9.2 shows dynamic linear viscoelastic data for a virgin metallocene-LLDPE (Exact® 3128) and m-LLDPE *with* three different levels of BN

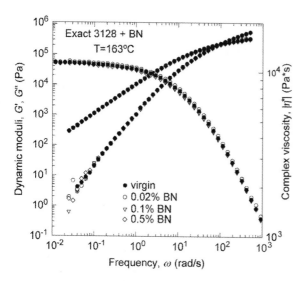

**FIGURE 9.2** Linear viscoelastic data for PE Exact 3128 at 163°C with and without BN (type CTF5). (From Ref. 23.)

(21–23). No significant difference was found in the linear viscoelastic behavior of pure and filled resins as long as the loadings were kept at relatively low levels (up to 0.5 wt.% loadings were examined). Similar observations were reported by Lee et al. (24,25). A variety of different BN powders were also examined by Yip (26) and Yip et al. (27–29) and it was again reported that use of BN up to 0.5 wt.% loadings have no effect on the linear viscoelasticity of polyolefins.

### 9.5.1.2. Nonlinear Rheology

Rozenbaum et al. (18,21,30,31) have studied in detail the effects of the addition of boron nitride on the nonlinear rheology of metallocene poly-ethylenes. Using steady shear experiments in a sliding plate rheometer, it was found that the addition of BN (up to 0.1 wt.%) has no effect on the flow curve of metallocene polyethylenes (Exact® 3128 and Exceed® 116).

Another series of experiments with these resins involved relaxation after cessation of steady shear and it was carried out in a sliding plate rheometer (21). The shear stress decay has shown to increase at small shear rates with the addition of BN (see Fig. 9.3). The addition of BN seems to slow down the relaxation of the polymer. This is characteristic for cross-linked and phase-separated polymers (32,33). This is because of form relaxation that occurs when the motion of a polymer matrix is slower than that of the BN particles. Thus longer relaxation times are seen in the latter case. However, at higher shear rates ($> 5$ s$^{-1}$), the curves are indistinguishable, which means that

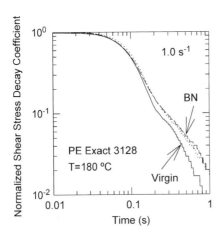

**FIGURE 9.3** Shear stress decay coefficient, $\eta^-(t,\dot{\gamma})/\eta(\dot{\gamma})$, for metallocene PE Exact 3128 as a function of time (s) at 180°C. Solid lines correspond to the virgin resin, dashed lines to the resin with 0.05 wt.% BN, and dotted lines to 0.5 wt.% BN. At higher rates, these differences disappear (21). (From Ref. 30.)

the resin relaxes in the same way regardless of the presence of BN. Similar results were obtained in elongational experiments (21). Small differences at small extensional rates were observed. These disappear at higher rates. Rozenbaum (21) has concluded that such small differences in linear and nonlinear rheology cannot explain the dramatic effects that have been observed regarding the elimination of gross melt fracture (reviewed below in this chapter).

### 9.5.1.3. Capillary Experiments of Boron Nitride Filled Polymers

Based on the effects of the addition of BN into polymers on the rheology of polymers, it would be expected that similar effects (no significant changes in the flow curve) would be observed in capillary experiments. Rozenbaum (21,30) performed capillary experiments to examine such possible effects. Figure 9.4 depicts representative results. The flow curves of the virgin and filled Exact® 3128 with 0.1 wt.% BN (CTF5, one of the most effective BN powders in eliminating gross melt fracture) obtained using the capillary

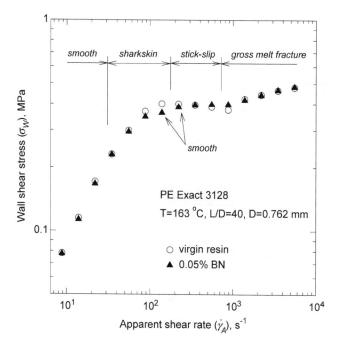

**FIGURE 9.4** The flow curves for PE Exact 3128 without and with boron nitride obtained in a rheometer with a capillary die having $L/D = 40$ and $D = 0.762$ mm at 163°C. (From Ref. 30.)

rheometer with a capillary die having a diameter of 0.762 mm and a length-to-diameter ratio of 40 at $T = 163°C$. The flow curves almost coincide. The addition of 0.1 wt.% has no effect on the flow curve of this metallocene polyethylene. However, it was found to have an effect on the extrudate appearance, eliminating extrudate distortions in the range of shear rates corresponding to transition from sharkskin to stick–slip melt fracture (200–250 $s^{-1}$). Therefore, the addition of BN improves the processability of the resin by eliminating sharkskin and oscillating melt fracture. Once in the gross melt fracture regime, the presence of BN may eliminate gross distortions but fails in eliminating surface distortions at such high rates. In other words, gross distorted extrudates of virgin resin appear as sharkskinned in the presence of BN at rates well within the gross melt fracture region. This can also be seen from the experiments of Lee (24,25) who performed capillary experiments using tungsten carbide and BN hot-pressed capillary dies. It was reported that sharkskin melt fracture can be eliminated when BN hot-pressed dies and/or BN as an additive to the base resin are used. Carefully examining the photos published by Lee (24,25), one can make an additional important observation (see Fig. 9.5). It can be seen that the addition of BN can indeed eliminate gross melt fracture at high shear rates. In fact, the gross distortions obtained in the

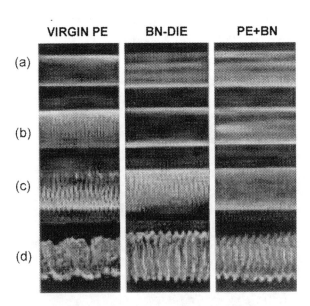

**FIGURE 9.5** The effect of the addition of BN on the melt fracture behavior of polyethylene at various shear rates in capillary experiments (a) 115.4 $s^{-1}$; (b) 224.1 $s^{-1}$; (c) 851.3 $s^{-1}$;and (d) 2007.4 $s^{-1}$. (From Ref. 25.)

extrusion of virgin resin are substituted by sharkskin melt fracture in the case of BN-filled resin. For example, the gross distortions (irregular and large distortions) obtained in the extrusion of PE through the tungsten carbide die at $2007\,s^{-1}$ (Fig. 9.5d), in fact, become surface distortions (relatively small and regular) when BN is added into the base resin. It is noted that surface and gross melt fracture are two separate phenomena. The former originates at the exit, while the latter at the entrance. Therefore at high rates, one may easily see the gross distortions and ignore the small scale ones that possibly imply sharkskin. Clearly BN can eliminate the gross and irregular distortions but not the surface and regular ones at high shear rates. The latter originate at the exit and lubrication is needed to eliminate them. As will be discussed below, a careful selection of the surface chemistry of the BN might lead to the elimination of surface distortions at high shear rates as well.

### 9.5.2 Effects of Boron Nitride on Polyethylene Processing

The effect of the boron nitride addition on the processability of PEs has been tested in continuous extrusion by means of a cross head die that simulated the wire coating process (30,31), in blow molding (26,34,35), film blowing (36), and fiber spinning (37). In all these cases, the results indicate that the use of BN is beneficial in these operations. The use of BN not only can eliminate surface melt fracture but also can postpone the gross melt fracture at high shear rates.

Rozenbaum et al. (30) examined the effect of BN addition on the processability of a metallocene linear low-density PE in continuous extrusion by using a crosshead die. They have reported that the virgin resin starts exhibiting melt fracture at about $50\,s^{-1}$ at $163\,^{\circ}C$. The addition of 0.1 wt.% BN (CTF5) virtually eliminates melt fracture in the broad range of apparent shear rates up to approximately $900\,s^{-1}$. This value falls within the gross melt fracture region. Furthermore, they have reported that while the effect on the melt fracture is significant, there is no effect on the flow curve. This is similar to results summarized in Sec. 10.5.1.3, where the effects of BN on the processability of polymers in capillary flow were discussed.

Further experiments in the extruder (21,22,30,31) have shown that the maximal shear rate yielding a smooth extrudate is sensitive to the BN (CTF5) content. In Fig. 9.6, one can see that optimal performance was obtained with the BN content of 0.01 wt.%; this composition resulted in a maximum shear rate for the onset of distortions of about $1000\,s^{-1}$. Increase in the BN concentration above 0.01 wt.% resulted in gradual decrease in the limiting shear rate for the onset of flow instabilities. Moreover, the highest concentration of BN (0.5 wt.%) did not completely eliminate sharkskin at low shear rates. It seems that higher concentrations facilitate the presence of agglom-

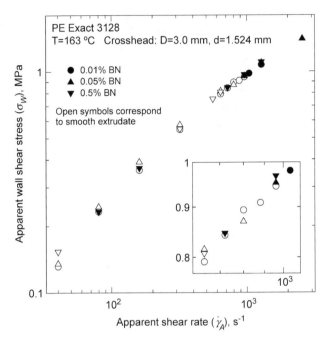

**FIGURE 9.6** The effect of the boron nitride concentration on the processability of PE Exact® 3128 in an Entwistle extruder with Nokia Maillefer crosshead having 3.00-mm die and 1.52-mm tip at 163°C. (From Ref. 30.)

erated particles, which has a detrimental effect on the BN performance as a processing aid (see Sec. 9.5.4). Figure 9.7 clearly shows the effect of BN concentration. While the addition of 0.02 or 0.1 wt.% of BN eliminates gross melt fracture, the addition of 0.5 wt.% cannot eliminate it. In fact, in the case of the 0.5 wt.% BN addition, the melt fracture rather appears as surface melt fracture. Therefore, instabilities at the entrance (gross melt fracture) have been eliminated to a certain extent but not at the exit (surface melt fracture). This effect is similar to that discussed in reference to the photographs of Fig. 9.5. As will be discussed later, the addition of a traditional processing aid such as a fluoroelastomer used in combination with BN may result the best processing aid eliminating both types of instabilities (surface and gross fracture).

Yip (26) and Yip et al. (34) have examined the effect of BN on the processability of two high-density polyethylenes in blow molding. They have reported that the addition of BN at levels as low as 0.05–0.1 wt.% significantly improves the processability of such polymers. Induction times of the order of 10 min were reported before melt fracture could be eliminated. Induction

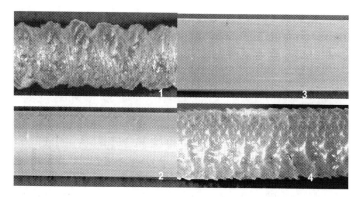

**FIGURE 9.7**  The extrudate samples to illustrate the effect of BN (CTF5) on the extrusion of m-LLDPE Exact® 3128 obtained at 617 s$^{-1}$ and 163°C: (1) pure resin; (2) 0.02% BN; (3) 0.1% BN; (4) 0.5% of BN (CTF5). (From Ref. 27.)

times were also reported by Rozenbaum (21). He has reported that after extruding polyethylene with BN where sharkskin was eliminated, switch to extrusion of pure PE under the same conditions does not exhibit sharkskin immediately but only after a few minutes. This implies the presence of BN on the die surface and certain time is needed for BN to be washed off the surface, an effect similar to that seen in the case of fluoroelastomer processing aids. Based on these experimental observations on the induction time, one may hypothesize for the case of BN a similar surface melt fracture elimination mechanism with that believed for the case of fluoroelastomers. The gradual and continuous accumulation of BN at the polymer–wall interface (particularly and most importantly at the exit) gradually changes the local stick–slip boundary condition (Schallamach waves) to a continuous constant slip velocity one. This hypothesis also explains the no-pressure reduction observed in the extrusion of polyolefins containing BN.

Temperature effects are also enhanced in the presence of BN with the elimination of melt fracture to be favored at a higher temperature (26,34). The addition of BN was also reported to slightly decrease the head extrusion pressure. For example, in Ref. (35), for one high-density polyethylene that exhibits surface melt fracture at a critical rate of about 1000 s$^{-1}$, this becomes about 1850 s$^{-1}$ with the addition of 0.1 wt.% of BN. It is noted that the addition of 0.1 wt.% of fluoroelastomer into the same resin results into a critical shear rate of only 1300 s$^{-1}$.

Pruss et al. (36) have examined the use of boron nitride based processing aids (possibly combination of BN with fluoroelastomers because a throughput increase is reported) in the film blowing process of metallocene LLDPEs.

Extrusions with the use of BN have resulted in an over twofold increase in volumetric flow rate compared to the rate obtained by using a conventional fluoropolymer. In addition, they have reported improvements in film quality, such as better caliper control, mitigation of film streaking, control of coefficient of friction, enhanced gloss, reduced haze, and enhanced heat sealability when BN is used as a process aid. On the other hand, mechanical properties such as yield strength and Graves tear remain the same. They have concluded that this unique combination of both process and product improvements, at equivalent to fluoroelastomer process aid costs, demonstrate that BN is the next generation of polymer process aids.

Vogel et al. (37) have examined the melt spinning process of high-molecular weight metallocene LLLPEs. Without the use of a BN-based process aid, the melt spinning of PEs is only possible up to a critical molecular weight. This poses a limitation to the mechanical properties of the melt spun fibers as these improve dramatically with increase of molecular weight. They have reported that this limitation can be overcome with the use of a BN-based process aid. The use of BN causes an increase of the physical break stress of the spun fibers. They have speculated that the well-bonded boron nitride particles into the PE matrix could generate a large number of small, stable crazes instead of a large one when stressed. As a consequence, a more ductile failure of the fiber occurs, which causes the increase of the physical break stress.

### 9.5.3 Effects of Boron Nitride on Fluoropolymer Processing

Fluoropolymers exhibit all different types of melt fracture phenomena as discussed above (18,38). Therefore to process them at rates that can be economical from the industrial point of view, processing aids must be used. The addition of a small amount of polyethylene (typical 0.1 wt.%) may eliminate sharkskin melt fracture in the capillary extrusion of an FEP resin (tetrafluoroethylene–hexafluoropropylene polymer) (18,38). However, the addition of PE cannot eliminate gross melt fracture. It has been discovered that the addition of BN into an FEP resin may eliminate gross melt fracture and postpone it to very high shear rates (10). This was examined into more detail by Rozenbaum (21,30,38).

The rheological and processing performance of FEP 100 resin was studied using a crosshead die mounted into a capillary rheometer. The maximum shear rate limit for smooth surface of virgin FEP 100 resin was found to be about 90 s$^{-1}$. With the addition of 0.25 wt.%, smooth extrudates can be obtained in the whole range of shear rates up to 4000 s$^{-1}$, which is well beyond the onset of gross melt fracture. Decrease in BN content to 0.05 wt.% resulted in a drop in the maximum shear rate for the onset of flow instabilities (to about 3200 s$^{-1}$). Figure 9.8 shows photos of the extrudate samples of FEP 100 obtained in an extruder at 371°C.

Other experiments with FEP 4100 to assess their processability with the use of BN exhibited similar results. FEP 4100 filled with 0.17% of BN and dried overnight at 150°C yielded smooth extrudates at shear rates as high as 6000 s$^{-1}$. At 0.035% BN content, the maximum shear rate was 3200 s$^{-1}$. A similar performance was observed with the FEP 3100 and 5100 as well as with PFA resins (21). The results with all fluoropolymers are summarized in Table 9.1. Similar results were also reported by Yip (26). Induction time effects using BN in a fluoropolymer matrix has also been reported (21,26–30).

### 9.5.4  Effects of Particle Size and Dispersion

The effects of particle size and the dispersion quality of BN into the processability were also studied in detail (26). Several BN powders were examined. It had been reported that the optimum particle size is in the range of 5–20 μm. In addition, the absence of agglomerated particles is a requirement for optimum performance of BN as a process aid. Possibly the presence of agglomerated particles (100–400 μm) cannot provide the necessary lubrication at the die exit (change from a local stick–slip to a continuous smooth boundary condition) to eliminate sharkskin melt fracture.

The quality of the dispersion is also a requirement for the optimum performance of BN as a process aid. For example, Rozenbaum (21) mixed several types of BN powders with PE pellets in dry form. Extrusion of such blends resulted into a poor performance of BN as process aid. In certain cases,

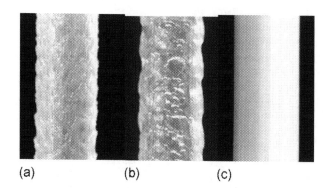

(a)                    (b)                    (c)

FIGURE 9.8   The extrudate samples of Teflon® FEP 100 at 371°C: (a) sharkskin for pure FEP 100 at $\dot{\gamma}_A = 320$ s$^{-1}$; (b) gross melt fracture for pure FEP 100 at $\dot{\gamma}_A = 4000$ s$^{-1}$; (c) smooth extrudate for FEP 100 with 0.01% BN at $\dot{\gamma}_A = 4000$ s$^{-1}$. (From Ref. 30.)

TABLE 9.1 Influence of Boron Nitride Concentration on Tubular Extrudate Surface Smoothness in the Extrusion of Fluoropolymers (extrusion tests in the Entwistle extruder with Nokia Maillefer crosshead 3.0-mm die and 1.52-mm tip)

| $T$ (°C) | BN concentration (mass%) | Maximum shear rate to yield smooth extrudate (s$^{-1}$) |
|---|---|---|
| FEP 100 | | |
| 371 | 0 | 40 |
| | 0.05 | 3200 |
| | 0.25 | 4000 |
| FEP 4100 (FEP 5100) | | |
| 371 | 0 | 350 |
| | 0.035 | 3200 |
| | 0.055 | 4000 |
| | 0.17 | 5600 |
| FEP 3100 | | |
| 371 | 0 | 250 |
| | 0.25 | 6000 |
| PFA 340 | | |
| 385 | 0 | 50 |
| | 0.25 | 2000 |

sharkskin melt fracture could be eliminated to a certain extent while gross melt fracture still persisted. This indicates that gross melt fracture is a bulk phenomenon and the presence of well-dispersed BN is needed to eliminate it. To attain an ideal dispersion of BN into a polymer, a concentrate should be first prepared (5 or 10 wt.%) either by using a twin-screw extruder or a single-screw extruder if the polymer is available in powder form. Consequently, final dilution should be followed to attain the final BN concentration. Scanning electron microscopic photography was used to assess the quality of the dispersion attained by these two methods (26). It has been reported that preparation of a concentrate properly disperses the BN particles into the resin and as a result BN results to be an ideal processing aid for the extrusion of polymers.

## 9.5.5 Effects of Surface Energy

Surface energy of BN powders plays a significant role in eliminating melt fracture phenomena in the extrusion of polymers (22,26,39–41). In particular,

Yip (26) and Seth (41) performed experiments with two types of BN having the same particle size, degree of agglomeration, and degree of dispersion in a LLDPE. It was found that BN having a small amount of boron oxides could not effectively eliminate melt fracture phenomena and failed at small shear rates. This was attributed to the effect of surface energy.

Seth (41) and Seth et al. (39,40) examined a number of BN powders as processing aids by measuring their surface energy. A capillary rise technique based on Washburn's equation was used (42). Both the dispersive and nondispersive components of surface energy were determined. The results of the surface energy of BN powders has been found to correlate well with the critical shear rate for the onset of melt fracture, indicating the importance of surface energy in the procedure of selecting an effective processing aid. In particular, the critical shear rate for the onset of melt fracture was found to increase with the ratio of dispersive component to nondispersive component of the surface energy. This trend was confirmed by Rathod et al. (43) by using a sessile drop method technique to assess the surface energy of several BN powders. He found that the presence of oxides increases the nondispersive component (NDC) of the surface energy of polymers whereas leaving the dispersive (DC) to be about the same ($36.5 \pm 0.6 \, mJ/m^2$). Then, they identified that the critical shear rate for the onset of melt fracture correlates well with the nondispersive component of the surface energy. A nondispersive component of $8{-}12 \, mJ/m^2$ is ideal for optimizing the performance of BN as process aid in polyolefins. Higher values ($23{-}27 \, mJ/m^2$) were found to have a detrimental effect in the performance of BN as a process aid. Thus it can be concluded that the higher the NDC of the surface energy, the lower the critical shear rate for the onset of melt fracture.

Using the Young and Good-Girifalco equation, Rathod et al. (43) have evaluated the affinity of the various additives (fluoroelastomers, PTFE, BN) with the die wall and the polymer in an attempt to explain the critical factors in elimination of surface and gross melt fracture phenomena. As will be discussed in the next section, a good affinity of the additive with the polymer can eliminate instabilities (gross melt fracture) within the polymer itself by healing (lubricating) fractured surfaces. On the other hand, good affinity of the additive with the die wall will lubricate the flow at the exit and as a result sharkskin melt fracture can be eliminated. Based on these findings, Rathod et al. (43) argued that fluoroelastomers and PTFE have a high affinity with the wall and as a result can only eliminate sharkskin fracture by coating the die land (particularly the exit). On the other hand, BN with a high nondispersive component have a good affinity with the polymer itself (in the present case, polyethylene) but not with the wall. Thus they can eliminate gross fracture but not sharkskin. This is the reason that gross melt fracture is eliminated although sharkskin still appears with the use of BN in the extrudate photos shown in Fig. 9.5d. On the other hand, a BN with a low NDC can eliminate both sharkskin and gross melt fracture because its affinity is relatively high for both the wall and polymer. As will be explained later, a combination of BN with a fluoroelastomer would result an ideal processing aid for polyolefins, as BN can effectively eliminate gross fracture whereas the fluoroelastomer can effectively lubricate the die exit to eliminate sharkskin.

### 9.5.6 Mechanisms

The mechanism by which the addition of BN eliminates gross melt fracture was studied by Kazatchkov et al. (44) by using visualization experiments. They have used a transparent die made out of fused quartz to visualize the change in the flow patterns at the entrance to a capillary die (site of initiation of gross melt fracture). Long-time exposure photographs were taken at the capillary entry region by using a Nikon FM-2 35mm photographic camera attached to a microscope (Nikon SMZ-2T). Figure 9.9 shows the entrance flow of a virgin polypropylene resin (no BN) at three different shear rates (32.4, 324, and $650\ s^{-1}$) at $200\,°C$ (44,45). The key features can be summarized as follows. At low shear rates, the flow enters the capillary at a higher entry angle, and the vortex in the corner of the reservoir is larger. As the shear rate increases, the streamlines bend more near the entrance to the capillary. It is noted that the vortex also becomes slightly smaller in size because it is suppressed toward the corner by the higher pressure. At the highest shear rate ($650\ s^{-1}$), gross melt fracture is observed, and the streamlines are no longer smooth. It is not obvious from the photo, but it is clearly visible in the viewfinder that these "zigzag" streamlines lines are caused by discontinuous motion of parts of the bulk material.

A schematic is drawn in Fig. 9.10 to explain this flow pattern development (26,44,45). The flow in the entry region appears to be broken into several layers, and each layer moves with its own velocity. At regular time intervals, different in each layer, the motion stops for a brief period. The closer the layer is toward the center of the stream, the larger and more frequent are the jumps and stops inside it.

An enlarged segment of Fig. 9.9 (marked by a rectangle) is shown in Fig. 9.11. It can be seen that the series of dots produced by illuminated (at a frequency of 85 Hz) tracer particles are interspaced by larger, brighter spots. These brighter spots are the result of the above-described intermittent stops of the flow. This flow pattern closely resembles the stick–slip phenomenon observed in the capillary flow of linear polymers. The difference is that the former occurs at multiple polymer surfaces within the bulk material, while the latter occurs only close to the polymer–wall interface.

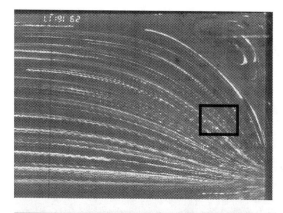
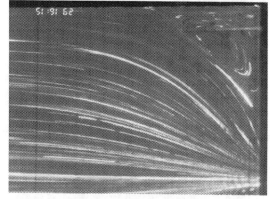
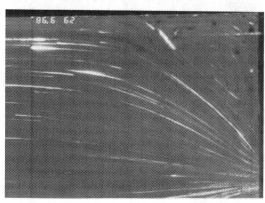

**FIGURE 9.9** Pictures of the flow of polypropylene at various apparent shear rates (left 32.4 s$^{-1}$, middle 324 s$^{-1}$, and right 650 s$^{-1}$) at 200°C. (From Ref. 44.)

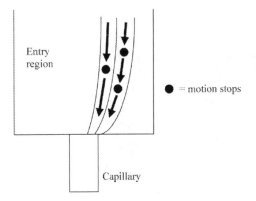

Entry region

● = motion stops

Capillary

**FIGURE 9.10** A schematic illustrating the unstable flow development at the entrance region of the capillary at high apparent shear rates (gross melt fracture) in the absence of boron nitride. (From Ref. 44.)

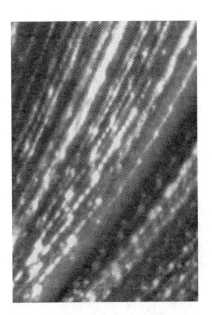

**FIGURE 9.11** Enlarged segment of Fig. 9.9 (right) in the entry region of the capillary showing the flow instabilities in the bulk flow of polypropylene. (From Ref. 44.)

Figure 9.12 shows the flow of polypropylene with and without the addition of 0.1% BN (CTF5) at the shear rate of 650 s$^{-1}$ at 200°C (26,44,45). It can be seen that the addition of 0.1% BN eliminates the discontinuous streamlines even at the highest shear rate of 650 s$^{-1}$. The streamlines are now smooth, the flow seems to be more organized, and the extrudate is free of any distortions. This suggests that BN is a suitable processing aid for eliminating and postponing the onset of gross melt fracture to higher shear rates. It seems that the presence of the BN over the surfaces within the bulk material where stick–slip was observed in its absence, provides proper lubrication between these singular layers (change of stick–slip or intermittent fluid layer motion to a continuous slippage between those fluid layers), thus eliminating stick–slip phenomena within the bulk of the polymer. The motion is rather continuous in the presence of BN. It is noted that BN is a solid lubricant and such action can somehow be expected.

The presence of BN in the bulk of the polymer seems to eliminate always and postpone the onset of gross melt fracture at higher shear rates. However, elimination of surface melt fracture seems to be more complex in the case of BN. Boron nitride should have a low nondispersive surface energy component to have a good affinity with the wall. Good affinity with the wall means that it

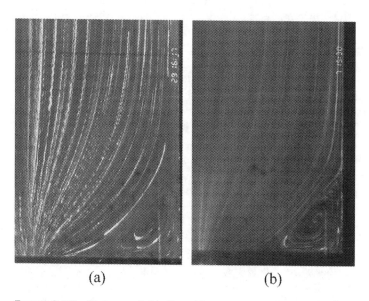

(a)                                                    (b)

FIGURE 9.12  Pictures of the flow of virgin polypropylene (a) and polypropylene filled with 0.1 wt.% BN-CTF5 (b) at the shear rate of 650 s$^{-1}$ at 200°C. (From Ref. 44.)

could provide lubrication at the die exit where surface melt fracture occurs. Having these in mind, one might naturally argue that a combination of BN with a process aid that is capable of eliminating sharkskin would be an ideal process aid. Such combinations are summarized below in Sec. 9.6.

## 9.6 COMBINATIONS OF BORON NITRIDE WITH FLUOROELASTOMERS

As discussed before, it is reasonable to assume that, because the mechanisms of the action of fluoropolymers as a processing aid and BN are essentially different, they might supplement each other if they were used together. Fluoropolymer could work in the sharkskin and stick–slip melt fracture regions (this volume, Chapter 8 by Kharchenco, Hatzikiriakos, and Migler), eliminating melt fracture and reducing the pressure drop. On the other hand, BN works in the gross melt fracture region, where fluoropolymers have no effect, thus delaying the onset of melt fracture to even higher shear rates. The examples below demonstrate the synergistic effects resulting from combining BN with conventional processing aids.

Rozenbaum (21) first combined a Teflon® processing aid (Teflon® APA) with BN (CTF5) in the processing of a metallocene polyethylene (Exact® 3128). It reported that the presence of the BN particles has only a small effect on the flow curve (seen also before). However, the onset of melt fracture with the addition of the BN can be postponed from 60 to 1850 s$^{-1}$. The onset of gross melt fracture for virgin Exact® 3128 was reported to be at about 300–400 s$^{-1}$. The addition of Teflon® particles decreases the shear stress practically over the whole range of apparent shear rates up to those in the gross melt fracture region regardless of the presence of BN. More important and surprising is the effect of the combination of the two processing aids (Teflon® APA and BN) on the onset of melt fracture. For this case, the critical shear rate was found to be 2250 s$^{-1}$. Similar results have been reported by other reports that combine BN with Teflon® process aids or fluoroelastomers such as Viton® and Dynamar® (21,30,34,37,46–48).

In another example, the effect of the addition of the combination (BN and Teflon®) on the extrusion of a m-LLDPE (Exceed® 116) at 204°C was examined. The addition of only 0.1 wt.% of BN (CTF5) on the resin increases the critical shear rate for the onset of melt fracture from about 100 s$^{-1}$ for the virgin resin to almost 1000 s$^{-1}$. The combined processing aid (BN and Teflon® APA) extends the critical shear rate to about 2000 s$^{-1}$.

In most of these cases where the effects of the combined processing aid is examined, no particular reference is made to the properties of BN used. In fact, particle size or degree of agglomerations is not a factor when the combined process aid is used. These characteristics are important for BN when used alone. In such cases, BN should possess these properties to eliminate surface melt fracture. It is speculated that such characteristics are not important for the elimination of gross melt fracture. As a combination, the best BN should be the one having a high nondispersive surface energy component (43).

## 9.7 CONCLUSIONS

Boron nitride has been found to be an effective processing aid in the extrusion of both fluoropolymers and polyolefins. It was shown in certain cases to eliminate not only sharkskin and stick–slip (oscillating) melt fracture, but also to postpone gross melt fracture to significantly higher shear rates. For BN to be an effective processing aid, the following requirements apply: (1) to have a fine particle size in the range of 5–20 μm; (2) to have no agglomerated particles; (3) to be finely dispersed into the resin under processing by appropriately preparing a concentrate first before final dissolution; (4) to be used at its optimum concentration; and (5) to have proper surface chemistry with minimum nondispersive surface energy component if BN is to be used alone (essentially no boron oxides on the surface).

Boron nitride is capable of eliminating both surface and gross melt fracture. If the nondispersive surface energy component of BN is minimum, its affinity with the wall increases and this facilitates suppression of sharkskin at the exit of the die (essentially suppressing tearing at the exit of the die by appropriately lubricating the flow). The good affinity with the metal wall of the die was demonstrated by the continuous extrusion experiments performed by Rozenbaum (21). After extruding polyethylene with BN where sharkskin was eliminated, switch to extrusion of pure PE under the same conditions does not exhibit sharkskin immediately but after a few minutes. This implies the presence of BN on the die surface and certain time is needed for BN to be washed off the surface, an effect similar to that seen in the case of fluoroelastomer processing aids.

Addition of BN eliminates gross melt fracture by eliminating the upstream flow instabilities at the entrance of the die. Singular surfaces that exhibit local stick–slip within the bulk material, are healed by the presence of BN that provides proper lubrication over these singular surfaces.

Finally, it was demonstrated that combinations of BN with conventional processing aids result in enhanced processing aids that have certain advantages over their constituents when they are used as processing aids independently. For example, enhanced processing aids can be produced by using a combination of BN with fluoroelastomers or fluoropolymers in the extrusion of polyolefins. These combinations result in enhanced processing of polyolefins at high rates.

## REFERENCES

1. Pearson, J.R.A. *Mechanics of Polymer Processing*, 1st Ed.; Elsevier: New York, 1986; 184–196.
2. Larson, R.G. Instabilities in viscoelastic flows. Rheol. Acta 1992, *31*, 213–263.
3. Tordella, J.P. Unstable flow of molten polymers. In *Rheology, Theory and Applications*; Eirich, F.R., Ed.; New York: Academic Press, 1969; 57–92.
4. Kurtz, S.J. The dynamics of sharkskin melt fracture: effect of die geometry. Proc. XIth Int. Congr. on Rheology. In *Theoretical and Applied Rheology*; Moldenaers, P., Keunings, R., Eds.; Elsevier Science Publishers: Brussels, 1992; 377–379.
5. Migler, K.B.; Lavallee, C.; Dillon, M.P.; Woods, S.S.; Gettinger, C.L. Visualizing the elimination of sharkskin through fluoropolymer additives: Coating and polymer–polymer slippage. J. Rheol. 2001, *45*, 565–581.
6. Migler, K.B.; Son, Y.; Qiao, F.; Flynn, K. Extensional deformation, cohesive failure and boundary conditions during sharkskin melt fracture. J. Rheol. 2002, *46*, 383–400.
7. Piau, J.M.; El Kissi, N.; Tremblay, B. Influence of upstream instabilities and wall slip on melt fracture and sharkskin phenomena during silicones extrusion through orifice dies. J. non-Newtonian Fluid Mech. 1990, *34*, 145–180.
8. Vinodradov, G.V.; Malkin, A.Y. *Rheology of Polymers*, 1st Ed.; Mir Publishers: Moscow, Springer, Berlin, 1980; 140–150.
9. Leonov, A.L.; Prokunin, A.N. *Nonlinear Phenomena in Flows of Viscoelastic Polymer Fluids*; Chapman and Hall: New York, 1994; 356–396.
10. Buckmaster, M.D.; Henry, D.L.; Randa, S.K. U.S. Patents 5,688,4 and 575,945,478 issued to E.I. DuPont de Nemours & Co Inc, 1997.
11. Achilleos, E.C.; Georgiou, G.; Hatzikiriakos, S.G. The role of processing aids in the extrusion of molten polymers. J. Vinyl Addit. Technol. 2002, *8*, 7–24.
12. Hatzikiriakos, S.G.; Hong, P.; Ho, W.; Stewart, C.W. J. Appl. Polym. Sci. 1995, *55*, 595–603.
13. Kalika, D.S.; Denn, M.M. Wall slip and extrudate distortion in linear low-density polyethylene. J. Rheol. 1987, *31*, 815–834.
14. Ramamurthy, A.V. Wall slip in viscous fluids and influence of materials of construction. J. Rheol. 1986, *30*, 337–357.
15. Vinogradov, G.V. Flow and rubber elasticity of polymeric systems. Pure Appl. Chem. 1971, *26*, 423–449.
16. Vinogradov, G.V. Fundamental problems concerning the interrelation of the structure of polymers and their rheological properties in the fluid state. Pure Appl. Chem. 1974, *39*, 115–143.
17. Vinogradov, G.V. Ultimate regimes of deformation of linear flexible chain fluid polymers. Polymer 1977, *18*, 1275–1283.
18. Rozenbaum, E.; Hatzikiriakos, S.G.; Stewart, C.W. Flow implications in the processing of Teflon® resins. Int. Polym. Process. 1995, *X*, 204–212.
19. El Kissi, N.; Piau, J.M. Stability phenomena during polymer melt extrusion. In *Rheology for Polymer Melt Processing*; Piau, J.M., Agassant, J.F., Eds.; Elsevier: Amsterdam, 1996; 389–420.

20.  El Kissi, N.; Piau, J.M.; Toussaint, F. Sharkskin and cracking of polymer melt extrudates. J. non-Newtonian Fluid Mech. 1997, 68, 271–290.
21.  Rozenbaum, E.E. Rheology and processability of FEP resins for wire coating. Ph.D. dissertation, The University of British Columbia: Vancouver, BC, 1998.
22.  Yip, K.; Rozenbaum, E.E.; Randa, S.K.; Hatzikiriakos, S.G.; Stewart, C.W. The effect of boron nitride type and concentration on the rheology and processability of molten polymers. Proc. ANTEC, Tech. Pap.; Society of Plastic Engineers: New York, NY, USA, 1999; Vol. 45, 1223–1227.
23.  Yip, F.; Rozenbaum, E.; Hatzikiriakos, S.G. Boron nitride powders: new processing aids for molten polymers. J. Plast. Film Sheeting 2000, 16, 16–32.
24.  Lee, S.M.; Kim, J.G.; Lee, J.W. The effect of boron nitride on the processability of metallocene based LLDPE. Proc. ANTEC, Tech. Pap.; Society of Plastic Engineers: New York, NY, USA, 2000; Vol. 46, 2862–2866.
25.  Lee, S.M.; Lee, J.W. The effect of boron nitride on the rheological properties and processability of polyethylene. Proc. ANTEC, Tech. Pap.; Society of Plastic Engineers: New York, NY, USA, 2001; Vol. 45, 1223–1227.
26.  Yip, F. The effect of boron nitride on the rheology and processability of polymers. In M.A.Sc. Thesis; The University of British Columbia, Vancouver, BC, 1999.
27.  Yip, F.; Hatzikiriakos, S.G.; Clere, T.M. A new processing aid for the extrusion of polyolefins. J. Vinyl Addit. Technol. 2000, 6, 113–118.
28.  Yip, F.; Hatzikiriakos, S.G.; Clere, T.M. A new processing aid for the extrusion of polyolefins. Proceedings of Polyolefins, Houston, TX, 2000; 563–570.
29.  Yip, F.; Hatzikiriakos, S.G.; Clere, T.M. New boron nitride processing aids for the extrusion of molten polymers. Proc. ANTEC, Tech. Pap.; Society of Plastic Engineers: Orlando, FL, 2000; Vol. 46, 2852–2857.
30.  Rozenbaum, E.E.; Randa, S.; Hatzikiriakos, S.G.; Stewart, C.W. Boron nitride as a processing aid for the extrusion of polyolefins and fluoropolymers. Polym. Eng. Sci. 2000, 40, 179–190.
31.  Rozenbaum, E.E.; Randa, S.K.; Hatzikiriakos, S.G.; Stewart, C.W.; Henry, D.L.; Buckmaster, M. A new processing additive in the extrusion of fluoropolymers. Proc. ANTEC, Tech. Pap.; Society of Plastic Engineers: Atlanta, GA, 1998; Vol. 44, 952–956.
32.  Kapnistos, M.; Vlassopoulos, D.; Anastasiadis, S.H. Determination of both the bimodal and spinodal curves in polymer blends by shear rheology. Europhys. Lett. 1996, 34, 513–518.
33.  Vlassopoulos, D. Rheology of critical LCST polymer blends. Rheol. Acta 1996, 35, 556–566.
34.  Yip, F.; DiRaddo, R.; Hatzikiriakos, S.G. The effect of combining boron nitride with fluoroelastomer on the melt fracture of HDPE in extrusion blow molding. J. Vinyl Addit. Technol. 2000, 6, 196–204.
35.  Seth, M.; Yip, F.; Hatzikiriakos, S.G. Combining boron nitride with a fluoroelastomer: An enhanced polymer processing additive. Proc. ANTEC, Tech. Pap.; Society of Plastic Engineers: Dallas, TX, 2000; Vol. 47, 2649–2653.
36.  Pruss, E.A.; Randa, S.K.; Lyle, S.S.; Clere, T.M. Properties of m-LLDPE

blown films extruded utilizing boron nitride based polymer process aids. Proc. ANTEC, Tech. Pap.; Society of Plastic Engineers: San Francisco, CA, 2002; Vol. 46, 2864–2868.

37. Vogel, R.; Hatzikiriakos, S.G.; Brunig, H.; Tandler, B.; Golzar, M. Improved spinnability of metallocene polyethylenes by using processing aids. Int. Polym. Process. 2003, *XVIII*, 67–74.

38. Rozenbaum, E.E.; Hatzikiriakos, S.G.; Stewart, C.W. The melt fracture behavior of Teflon® resins in capillary/slit extrusion. Proc. ANTEC, Tech. Pap.; Society of Plastic Engineers: Boston, MA, 1995; Vol. 41, 1111–1115.

39. Seth, M.; Hatzikiriakos, S.G.; Clere, T. Gross melt fracture elimination: the role of surface energy of boron nitride powders. Polym. Eng. Sci. 2002, *42*, 743–752.

40. Seth, M.; Hatzikiriakos, S.G.; Clere, T. The effect of surface energy of boron nitride powders on gross melt fracture elimination. Proc. ANTEC, Tech. Pap.; Society of Plastic Engineers: Dallas, TX, 2001; Vol 47, 2634–2638.

41. Seth, M. The role of surface energy of boron nitride on the gross melt fracture elimination of polymers. In *M.A.Sc. Thesis*; The University of British Columbia: Vancouver, BC, 2001.

42. Laskowski, J.S. *Coal Flotation and Fine Coal Utilization*, 1st Ed., Elsevier: New York, 2001.

43. Rathod, N.; Hatzikiriakos, S.G. Polymer processing aids and surface energy. Proc. ANTEC, Soc. Plast. Eng., Tech. Pap. 2003, *49*. (CD-ROM, Nashville, TN).

44. Kazatckov, I.B.; Yip, F.; Hatzikiriakos, S.G. The effect of boron nitride on the rheology and processing of polyolefins. Rheol. Acta 2000, *39*, 583–594.

45. Hatzikiriakos, S.G. Gross melt fracture elimination during the extrusion of polyolefins. Proceedings of the XIIIth International Congress on Rheology, Cambridge, UK 2000; Vol. 3, 143–145.

46. Seth, M.; Hatzikiriakos, S.G. Combining boron nitride with fluoroelastomer: An enhanced polymer processing additive. J. Vinyl Addit. Tech. 2001, *7*, 90–97.

47. Randa, S.; Stewart, C.W.; Rozenbaum, E.E.; Hatzikiriakos, S.G. Extrusion Aid Combination. US patent published December, 2001. No:0048179 (UBC-DuPont & de Nemours Co.).

48. Randa, S.; Stewart, C.W.; Rozenbaum, E.E.; Hatzikiriakos, S.G. Extrusion Aid Combination. PCT publication (WO01/46313A1), 2001.

# 10

## Draw Resonance in Film Casting

**Albert Co**
University of Maine, Orono, Maine, U.S.A.

### 10.1 INTRODUCTION

In the film casting process, polymer melt is extruded through a slit die to form a molten film. The film is then taken up by a chill roll, which quenches the molten film and solidifies it for windup. The process is shown schematically in Fig. 10.1. The distance $L$ between the die exit and the contact point at the chill roll is referred to as the *drawing length*. The ratio of the take-up velocity $v_L$ at the chill roll to the extrusion velocity $v_0$ at the die exit is known as the *draw ratio* $D_R$. Since the draw ratio is greater than unity, the molten film is stretched and drawn down in its cross section. The reduction in the film cross section involves the decrease of the film thickness and the reduction of the film width, usually referred to as *neck-in*. In addition, beads or regions of greater thickness are formed at the film edges. Typically, the die opening is in the order of 1 mm and the final film thickness ranges from 25 to 100 μm.

When the draw ratio is increased beyond a critical value, the film-casting process can become unstable. This flow instability is characterized by a sustained oscillation in the film thickness and is referred to as *draw resonance*. It occurs in spite of constant extrusion speed and constant take-up speed. Examples of the thickness oscillations along the length of solidified films for a linear low-density polyethylene (LLDPE) film (1) are shown in Fig. 10.2. At

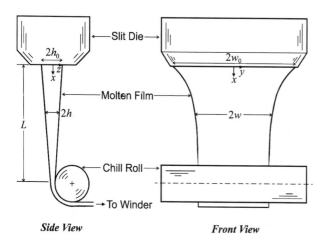

Side View          Front View

FIGURE 10.1 The film casting process.

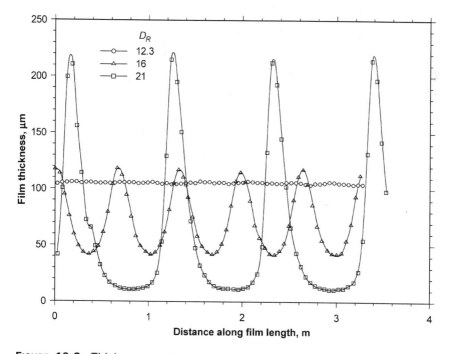

FIGURE 10.2 Thickness measurements along the length of solidified LLDPE films produced at different draw ratios. The measurements are made at the center of the films. (From Ref. 1.)

$D_R = 12.3$ (below the critical draw ratio), the process was stable and the thickness showed only very small random noises. At $D_R = 16$ (near the critical draw ratio), the thickness showed sinusoidal oscillations with wavelength in the order of a meter. At $D_R = 21$ (beyond the critical draw ratio), the thickness oscillations consist of sharp peaks and thin troughs with longer wavelength.

Aside from the thickness fluctuations, one also observes width fluctuations of the solid film. Figure 10.3 shows the width and thickness fluctuations along the length of the film obtained by Barq et al. (2) for a polyethylene terephthalate (PET) film. Here we see that the maxima of the width oscillations occur at positions where the thickness oscillations have minima. However, the amplitude of the width oscillations is relatively smaller than that of the thickness oscillations. Similar behaviors in width and thickness oscillations were reported by Bian (3). He also observed oscillations in measured film tension, at comparable frequency as the thickness oscillations.

In the limit of one-dimensional flow, the kinematics of film casting and fiber spinning (see this volume, Chap. 11 by Hyun) are similar. However, the

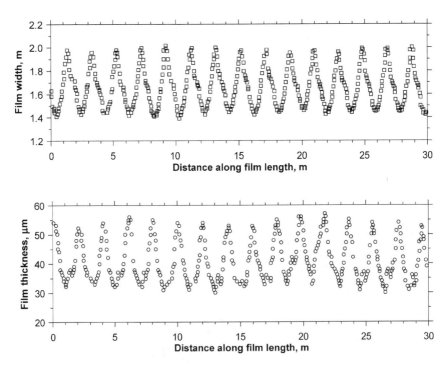

FIGURE 10.3 Width and thickness fluctuations along the length of a PET film produced at $D_R = 28.4$. The thickness measurements are made at the center of the film. (From Ref. 2.)

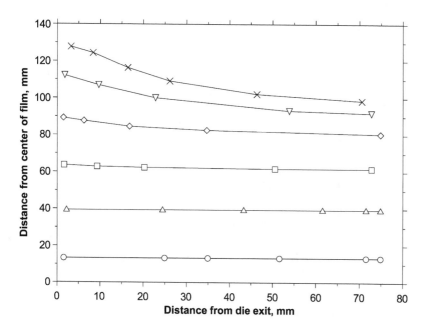

**FIGURE 10.4** Streamlines from tracer measurements for a LDPE film drawn at $D_R = 12$. (From Ref. 5.)

similarity ends when one investigates the two and three dimensionalities of film casting. The occurrence of neck-in and edge beading in film casting has no equivalence in fiber spinning. Due to the neck-in phenomena, the central region of the drawn film undergoes planar elongational flow and the region near the edge undergoes uniaxial elongational flow. In fiber spinning, the drawn filament undergoes uniaxial elongation flow only. The presence of two flow regions in cast films was first pointed out by Dobroth and Erwin (4) and is further demonstrated in Fig. 10.4, which shows the streamlines created by tracers in the film casting of a low-density polyethylene (LDPE) film (5). In the central region, the streamlines are parallel, indicating that the fluid elements are not deformed in the width direction (planar elongational flow). Near the film edges, the streamlines are converging, indicating that the fluid elements are deformed in both width and thickness direction (uniaxial elongational flow).

## 10.2 MODEL DIMENSIONALITY

In analyzing the film-casting problem, various researchers have made certain assumptions on the velocity flow field in the stretching film. The resulting models range from the simpler one-dimensional models to the more complicated two-dimensional model. In addition, the models that were used to investigate stability in film casting to date considered the process to be *isothermal*. Although recent works (6–11) had investigated the nonisothermal effects in film casting, these studies dealt with the steady-state behaviors and did not investigate flow instability. In industrial operations where the slit die is typically more than a meter wide and the drawing length is between 25 and 65 mm, effects of temperature variations in the drawing zone on steady-state performances are usually considered to be small.

### 10.2.1  Two-Dimensional Model

Since the film thickness is much smaller than the drawing length and the film width, one can assume that the velocity component $v_x$ in the drawing direction and the velocity component $v_y$ in the width direction are uniform across the film thickness; that is, they are independent of $z$. Hence, the velocity field in the drawn film can be described by:

$$v_x = v_x(x, y, t) \tag{10.1}$$

$$v_y = v_y(x, y, t) \tag{10.2}$$

$$v_z = -z\left(\frac{\partial v_x}{\partial x} + \frac{\partial v_y}{\partial y}\right) \tag{10.3}$$

Here $v_z$ is the velocity component in the thickness direction. Equation (10. 3) follows from the continuity equation for an incompressible fluid.

The continuity equation for an incompressible fluid can be integrated across the film thickness to get:

$$\frac{\partial h}{\partial t} + \frac{\partial(hv_x)}{\partial x} + \frac{\partial(hv_y)}{\partial y} = 0 \tag{10.4}$$

Here $h(x,y,t)$ is the half-thickness of the film at a specific position and time. Derivation of Eq. (10.4) makes use of the free surface kinematic condition on the film surface:

$$\text{At } z = h(x, y, t): \qquad \frac{\partial h}{\partial t} + v_x\frac{\partial h}{\partial x} + v_y\frac{\partial h}{\partial y} - v_z = 0 \tag{10.5}$$

Similarly the $x$- and $y$-component of the equation of motion can be integrated across the film thickness to yield:

$$\frac{\partial[h(\tau_{xx} - \tau_{zz})]}{\partial x} + \frac{\partial(h\tau_{yx})}{\partial y} = 0 \tag{10.6}$$

$$\frac{\partial[h(\tau_{yy} - \tau_{zz})]}{\partial y} + \frac{\partial(h\tau_{yx})}{\partial x} = 0 \tag{10.7}$$

Here $\tau_{ij}$ is the $ij$ component of the extra stress tensor. In deriving the above equations, fluid inertia, gravity, surface tension, and aerodynamic drag are neglected. In addition, it is assumed that the film curvature is small, that is, $\partial h/\partial x \ll 1$ and $\partial h/\partial y \ll 1$.

Equations (10.4), (10.6), and (10.7) were first derived by Yeow (12) in his study of two-dimensional disturbances in film casting of Newtonian fluids. These were also used in the two-dimensional disturbance analyses of Aird and Yeow (13) for power-law fluids and of Anturkar and Co (14) and Iyengar and Co (15) for viscoelastic fluids. These equations, with the time derivatives dropped, were used in the steady-state two-dimensional analyses of d'Halewyu et al. (16) for Newtonian fluids and of Debbaut and Marchal (17) for viscoelastic fluids.

### 10.2.2 One-Dimensional Model with Varying Film Width

An approach to reduce the dimensionality of the problem and to retain the concept of varying width in the drawn film is to assume that the velocity field in the drawn film can be approximated by:

$$v_x = v_x(x, t) \tag{10.8}$$

$$v_y = yf(x, t) \tag{10.9}$$

The continuity equation then leads to:

$$v_z = -zg(x, t) \tag{10.10}$$

With this approximated velocity field, Eqs. (10.4) and (10.6) can be integrated across the film width to yield the following simplified form of the mass and momentum balances:

$$\frac{\partial (hw)}{\partial t} + \frac{\partial (hwv_x)}{\partial x} = 0 \tag{10.11}$$

$$\frac{d[hw(\tau_{xx} - \tau_{zz})]}{dx} = 0 \tag{10.12}$$

In the derivation of Eq. (10.11), the free surface kinematic condition at the film edge [see Eq. (10.17)] is used.

This model implies that $h = h(x,t)$ and therefore cannot predict the variation in film thickness due to edge beading. Moreover, the assumed form for $v_y$ in Eq. (10.9) may not be valid for viscoelastic fluids. Two-dimensional numerical simulation for viscoelastic fluids by Debbaut and Marchal (17) and available experimental data, as shown in Fig. 10.4, show that the streamlines are parallel in the central region of the film. This observation is inconsistent with the assumption that $v_y$ varies linearly with $y$, as given in Eq. (10.9).

Another shortcoming of the one-dimensional model is that it predicts a more pronounced neck-in of the film width. Figure 10.5 shows the film width profiles reported by Silagy et al. (18) for one- and two-dimensional models of Newtonian and convected Maxwell fluids. For both fluids, the one-dimensional model predicts larger film width reductions than the two-dimensional model. These profiles also show that viscoelastic fluids exhibit less film width reduction than Newtonian fluids. Similar observation was reported by Debbaut and Marchal (17).

This one-dimensional varying film width model was used by Silagy et al. (18,19) in their stability analysis of Newtonian and viscoelastic films. It was also used by Lee et al. (20,21) in their stability analysis of film casting for viscoelastic fluids.

### 10.2.3   One-Dimensional Model of Infinite Film Width

Another approach to reduce the dimensionality of the problem is to assume that $v_x = v_x(x,t)$ and $v_y = 0$, which is the flow field for a fluid undergoing planar elongational flow. This assumed flow field implies that the planar-elongational central region of the film is much wider than the uniaxial-elongational

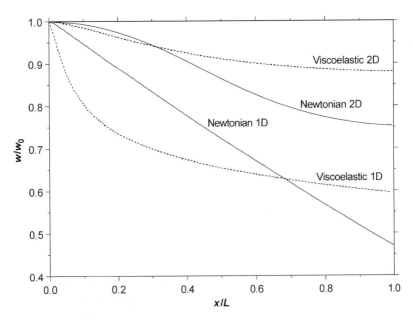

FIGURE 10.5 Width profiles from one- and two-dimensional models of Newtonian and convected Maxwell fluids. (From Ref. 18.)

film edge region, or, in other words, the film is infinitely wide. Hence this model cannot predict the reduction of film width.

With this approximated flow field, the mass and momentum balances in Eqs. (10.4) and (10.6), respectively, are simplified by dropping the $y$-derivative terms. This one-dimensional model was used by Yeow (12), Aird and Yeow (13), Anturkar and Co (14), and Iyengar and Co (15) to obtain the steady-state solutions in their stability analyses of film casting.

## 10.3 BOUNDARY CONDITIONS

The mass and momentum balances discussed in the previous section are to be solved with an appropriate constitutive equation using the following boundary conditions at the die exit and the take-up point on the chill roll.

At $x = 0$ :

$$v_x = v_0 \tag{10.13}$$

$$h = h_0 \tag{10.14}$$

$$w = w_0 \tag{10.15}$$

At $x = L$ : $\qquad v_x = v_L$ \hfill (10.16)

Along the film edge, the free-surface kinematic condition gives:

At $y = w(x, t)$ : $\qquad \dfrac{\partial w}{\partial t} + v_x \dfrac{\partial w}{\partial x} - v_y = 0$ \hfill (10.17)

and the free-surface stress condition leads to:

At $y = w(x, t)$ : $\qquad \tau_{yx} = \left(\dfrac{\partial w}{\partial x}\right)(\tau_{xx} - \tau_{zz})$ \hfill (10.18)

$$\tau_{yy} - \tau_{zz} = \left(\dfrac{\partial w}{\partial x}\right)^2 (\tau_{xx} - \tau_{zz}) \tag{10.19}$$

For viscoelastic fluids, the stress conditions at $x = 0$ need to be specified. The stresses at the die exit depend on the flow conditions inside the die, as well as the conditions downstream. The accurate determination of the stresses at the die exit would require extensive numerical calculations involving both the flows in the die and in the drawn film. In the literature, several approaches have been used to approximate the stress conditions at the die exit.

Anturkar and Co (14) estimated the value of $\tau_{xx}$ at the die exit with the mean stress value for fully developed slit flow in an infinitely wide die. Silagy et al. (18,19) extended this concept and proposed that $\tau_{xx}$ at the die exit could range from zero (i.e., the fluid was totally relaxed due to the die swell phenomena) to the mean stress value in infinite flat die. They found that the stress conditions at the die exit had a rather weak influence on the solution. Instead of specifying the stress value, Iyengar and Co (22) specified the stress ratio $\tau_{zz}/\tau_{xx}$ at the die exit, with values ranging from that for planar elongational flow to that for slit flow in the die. The stress values were then calculated using the specified stress ratio and a macroscopic force balance between the die exit and the take-up point on the chill roll. Iyengar (1) found that the stress ratios corresponding to the two limiting cases gave essentially the same velocity and stress profiles.

In two-dimensional analysis, the value of the stress component $\tau_{yy}$ at the die exit also needs to be specified. In their work, Silagy et al. (18,19) assumed the value of $\tau_{yy}$ for a viscoelastic fluid to be the same as that for a Newtonian fluid.

## 10.4  LINEAR STABILITY ANALYSIS

In linear stability analysis, we study the system response to an *infinitesimal* flow disturbance about the steady-state condition. Various researchers had applied this type of analysis to investigate the behavior of disturbances about one-dimensional steady-state solutions in film casting. For two-dimensional flow disturbance, the perturbation about a one-dimensional steady-state solution can be represented by:

$$f(x, y, t) = f_s(x) + \text{Re}\left\{ f^*(x) \exp\left( \Omega \frac{t v_0}{L} + iA \frac{y}{L} \right) \right\} \tag{10.20}$$

Here $f$ represents the film thickness, a component of the velocity vector, or a component of the stress tensor. The first term $f_s(x)$ is the steady-state solution and the second term is the disturbance quantity. In the disturbance quantity, $\text{Re}\{\}$ represents the real part of a complex variable and $f^*(x)$ is a complex function. Since the disturbances are considered infinitesimal, all nonlinear terms of the disturbance quantities or their derivatives are neglected in the resulting equations used to determine the eigenvalues.

In Eq. (10.20), the frequency $\Omega$ ($\equiv \Omega_r + i\Omega_i$) is complex and the wave number $A$ is real. The amplitude of the disturbance decays exponentially for $\Omega_r < 0$, remains constant for $\Omega_r = 0$, and grows exponentially for $\Omega_r > 0$. The conditions for $\Omega_r = 0$ to occur establish the stability limits of the process. In the literature, the exponential term in Eq. (10.20) is sometimes written as $\exp[i(-\Omega t v_0/L + Ay/L)]$, for example, in the works of Yeow and coworkers (12,13). Using this other convention, the growth or decay of the disturbances is then determined by the imaginary part of $\Omega$, with positive value for growth and negative value for decay.

For nonzero values of $A$, Eq. (10.20) implies that the disturbance can occur as a periodic wave along the width with wave number $A$. When $A = 0$, there is no disturbance across the width and the flow disturbance is only one-dimensional.

This standard approach of linear stability analysis was described by Chandrasekhar (23) and by Drazen and Reid (24). Analysis of two-dimensional disturbances was used by Yeow (12) and Aird and Yeow (13) for film casting of inelastic fluids and by Anturkar and Co (14) and Iyengar and Co (15) for viscoelastic fluids. Analysis of one-dimensional disturbances was used by Silagy et al. (18,19) for Newtonian and viscoelastic fluids.

## 10.5 NONLINEAR STABILITY ANALYSIS

The stability limits determined from linear stability analysis are found for infinitesimal disturbances. Nonlinear stability analysis is usually performed to ascertain that the stability limits for infinitesimal disturbances still apply for finite-amplitude disturbances and to investigate the nonlinear dynamics of the disturbances.

In nonlinear stability analysis, the disturbance over the steady-state solution is introduced by changing one of the variables, for example, the take-up velocity. In this case, the change is analogous to a typical experimental procedure in which the extrusion rate is held constant and the take-up velocity at the chill roll is varied. The disturbance introduced can be either a step disturbance of certain percentage change or a ramp disturbance spread out over a period of time. Transient calculations are then conducted to study the response of the system.

This approach was used by Barq et al. (2) for Newtonian fluids, by Iyengar and Co (15) for Newtonian and viscoelastic fluids, by Silagy et al. (18) for Newtonian and viscoelastic fluids, and by Lee et al. (20,21) for viscoelastic fluids.

## 10.6 STABILITY OF PURELY VISCOUS FILMS

In the literature, studies of film casting of purely viscous liquids include Newtonian fluids and power-law fluids. For Newtonian fluids, both infinite film width and varying film width have been considered. For power-law fluids, only the case of infinite film width has been investigated.

### 10.6.1 Newtonian Films of Infinite Film Width

The draw resonance in isothermal, Newtonian film casting was first investigated theoretically by Yeow (12). He neglected edge effects, fluid inertia, surface tension, gravitational forces, and aerodynamic drag. He also assumed small film curvature and uniform axial stress across the film thickness. On the basis of these assumptions, he derived an analytical steady-state solution (denoted by the subscript s) for one-dimensional Newtonian model of infinite film width:

$$v_{x,s} = v_0 \exp(\beta x/L) \tag{10.21}$$

$$v_{z,s} = -v_0 \beta (z/L) \exp(\beta x/L) \tag{10.22}$$

$$h_s = h_0 \exp(-\beta x/L) \tag{10.23}$$

Here $\beta = \ln D_R = FL/(4\mu Q)$, where $F$ is the axial drawing force per unit width, $\mu$ is the fluid viscosity, and $Q$ is the volumetric flow rate.

He then introduced infinitesimal disturbances into the steady-state solution. The perturbed velocity components $v_x$, $v_y$ (with $v_{y,s} = 0$), $v_z$, and the thickness $h$ were described by expressions similar to Eq. (10.20). When these expressions were substituted into Eqs. (10.4), (10.6), and (10.7) and only the linear terms of the perturbed quantities were retained, he obtained a system of homogeneous first-order differential equations for $v_x^*$, $v_y^*$, and $h^*$. For the boundary conditions, he specified:

$$\text{At } x = 0, \qquad v_x^* = 0, \qquad v_y^* = 0, \qquad h^* = 0 \tag{10.24}$$

$$\text{At } x = L, \qquad v_x^* = 0, \qquad v_y^* = 0 \tag{10.25}$$

These are appropriate for the common practice of holding the take-up speed at the chill roll constant. In his study, he also specified a second set of boundary conditions for constant tension operations.

The set of differential equations and the homogeneous boundary conditions constitute an eigenvalue problem, for which the parameters are $D_R$, $A$, and $\Omega$. For a given $D_R$ and $A$, there exist various $\Omega$'s corresponding to the nontrivial solutions for this eigenvalue problem. Inspection of the differential equations and the boundary conditions indicates that if the set consisting of $D_R$, $A$, and $\Omega$ is an eigencombination, then the set consisting of $D_R$, $A$, and the complex conjugate of $\Omega$ is also an eigencombination. Hence only $\Omega_i \geq 0$ need to be considered for disturbances of the form of Eq. (10.20).

To find the eigencombinations of $D_R$, $A$, and $\Omega$, Yeow used a numerical scheme to locate the first three of the infinite number of $\Omega$'s for a given $D_R$ and $A$. Then neutral stability curves (contour lines of $\Omega_r = 0$) corresponding to the first three modes of disturbance were generated in the $A$–$D_R$ plane. The neutral stability curves separated the plane into stable and unstable region. For operations with constant take-up velocity, he found that the eigenvalue corresponding to the first mode was dominant and that the lowest $D_R$ for transition from stable to unstable occurred at $A = 0$ and had a value of 20.21. This value for the critical draw ratio is identical to that derived for isothermal, Newtonian fiber spinning with constant take-up velocity (see this volume, chapter by Hyun). For operations with constant tension, he found that the most unstable disturbance had a nonzero $A$ and this disturbance became unstable at $D_R = 2.25 \times 10^3$, not likely to be attained in practice.

The nonlinear stability analysis of isothermal Newtonian film casting was first carried out by Barq et al. (2). In their numerical calculations, they considered the one-dimensional model of infinite film width and imposed three types of perturbations: (1) a 10% stepwise increase of the take-up velocity, (2) a sinusoidal oscillation of the take-up velocity, and (3) random fluctuations of the take-up velocity. For the case of stepwise increase in take-up velocity, they obtained critical draw ratio similar to that obtained from linear stability analysis (12). For a final draw ratio below the critical draw ratio, the amplitude of the oscillation decayed with time and a new equilibrium state was eventually attained. For a final draw ratio near the critical draw ratio, the amplitude of the oscillation remained constant. For a final draw ratio above the critical draw ratio, the amplitude of the oscillation grew with time. For the case of sinusoidal take-up velocity, the calculated thickness exhibited sinusoidal oscillation with the same period and the same relative amplitude. For the case of random fluctuations in take-up velocity, the calculated thickness showed sinusoidal fluctuations, with time-varying amplitudes.

Similar nonlinear behavior for stepwise increase of take-up velocity in Newtonian film casting was also reported by Iyengar and Co (15). In addition to the agreement between the critical draw ratios for infinitesimal disturbances and for finite-amplitude disturbances, they also found that the initial growth or decay rate and the frequency of oscillation during the initial growth or decay of a finite-amplitude disturbance agreed with those predicted by linear stability analysis.

Barq et al. (2) also reported experimental data for a polyethylene terephthalate (PET) film. They considered the PET melt to be Newtonian because the viscosity of the PET melt was constant over the shear rate range corresponding to the rate of deformation of the melt between the die and the chill roll. In addition, the primary normal stress difference was low. Their thickness measurements showed that the critical draw ratio is about 20, which agreed with the prediction for Newtonian film casting. They also measured the fluctuations in the width of the solidified film and observed the coupling of the thickness oscillations and the width oscillations, as shown earlier in Fig. 10.3. The coupling of the thickness and width fluctuations was further substantiated when their measured maximum and minimum values for film width was plotted against the draw ratio, as shown in Fig. 10.6. Here we see that the fluctuations in film width started at a draw ratio of about 20, similar to that for thickness fluctuations.

Earlier experimental works on draw resonance in film casting had been reported in the literature. Kase (25) carried out his experiments for a polypropylene film at only one draw ratio, $D_R = 33$, but at drawing length

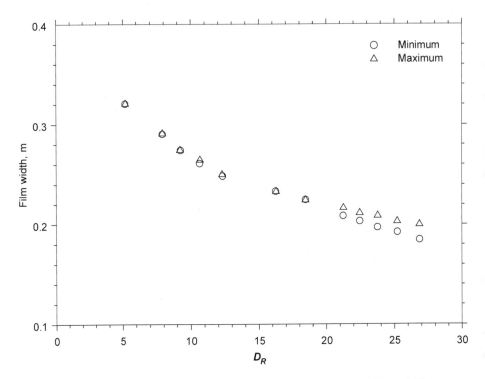

**FIGURE 10.6** The dependence of the minimum and maximum film widths on draw ratio for a PET film. (From Ref. 2.)

ranging from 50 to 200 mm and take-up speed ranging from 10 to 60 m/min. He found the film to be stable at $D_R = 33$, which is higher than the critical draw ratio for isothermal, Newtonian fluids. He attributed the improved stability to the cooling of the film in the drawing zone during his experiments. However, he did find that the oscillation periods obtained by subjecting the films to disturbances agreed with predicted values for isothermal, Newtonian films. Bergonzoni and DiCresce (26) reported draw resonance in their water-quenched extrusion of thick polypropylene ribbons. However, their ribbons were approximately 1 mm thick and 5 mm wide at the take-up. This makes their study closer to melt spinning, rather than film casting, in which the thickness to width ratio is much less than unity. To improve performance in commercial operations, Lucchesi et al. (27) described a device that could reduce draw resonance in melt embossing and extrusion coating of various linear low-density polyethylene (LLDPE) resins. They achieved this by blowing a curtain of cooling air at the midpoint between the die and chill roll.

The one-dimensional Newtonian film casting model was extended by Ramos (28) to investigate the effects of fluid inertia and gravity. He investigated the nonlinear dynamics of the process subjected to a perturbation in the take-up velocity at low Reynolds and Froude numbers. His numerical results showed that the film thickness at the take-up had very sharp spikes and the separation between the spikes depended on the Reynolds and Froude numbers, as well as the amplitude and frequency of the imposed velocity perturbations.

### 10.6.2 Newtonian Films with Varying Film Width

The one-dimensional model with varying film width was considered by Silagy et al. (19) to investigate the effect of aspect ratio $L/w_0$ on the stability of film casting. Their neutral stability curve generated from linear stability analysis for Newtonian fluids is shown as the solid line in Fig. 10.7. At $L/w_0 = 0$, the neutral stability curve gives a critical draw ratio of 20.2, which is identical to that for one-dimensional model of infinite film width. Within the range of typical values of aspect ratio ($0.1 \leq L/w_0 \leq 1.2$), the critical draw

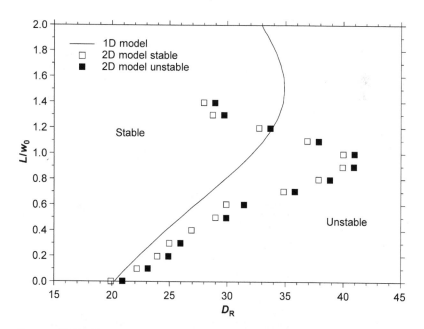

FIGURE 10.7  The effect of the aspect ratio on the critical draw ratio from one- and two-dimensional Newtonian models. The open square at each aspect ratio represents the highest draw ratio that is stable and the solid square represents the lowest draw ratio that is unstable. (From Refs. 18 and 19.)

ratio increases (more stable) as the aspect ratio $L/w_0$ increases. Their model also predicts that the maximum critical draw ratio for Newtonian film casting is about 35, at an aspect ratio of 1.55.

Later, Silagy et al. (18) expanded their study by considering the two-dimensional model. In their nonlinear stability analysis, they obtained for a given aspect ratio the highest draw ratio that is stable and the lowest draw ratio that is unstable. The critical draw ratio lies between these two draw ratios. Their calculated results for a Newtonian fluid are represented by the open (highest draw ratio that is stable) and solid (lowest draw ratio that is unstable) squares in Fig. 10.7. The two-dimensional model also predicts that a higher aspect ratio leads to a more stable process. This stabilizing effect is more pronounced in the two-dimensional model than in the one-dimensional model. The two-dimensional model predicts that the maximum critical draw ratio for Newtonian fluids is about 45, at an aspect ratio of unity. The numerical results of Silagy et al. for the two-dimensional model also showed that the thickness fluctuations at the central part and at the periphery of the film were out-of-phase. Their calculated width fluctuations were also out-of-phase with the thickness fluctuations at the central part of the film.

### 10.6.3 Power-Law Films of Infinite Film Width

The film casting of power-law fluids was investigated by Aird and Yeow (13). The power-law model can be expressed as:

$$\tau_{ij} = -m\dot{\gamma}^{n-1}\dot{\gamma}_{ij} \tag{10.26}$$

Here $\tau_{ij}$ is the $ij$ component of the stress tensor $\tau$, $\dot{\gamma}_{ij}$ is the $ij$ component of the rate-of-deformation tensor $\dot{\gamma}$ [$\equiv \nabla v + (\nabla v)^+$], and $\dot{\gamma}$ is the magnitude of the rate-of-deformation tensor, i.e., $\dot{\gamma} = \sqrt{(1/2)\dot{\gamma}:\dot{\gamma}}$. The parameters of the model are the consistency index $m$ and the power-law index $n$. For $m = \mu$ and $n = 1$, the power-law model reduces to the Newtonian fluid.

Following the development of Yeow (12) for Newtonian fluids, Aird and Yeow first showed that the steady-state velocity and thickness profiles in the film casting of power-law fluids were given by:

$$\frac{v_{x,s}}{v_0} = \frac{h_0}{h_s} = \left[1 + \left(\frac{n-1}{n}\right)\beta^{1/n}\frac{x}{L}\right]^{n/(n-1)} \tag{10.27}$$

Then the draw ratio of the drawn film was given by:

$$D_R = \left[1 + \left(\frac{n-1}{n}\right)\beta^{1/n}\right]^{n/(n-1)} \tag{10.28}$$

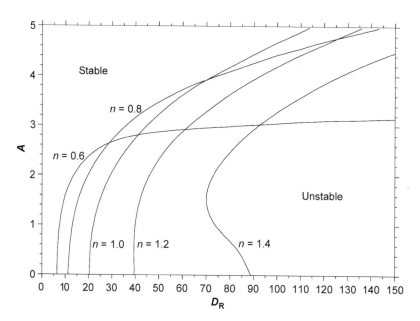

**FIGURE 10.8** Neutral stability curves for power-law fluids of different power-law indices. (From Ref. 13.)

Here $\beta = FL^n/2^{n+1}mQv_0^{n-1}$. All the variables are as defined for the Newtonian case.

Aird and Yeow then carried out a linear stability analysis with disturbances of the form given by Eq. (10.20). They obtained neutral stability curves on the $A-D_R$ plane for different values of $n$, as shown in Fig. 10.8. For $n<1.2$, the critical draw ratio for a given $n$ occurred at $A=0$ and the disturbances were uniform across the film width. The critical draw ratio was also identical to that for the isothermal fiber spinning of a power-law fluid. For $n>1.2$, the critical draw ratio occurred at $A>0$ and the disturbances were traveling waves across the width. The critical draw ratio was lower than that for fiber spinning. Moreover, their results showed that extensional thickening ($n>1$) stabilized, whereas extensional thinning ($n<1$) destabilized.

## 10.7 STABILITY OF VISCOELASTIC FILMS

Earlier works on the stability of film casting of viscoelastic liquids considered the film to be infinitely wide. Later investigations allowed for the variation of film width.

## 10.7.1  Viscoelastic Films of Infinite Film Width

The role of viscoelasticity on draw resonance in film casting was investigated by Anturkar and Co (14), who conducted a linear stability analysis using a modified convected Maxwell model. This model relates the stress tensor $\tau$ and the rate-of-deformation tensor $\dot{\gamma}$ as follows:

$$\tau + \lambda(\dot{\gamma})\tau_{(1)} = -\eta(\dot{\gamma})\dot{\gamma} \tag{10.29}$$

Here the convected derivative $\tau_{(1)}$ is given by:

$$\tau_{(1)} = \frac{\partial \tau}{\partial t} + v \cdot \nabla \tau - \left[ \tau \cdot \nabla v + (\nabla v)^+ \cdot \tau \right] \tag{10.30}$$

They used the Carreau viscosity function and a similar function for the fluid characteristic time. These are:

$$\eta(\dot{\gamma}) = \eta_0 \left( 1 + \lambda_v^2 \dot{\gamma}^2 \right)^{(n-1)/2} \tag{10.31}$$

$$\lambda(\dot{\gamma}) = \lambda_0 \left( 1 + \lambda_t^2 \dot{\gamma}^2 \right)^{(n'-1)/2} \tag{10.32}$$

The model parameters are the zero-shear rate viscosity $\eta_0$, the time constants $\lambda_v$, $\lambda_0$, and $\lambda_t$, and the dimensionless parameters $n$ and $n'$.

In their linear stability analysis, Anturkar ad Co considered the one-dimensional steady-state model with two-dimensional disturbances represented by Eq. (10.20). Their model predicted two critical draw ratios; the flow was stable below the *lower* critical draw ratio and above the *upper* critical draw ratio. Within the range of model parameters investigated, they found that the lower critical draw ratio had zero wave number ($A = 0$) while the upper critical draw ratio has nonzero wave number ($A > 0$) for a large number of cases. Their results indicated that a higher degree of viscosity shear thinning (smaller $n$) reduced the lower critical draw ratio and increased the upper critical draw ratio; hence, it expanded the unstable region. This is illustrated in Fig. 10.9, which shows the neutral stability curves for various $n$ in the $D_R$ versus $\lambda_0(2\lambda_t)^{n'-1}/(L/v_0)^{n'}$ (a dimensionless fluid characteristic time) plane. They also found that a higher degree of reduction of fluid characteristic time (smaller $n'$) did not affect the lower critical draw ratio; but it increased dramatically the upper critical draw ratio and therefore expanded the unstable region.

Although the modified convected Maxwell model used by Anturkar and Co can predict reasonably the shear behavior of many polymers, it cannot describe extensional thinning in elongational flow. Moreover, its elongational viscosity becomes infinite at the elongational rate of $1/2\lambda$. As explained later, this sharp rise of the elongational viscosity to infinity is responsible for the turn back of the neutral stability curves in Fig. 10.9.

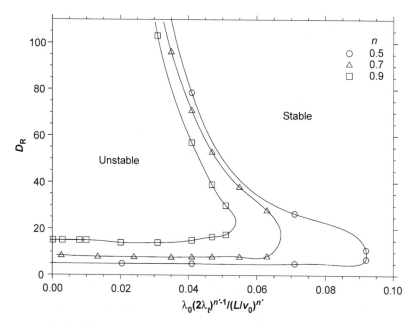

**FIGURE 10.9** The effect of the exponent $n$ of the viscosity function on the neutral stability envelopes for a modified convected Maxwell model. The other model parameters are $n' = 0.7$ and $\lambda_v v_0/L = \lambda_t v_0/L = 100$. (From Ref. 14.)

To further investigate the relationships of elongational properties and draw resonance, Iyengar and Co (15) considered a modified Giesekus model. In this model, the stress tensor $\tau$ is expressed as the sum of a Newtonian solvent contribution and a polymer contribution:

$$\tau = -\eta_s \dot{\gamma} + \tau_p \tag{10.33}$$

The polymer contribution $\tau_p$ is related to the rate-of-deformation tensor $\dot{\gamma}$ by the following equation (29):

$$
\left[ Z - \lambda_1 \frac{D \ln Z}{Dt} \right] \tau_p + \lambda_1 \tau_{p(1)} - \frac{\alpha Z}{nkT} \{ \tau_p \cdot \tau_p \}
$$
$$
= -nkT\lambda_1 \left[ \dot{\gamma} + \frac{D \ln Z}{Dt} \delta \right] \tag{10.34}
$$

The dimensionless parameter $Z$ is expressed as:

$$
Z = \frac{1}{b} \left[ b + 3 - \frac{\operatorname{tr} \tau_p}{nkT} \right] \tag{10.35}
$$

Here, $D(\ )/Dt$ [$\equiv \partial(\ )/\partial t + \mathbf{v}\cdot\nabla(\ )$] is the substantial derivative, $\boldsymbol{\tau}_{p(1)}$ is the convected derivative shown in Eq. (10.30), and $\boldsymbol{\delta}$ is the unit tensor. The parameters of the modified Giesekus model are the fluid relaxation time $\lambda_1$, the solvent viscosity $\eta_s$, the polymer number density $n$, the mobility factor $\alpha$, and the extensibility parameter $b$. The parameter $\alpha$ is related to the degree of anisotropy in the hydrodynamic drag on the polymer molecules; the drag is isotropic for $\alpha = 0$. The parameter $b$ describes the extensibility of the polymer chains; the chains are infinitely extensible for $b = \infty$. Instead of using $n$ and $\eta_s$, one can use the zero-shear rate viscosity $\eta_0$ and the retardation time $\lambda_2$ as two of the model parameters. These are given by:

$$\eta_0 = nkT\lambda_1 \frac{b+3}{b} \tag{10.36}$$

$$\lambda_2 = \lambda_1 \frac{\eta_s}{\eta_0} \tag{10.37}$$

This model can be reduced to several rheological models that have been used in the literature, as shown in a previous work of Iyengar and Co (22). It is capable of predicting a diverse range of rheological behaviors, including the elongational behavior of many polymers.

Iyengar and Co performed linear stability analysis for two-dimensional disturbances represented by Eq. (10.20). They considered four sets of model parameters: (1) $\alpha = 0.01$, $b = 10^5$; (2) $\alpha = 0.01$, $b = 10$; (3) $\alpha = 0.3$, $b = 10^5$; and (4) $\alpha = 0.3$, $b = 10$. In all cases, they considered the solvent viscosity to be negligible and set $\lambda_2 = 0$. For these model parameters, the modified Giesekus model predicts a wide range of rheological behavior in extensional flow. The elongational viscosity and shear viscosity for these sets of model parameters are shown in Fig. 10.10.

For each set of model parameters, they conducted linear stability analyses for $\lambda_1 v_0/L$ values in the range from 0.001 to 100. They found that the neutral stability curves for wave number $A = 0$ were the lowest for low $\lambda_1 v_0/L$. At higher $\lambda_1 v_0/L$, the curves for various $A$ values were indistinguishable. Hence, under the conditions they investigated, the neutral stability curves for $A = 0$ correspond to the critical draw ratio curves. These curves are shown in Fig. 10.11.

From comparison of the rheological behaviors in Fig. 10.10 and the neutral stability curves in Fig. 10.11, they related the variation of critical draw ratio with $\lambda_1 v_0/L$ to the behavior of the shear and elongational viscosity curves. Extensional thickening stabilizes flow and increases the critical draw ratio, whereas extensional thinning and shear thinning destabilize flow and decrease the critical draw ratio. The effect of model parameters on stability can be inferred from the elongational and shear viscosities. Fluids with larger

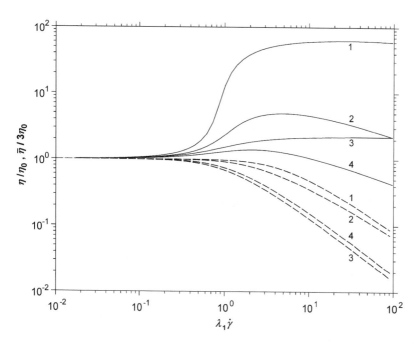

**FIGURE 10.10** The elongational viscosity curves (solid lines) and the shear viscosity curves (dashed lines) for a modified Giesekus fluid with different rheological parameters. The parameters are (1) $\alpha = 0.01$, $b = 10^5$; (2) $\alpha = 0.01$, $b = 10$; (3) $\alpha = 0.3$, $b = 10^5$; and (4) $\alpha = 0.3$, $b = 10$. (From Ref. 15.)

mobility factor $\alpha$ are more shear thinning and are therefore less stable, whereas fluids with larger chain extensibility $b$ exhibit greater extensional thickening and are therefore more stable.

As discussed previously, the presence of two critical draw ratios is predicted by the modified convected Maxwell model. This existence of two critical draw ratios is also exhibited by curve 1 in Fig. 10.11. Here the neutral stability curve has a high peak (trimmed off in Fig. 10.11) that tilts toward low $\lambda_1 v_0 / L$ values. Under these conditions, the average rates of deformation in the drawn film are equivalent to those at which the elongational viscosity curve (curve 1 in Fig. 10.10) shows a sharp rise. Similarly, the elongational viscosity curve of the convected Maxwell model also shows a sharp rise. It should be noted that the existence of the upper critical draw ratio has not been verified experimentally.

When the rise in elongational viscosity is moderate (curve 2 in Fig. 10.10), the height of the peak in the neutral stability curve is reduced (curve 2 in Fig. 10.11) and the turn back disappears. When the rise in elongational

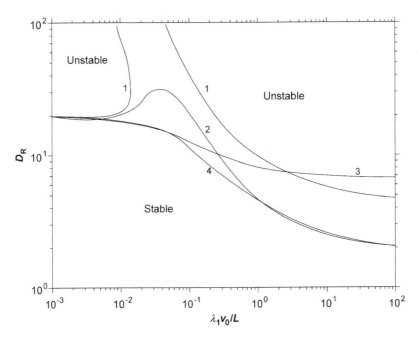

FIGURE 10.11 The neutral stability curves for a modified Giesekus fluid with different rheological parameters. The parameters are (1) $\alpha = 0.01$, $b = 10^5$; (2) $\alpha = 0.01$, $b = 10$; (3) $\alpha = 0.3$, $b = 10^5$; and (4) $\alpha = 0.3$, $b = 10$. (From Ref. 15.)

viscosity is small (curve 3 in Fig. 10.10), the neutral stability curve does not show any peak (curve 3 in Fig. 10.11). Hence the existence of two critical draw ratios is strongly linked to the steep rise in the elongational viscosity curve.

Iyengar and Co (15) also conducted nonlinear stability analysis for the modified Giesekus model. They considered the one-dimensional model of infinite film width and imposed a 10% step increase in the take-up velocity. They found that the stability limits from the nonlinear stability analysis were in agreement with the critical draw ratios predicted by the linear stability analysis. When an upper critical draw ratio was predicted by the linear stability analysis, the nonlinear analysis also indicated that the process was stable beyond the upper critical value.

They also found that initial growth or decay rate and the frequency of oscillation during the initial growth or decay of finite-amplitude disturbance agreed with those predicted by linear stability analysis. At draw ratios just above the critical value, the growth rate was low and thickness disturbance at the take-up was in the form of sinusoidal oscillation. At draw ratios further above the critical value, the growth rate was high and the thickness oscillation at the take-up exhibited narrow, sharp peaks alternating with wide, thin troughs. These behaviors are similar to the experimental observations for conditions near and beyond the critical draw ratio, as shown in Fig. 10.2.

In conjunction with his numerical simulations, Iyengar (1) also conducted film casting experiments for a low-density polyethylene (LDPE) melt, a linear low-density polyethylene (LLDPE) melt, and a polypropylene (PP) melt. A comparison of his prediction for the modified Giesekus model and his experimental data for LLDPE is shown in Fig. 10.12. The experimental data were represented by pairs of draw ratios: the lower value (open square) was the highest experimental draw ratio that maintained stable operation and the upper value (solid square) was the lowest experimental draw ratio that exhibited instability. The experiments were conducted with a drawing length of 7.0 cm and extrusion speeds ranging from 0.21 to 0.90 cm/sec. The model

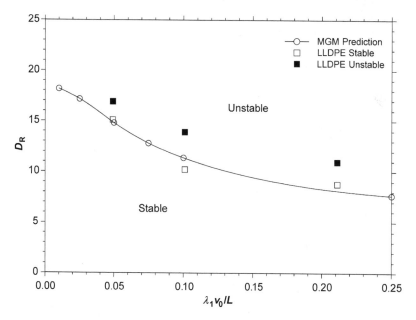

FIGURE 10.12 Comparison of the neutral stability curve for a modified Giesekus model and data for the film casting of a LLDPE melt. The model parameters are $\alpha = 0.3$, $b = 10$, and $\lambda_1 = 1.64$ sec. The experiments were conducted with a drawing length of 7.0 cm and extrusion speeds ranging from 0.21 to 0.90 cm/sec. The open squares represented the highest experimental draw ratios that maintained stable operation, and the solid squares represented the lowest experimental draw ratios that exhibited instability. (From Ref. 1.)

parameters used are $\alpha = 0.3$, $b = 10$, and $\lambda_1 = 1.64$ sec. He obtained these values by matching the experimental thickness oscillations at one draw ratio and the predicted thickness oscillations from nonlinear stability analysis of the modified Giesekus model. The elongational viscosity predicted by this set of values compared reasonably with the experimental data he obtained over a limited range of extensional rates (0.01–0.10 sec$^{-1}$), but its predicted shear viscosity showed a higher degree of shear thinning than his viscosity data in the shear rate range of 0.8–300 sec$^{-1}$. Nevertheless, the predicted neutral stability curve in the $D_R$ versus $\lambda_1 v_0/L$ plane did show good agreement with experimental data for LLDPE.

In his film casting experiments, Iyengar also observed different nonlinear behaviors for the LLDPE and PP melts. Plots of the ratio of maximum to minimum thickness at take-up versus the draw ratio for the two polymer melts are shown in Fig. 10.13. For the LLDPE melt, the thickness ratio increases drastically beyond the critical draw ratio. At a draw ratio of 34, the thickness ratio is about 37. For the PP melt, the rise of thickness ratio is fairly small. At a draw ratio of 37, the thickness ratio is only about 1.22. For the LDPE melt, Iyengar did not detect draw resonance in his experiments.

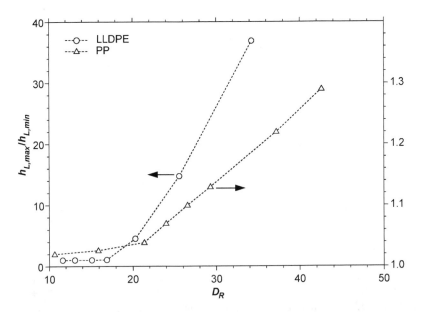

**FIGURE 10.13** The dependence of thickness ratio on the draw ratio for a LLDPE film and a PP film. The thickness measurements are made at the center of the film. (From Ref. 1.)

Another approach to simulate isothermal, one-dimensional film casting of viscoelastic fluid was described by Smith and Stolle (30). They utilized a material description of motion, rather than the usual spatial approach, and considered the occurrence of draw resonance as a response problem. The constitutive model they studied was similar to a Maxwell model but included some degree of compressibility. Their simulations suggested that a higher fluid relaxation time decreased the stability of the system. This behavior is opposite that predicted by linear stability of Maxwell fluids (14,19), which indicated that increasing fluid relaxation time enhanced system stability. They attributed the difference in prediction to the dissimilarity of the constitutive equations. They pointed out that their approach could be applied easily to more complex problems. However, they encountered difficulties with specifying the spatial boundary conditions exactly and with translating the constitution equations between the spatial and material formulations.

### 10.7.2  Viscoelastic Films with Varying Film Width

In their study of one-dimensional model with varying film width, Silagy et al. (19) also considered an upper-convected Maxwell fluid:

$$\boldsymbol{\tau} + \lambda \boldsymbol{\tau}_{(1)} = -\eta_0 \, \dot{\boldsymbol{\gamma}} \tag{10.38}$$

where the convective derivative $\boldsymbol{\tau}_{(1)}$ is defined in Eq. (10.30). They performed linear stability analysis to investigate the effects of Deborah number $\lambda v_0/L$ and aspect ratio $L/w_0$ on the stability of film casting. Figure 10.14 shows the neutral stability curves they obtained for aspect ratios ranging from $L/w_0 = 0$ (infinite film width) to $L/w_0 = 1$. The effect of aspect ratio on film stability is similar to that for Newtonian fluid shown in Fig. 10.7. At higher aspect ratio $L/w_0$, the critical draw ratio is higher and the unstable zone is smaller. They reported that the unstable zone almost disappeared when $L/w_0$ reached a value near 1.1 and then grew as $L/w_0$ was increased beyond 1.1. The increase in stability with increasing $L/w_0$ is consistent with the experimental observation of Iyengar (1). He reported that the critical draw ratio for the film casting of a LLDPE melt at an extrusion speed of 0.43 cm/sec was increased with increasing drawing length, as shown in Fig. 10.15. However, increasing the drawing length, aside from increasing the aspect ratio $L/w_0$, also decreases the Deborah number. In order to truly evaluate the effect of aspect ratio, experiments must be design to separate out the effect of Deborah number.

In a later paper, Silagy et al. (18) considered a two-dimensional model. They performed steady-state simulation using viscoelastic constitutive equations to study the effects of draw ratio, aspect ratio, and Deborah number on

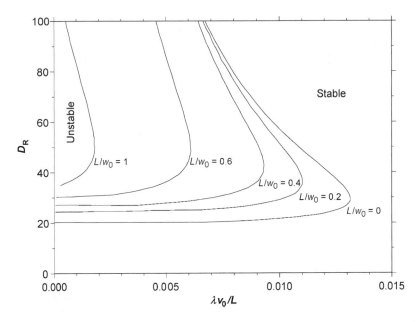

**FIGURE 10.14** The effect of the aspect ratio on the neutral stability curves for a convected Maxwell fluid. (From Ref. 19.)

the film geometry, as illustrated in Fig. 10.5. However, for two-dimensional nonlinear stability analysis, they reported only the results for Newtonian fluids, as discussed in Section 10.6.2.

Using the isothermal one-dimensional varying film width model developed by Silagy et al. (19), Lee et al. (20) studied the nonlinear dynamics of a Phan-Thien–Tanner (PTT) fluid in film casting. In this constitutive equation, the stress tensor $\tau$ is expressed as:

$$Z\tau + \lambda\tau_{(1)} + \frac{\xi}{2}\lambda\{\dot{\gamma}\cdot\tau + \tau\cdot\dot{\gamma}\} = -\eta_0\dot{\gamma} \tag{10.39}$$

$$Z = \exp(-\varepsilon\lambda\,\mathrm{tr}\,\tau/\eta_0) \tag{10.40}$$

Here $\tau_{(1)}$ is the convected derivative shown in Eq. (10.30) and $Z$ is closely related to the rate of destruction of junctions in network theory (31). The parameter $\xi$ can be determined from the shift between dynamic viscosity and shear viscosity. For polymer melts, it has a value in the order of 0.1. The value of $\varepsilon$ is determined from elongational viscosity data. It has a value in the order of 0.01. For weak flows such as viscometric flows, the model predictions are insensitive to the values of $\varepsilon$.

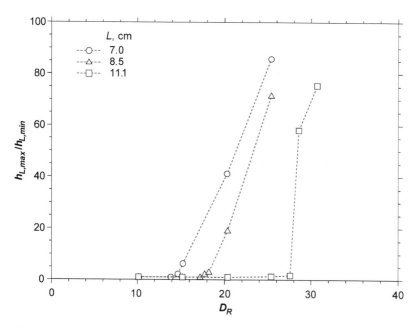

**FIGURE 10.15**   The effect of drawing length on the dependence of the thickness ratio on the draw ratio for a LLDPE film. The thickness measurements are made at the center of the film. The extrusion velocity is 0.43 cm/sec. (From Ref. 1.)

They obtained the transient simulation results for a PTT model with $\xi = 0.1$ and $\varepsilon = 0.015$ after a 5% step change in take-up velocity was imposed. They investigated the effects of Deborah number $\lambda v_0/L$ and the aspect ratio $L/w_0$ on the nonlinear transient behavior of film thickness and film width at a draw ratio of 50. Their results showed that as the Deborah number was increased from 0.003 to 0.01 at the same aspect ratio of 0.2, the magnitude of the film thickness oscillation was reduced, while that for the film width oscillation was increased. They also observed that the film width had a much smaller magnitude of oscillation than the film thickness, but the film width exhibited a much more complicated transient behavior. However, the one-dimensional varying width model that they used predicted narrower film width than the two-dimensional model, as shown in Fig. 10.5. In addition, as discussed in Section 10.2.2, the model assumed that $v_y$ varied linearly with $y$, which was inconsistent with the predictions of two-dimensional simulations of viscoelastic fluids. The complex behavior of the film width they observed maybe related to these limitations of the one-dimensional model. This conjecture can be evaluated with nonlinear stability analysis of two-dimensional model for viscoelastic fluids, which will be a challenging task.

As the aspect ratio was increased from 0.2 to 1.5 at a Deborah number of 0.01, their simulations showed that the oscillations of the film width became more severe, as indicated by the increase of the ratio of maximum film width to minimum film width. Their simulations also showed that the maximum film thickness increased with aspect ratio. However, the severity, expressed as the ratio of maximum film thickness to minimum film thickness, cannot be evaluated from their plots of film thickness versus time since these plots showed the minimum film thickness to be nearly zero. In a subsequent paper, Lee et al. (32) showed neutral stability curves similar to those from the linear stability analysis of Silagy et al. (19), as shown in Fig. 10.14. These results show that stability can be improved by increasing the aspect ratio.

Lee and Hyun (21) extended their previous work by investigating the effects of the extensional behavior of the fluids. They found that at an aspect ratio of 0.5, increasing the Deborah number enhanced the stability of extension-thickening fluids but reduced the stability of extension-thinning fluids. They also found that the stability responses for film thickness and film width were qualitatively the same.

In their nonlinear dynamic simulations, Lee et al. (20,21) also found that several kinematic waves traveled in the drawing zone due to the hyperbolic nature of the equation of continuity. Among these different waves, only the cross-sectional area ($hw$) waves and the unity-throughput ($vhw = v_0 h_0 w_0$) waves traveled the entire drawing length. They established that a relationship between the traveling times of these two waves was linked to the stability of the process. The relationship they found is given by:

$$2t_L + \frac{1}{2}T \geq 2\theta_L \text{ for } D_R \geq D_{R,\text{crit}} \tag{10.41}$$

where $t_L$ is the traveling time of the unity-throughput waves, $T$ is the period of oscillation, and $\theta_L$ is the traveling time of the cross-sectional area waves. They found the above relationship to hold for Newtonian fluids (20) and the viscoelastic PTT fluid (21). This relationship was identical to that observed in the earlier works of Kim et al. (33) and Jung et al. (34) on fiber spinning.

Using the equality in Eq. (10.41), Jung et al. (35) developed an approximate method to determine the critical draw ratio, without performing stability analysis. With certain approximations, they evaluated the terms in Eq. (10.41) using only the steady-state solution. They applied their approximate method on the modified convected Maxwell model studied by Anturkar and Co (14) and compared their neutral stability curves with those obtained from the linear stability analysis of Anturkar and Co. Their approximate stability curves had the same shape as those from linear stability analysis; however, the approximate method predicted a higher critical draw ratio.

## 10.8 STABILITY OF MULTICOMPONENT FILMS

In addition to films of single component, films consisting of multiple layers of different polymers are manufactured commercially. The layers of different polymers allow the end-use properties of the films to be tailor-made. In recent years, encapsulation dies have been used commercially to improve process-ibility. The encapsulation is achieved by extruding a second material posi-tioned along the outer edge of a core material in a slit die. In this section, we discuss the occurrence of draw resonance in these processes. Other types of instability such as the interfacial instability at the layer interfaces are not discussed.

### 10.8.1 Film Casting of Multilayer Films

In multilayer film casting, different polymers from separate extruders are fed into a feed block, where the separate streams are joined into a single multi-layer stream. This multilayer stream is then passed through a slit die to form a multilayer film. The stability of the film-casting process for multilayer films deviates from those for single-component films. The extent of the deviation depends on the rheological properties of the different polymers in the layers and the relative amount of the polymers.

Pis-Lopez and Co (36) considered a linear stability analysis of the multi-layer film casting of viscoelastic fluids. They used a modified Giesekus model, as shown in Eqs. (10.33)–(10.37), and followed the approach of Iyengar and Co (15). The overall governing equations were derived by integrating the equations for each layer across the thickness of that layer and summing up the resulting equations utilizing the interfacial conditions. The stress values at the die exit were calculated from stress ratios determined from planar elongational flow using the steady-state velocity gradient at the die exit. This steady-state velocity gradient was determined iteratively to satisfy the bound-ary condition at the take-up.

The results of their analysis indicate that the critical draw ratio in multi-layer films depends on the thickness fraction of the layers and the elonga-tional and shear viscosities of the fluid in each layer. Similar to single-layer films, extensional thickening stabilizes, whereas shear-thinning and exten-sional-thinning destabilize.

They found that the neutral stability curves for bilayer films of various thickness fractions were bounded by those for single-layer films of the two fluids. For cases in which the elongational viscosities of the two fluids were comparable, the neutral stability curves were evenly spaced between those for single-layer films and the critical draw ratio varied linearly with the thickness fraction. For cases in which one fluid had a much larger elongational viscosity than the other, the neutral stability curves were closer to that for a single-layer film of the fluid with the larger elongational viscosity. The critical draw ratio varied nonlinearly with the thickness fraction, with the values for most thickness fractions closer to the value for the single-layer film of the fluid with larger elongational viscosity. The dominating effect of the fluid with larger elongational viscosity remained even at small thickness fraction of that fluid. Hence if the dominating fluid is more stable, even a small amount of it can improve the stability of the bilayer film.

Bian (3) conducted experiments to investigate the stability of multilayer film casting. He studied one low-density polyethylene (LDPE) and two linear low-density polyethylene (LLDPE-1 and LLDPE-2) melts. The LDPE melt is extensional thickening. The extensional viscosities of the LLDPE melts are constant within the experimental range, with the LLDPE-1 melt having a higher extensional viscosity, but within the same order of magnitude, than LLDPE-2 melt. The LLDPE-1 melt is more shear-thinning than the LLDPE-2 melt. For a three-layer film of LLDPE-1/LLDPE-2/LLDPE-1 (a core layer of LLDPE-2 and even skin layers of LLDPE-1 on both sides), they found the critical draw ratio to vary linearly with the thickness fraction, as shown in Fig. 10.16. This is consistent with the theoretical predictions of Pis-Lopez and Co (36) for fluids of comparable elongational viscosities. For films that

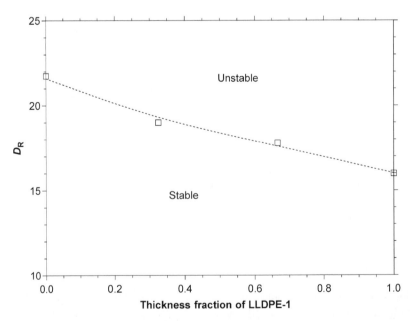

FIGURE 10.16 The dependence of the critical draw ratio on the thickness fraction for a three-layer film of LLDPE-1/LLDPE-2/LLDPE-1. (From Ref. 3.)

contained LDPE layer, they found that the process remained stable at high draw ratios. For example, for a three-layer film of LLDPE-1/LDPE/LLDPE-1 with a thickness fraction of 0.616 of LLDPE-1, they were able to operate at a draw ratio of 80 without the occurrence of draw resonance. This observation is consistent with the theoretical finding of Pis-Lopez and Co (36) on the stabilizing effect of adding an extensional-thickening layer. They also found from their experiments that switching the components of the skin layer and the core layer has no significant effect on process stability.

### 10.8.2 Film Casting with Encapsulation Die

In film casting with encapsulation die, the core material and the encapsulating material from separate extruders are fed into an encapsulation die via a feed block. The streams are redirected so that the encapsulating material is positioned along the outer edge of a core material. Frey et al. (37) conducted extrusion coating experiments to evaluate the effectiveness of encapsulation in processing linear polyolefin resins. For the core materials, they used LDPE/HDPE blends and LDPE/LLDPE blends of various concentrations. A LDPE resin was used as the encapsulating material. They found that encapsulation significantly reduced neck-in and oscillation of film width for the LDPE/HDPE blends with high concentrations of HDPE. For the LDPE/LLDPE blends, they observed only slight reduction in neck-in and film width oscillation with the use of encapsulation die.

In order to explain the effectiveness of encapsulation in stabilizing the film-casting process, Lee and Hyun (21) reported simulations they performed using a one-dimensional model with varying film width. The PTT model [Eqs. (10.39) and (10.40)] was used to describe the rheology of both the core and encapsulating materials. The shear terms of the rate of deformation tensors were included, and the shear effects at the interfaces of the two components were incorporated in the equation of motion. Using model parameters corresponding to a HDPE melt and a LDPE melt, they showed that adding more of the encapsulating LDPE at both film edges enhanced the stability of the films with HDPE as the core component.

### 10.9 CONCLUDING REMARKS

Various theoretical analyses and experimental investigations on draw resonance in film casting had been reported in the literature. Earlier works assumed the film to be of infinite width. In later studies, the film width was considered to vary in the process. For viscoelastic fluids, stability analyses indicated that the elongational behavior of the fluids played a significant role in determining the stability of the process. In general, extensional thickening has a stabilizing effect, whereas extensional thinning and shear thinning have destabilizing effects. These are consistent with reported experimental observations, e.g., those for LDPE and LLDPE melts. On the effect of aspect ratio on stability, simulation results and available experimental data indicated that increasing the aspect ratio improves stability. Investigations on multicomponent films indicate that the inclusion of a small amount of extensional-thickening material can improve the stability of the process.

To improve quantitative agreement between model predictions and experimental data in film casting, more extensive rheological data are needed. Elongational viscosity data are usually available only over limited range of extensional rates and data on planar elongational viscosity are scarce. Moreover, extensional rates and temperatures at which data are available are lower than those in film-casting operations. The availability of these data with wide range of extensional rates and temperatures will help in obtaining model parameters that are more representative of the fluids.

## REFERENCES

1.  Iyengar, V.R. Film Casting of Polymer Melts. Ph.D. dissertation, University of Maine, Orono, ME, 1993.
2.  Barq, P.; Haudin, J.M.; Agassant, J.F.; Roth, H.; Bourgin, P. Instability phenomena in film casting process. Int. Polym. Process. 1990, *V:4*, 264–271.
3.  Bian, B. Multilayer Film Casting of LLDPE and LDPE Melts. Ph.D. dissertation, University of Maine, Orono, ME, 1998.
4.  Dobroth, T.; Erwin, L. Causes of edge beads in cast films. Polym. Eng. Sci. 1986, *26*, 462–467.
5.  Canning, K.; Co, A. Edge effects in film casting of molten polymers. J. Plast. Film Sheeting 2000, *16*, 188–203.
6.  Acierno, D.; Di Maio, L.; Cuccurullo, G. Analysis of temperature fields in film casting. J. Polym. Eng. 1999, *19*, 75–94.
7.  Acierno, D.; Di Maio, L.; Ammirati, C.C. Film casting of polyethylene terephthalate: Experiments and model comparisons. Polym. Eng. Sci. 2000, *40*, 108–117.
8.  Smith, S.; Stolle, D. Nonisothermal two-dimensional film casting of a viscous polymer. Polym. Eng. Sci. 2000, *40*, 1870–1877.
9.  Lamberti, G.; Titomanlio, G.; Brucato, V. Measurement and modelling of the film casting process 1. Width distribution along draw direction. Chem. Eng. Sci. 2001, *56*, 5749–5761.
10. Lamberti, G.; Titomanlio, G.; Brucato, V. Measurement and modelling of the film casting process 2. Temperature distribution along draw direction. Chem. Eng. Sci. 2002, *57*, 1993–1996.
11. Lamberti, G.; Brucato, V.; Titomanlio, G. Orientation and crystallinity in film casting of polypropylene. J. Appl. Polym. Sci. 2002, *84*, 1981–1992.

12. Yeow, Y.L. On the stability of extending films: A model for the film casting process. J. Fluid Mech. 1974, *66*, 613–622.
13. Aird, G.R.; Yeow, Y.L. Stability of film casting of power-law liquids. Ind. Eng. Chem. Fundam. 1983, *22*, 7–10.
14. Anturkar, N.R.; Co, A. Draw resonance in film casting of viscoelastic fluids: A linear stability analysis. J. Non-Newton. Fluid Mech. 1988, *28*, 287–307.
15. Iyengar, V.R.; Co, A. Film casting of a modified Giesekus fluid: Stability analysis. Chem. Eng. Sci. 1996, *51*, 1417–1430.
16. d'Halewyu, S.; Agassant, J.F.; Demay, Y. Numerical simulation of the cast film process. Polym. Eng. Sci. 1990, *30*, 335–340.
17. Debbaut, B.; Marchal, J.M. Viscoelastic effects in film casting. Z. Angew. Math. Phys. 1995, *46*, S679–S698.
18. Silagy, D.; Demay, Y.; Agassant, J.F. Stationary and stability analysis of the film casting process. J. Non-Newton. Fluid Mech. 1998, *79*, 563–583.
19. Silagy, D.; Demay, Y.; Agassant, J.F. Study of the stability of the film casting process. Polym. Eng. Sci. 1996, *36*, 2614–2625.
20. Lee, J.S.; Jung, H.W.; Song, H.-S.; Lee, K.-Y.; Hyun, J.C. Kinematic waves and draw resonance in film casting process. J. Non-Newton. Fluid Mech. 2001, *101*, 43–54.
21. Lee, J.S.; Hyun, J.C. Nonlinear dynamics and stability of film casting process. Kor.-Aust. Rheol. J. 2001, *13*, 179–187.
22. Iyengar, V.R.; Co, A. Film casting of a modified Giesekus fluid: A steady-state analysis. J. Non-Newton. Fluid Mech. 1993, *48*, 1–20.
23. Chandrasekhar, S. *Hydrodynamic and Hydromagnetic Stability*; Oxford University Press: Oxford, 1961.
24. Drazen, P.G.; Reid, W.H. *Hydrodynamic Stability*; Cambridge University Press: London, 1981.
25. Kase, S. Studies on melt spinning: IV. On the stability of melt spinning. J. Appl. Polym. Sci. 1974, *18*, 3279–3304.
26. Bergonzoni, A.; DiCresce, A.J. The phenomena of draw resonance in polymer melts. Polym. Eng. Sci. 1966, *6*, 45–59.
27. Lucchesi, P.J.; Roberts, E.H.; Kurtz, S.J. Reducing draw resonance in LLDPE film resins. Plast. Eng. 1985, *41*, 87–90.
28. Ramos, J.I. Stability and nonlinear dynamics of planar film casting processes. Int. J. Eng. Sci. 2001, *39*, 1949–1961.
29. Wiest, J.M. A differential constitutive equation for polymer melts. Rheol. Acta 1989, *28*, 4–12.
30. Smith, S.; Stolle, D.F.E. Draw resonance in film casting as a response problem using a material description of motion. J. Plast. Film Sheeting 2000, *16*, 95–107.
31. Phan-Thien, N. A nonlinear network viscoelastic model. J. Rheol. 1978, *22*, 259–283.
32. Lee, J.S.; Jung, H.W.; Hyun, J.C. Frequency response of film casting process. Kor.-Aust. Rheol. J. 2003, *15*, 91–96.
33. Kim, B.M.; Hyun, J.C.; Oh, J.S.; Lee, S.J. Kinematic waves in the isothermal melt spinning of Newtonian fluids. AIChE J. 1996, *42*, 3164–3169.

34. Jung, H.W.; Song, H.-S.; Hyun, J.C. Draw resonance and kinematic waves in viscoelastic isothermal spinning. AIChE J. 2000, *46*, 2106–2111.
35. Jung, H.W.; Choi, S.M.; Hyun, J.C. Approximate method for determining the stability of the film-casting process. AIChE J. 1999, *45*, 1157–1160.
36. Pis-Lopez, M.E.; Co, A. Multilayer film casting of modified Giesekus fluids. Part 2. Linear stability analysis. J. Non-Newton. Fluid Mech. 1996, *66*, 95–114.
37. Frey, K.R.; Jerdee, G.D.; Cleaver, C.D. The use of encapsulation dies for processing linear polyolefin resins in extrusion coating. ANTEC 2001, 2001; 19–24.

# 11

## Fiber Spinning and Film Blowing Instabilities

**Hyun Wook Jung and Jae Chun Hyun**
Korea University, Seoul, South Korea

## 11.1 FIBER SPINNING PROCESS

### 11.1.1 Introduction

The subject of instability in fiber spinning and film blowing encompasses the entire gamut of instabilities leading to the process stoppage or unstable operations with deviations from desirable steady conditions. A host of factors ranging from various process disturbances to fundamental instabilities like draw resonance can possibly cause such stoppage or unstable operations. The category of process disturbances can cover those of input materials or of dynamic conditions, or even of geometric dimensions of the process equipment. The disturbances originating from the input materials are usually related to some sort of inhomogeneities in the input material characteristics such as inhomogeneous densities, molecular weights, etc. The group of process disturbances in dynamic conditions can include those in state variables such as process velocity, temperature, stress, etc. The final group of disturbances could be associated with imprecise geometric dimensions of the replaced parts of the equipment.

While this category of disturbances to the process undeniably has important implications in the steady operation of fiber spinning and film blowing, in this chapter, we rather focus on the other category of instabilities, namely, fundamental instabilities, which are caused by nonlinear dynamics of the processes. Particularly, we are most concerned with an intriguing instability phenomenon called draw resonance, which frequently occurs in extension deformation processes like fiber spinning, film blowing, and film casting when the drawdown ratio, defined as the ratio of the fluid velocities between the die exit and the take-up, exceeds certain critical values despite the velocity being maintained constant at both boundaries. (A thorough exposition on the instability in film casting process has been presented in Chapter 10 by Co, and also other instabilities like melt fracture exhibiting the surface irregularities on the extrudates at high extrusion rates are investigated in depth in Chapters 5–7 by Migler, Georgiou, and Dealy, respectively.) This instability is quite important industrially as well as academically because it is directly related to the productivity/profitability issue of the plants and also its theoretical analysis invariably involves fundamental understanding of the nonlinear dynamics of the processes.

In this chapter, we take up the instabilities in fiber spinning first and then in film blowing. While most of the subjects are similar in both processes, there are some distinct differences between them. First, uniaxial extension by fiber spinning is contrasted to the biaxial extension by film blowing. Moreover, numerically, transient solutions, i.e., temporal behavior of the state variables, are rather easy to obtain for fiber spinning, whereas they are exceedingly difficult to come by for film blowing largely due to the extensive nonlinearity involved in the film blowing model, resulting in no reported transient solutions in the literature until quite recently (1). In addition, there exist multiple steady states for film blowing, whereas usually single unique solutions exist for fiber spinning.

### 11.1.1.1. Description of Three Fiber Spinning Processes

Historically, there have been three different spinning methods to produce synthetic fibers: melt spinning, dry spinning, and wet spinning. Among these, melt spinning is the most important simply because it produces by far the largest volume of fibers such as nylon, polyester, and polyolefins. Acrylic fiber is probably the only major fiber of substantial volume produced by either dry or wet spinning method, which is otherwise employed primarily for producing specialty fibers such as Kevlar fibers.

For these reasons, here we discuss the melt spinning process only, in terms of its nonlinear stability dynamics and industrial implications for process productivity and product quality. It suffices for this purpose to state that in both dry and wet spinning processes, additional material balance equations are included in the typical set of governing equations of melt spinning consisting of continuity, momentum balances, and energy equations along with a suitable rheological constitutive equation for the material. The schematic diagrams of the three processes are shown in Fig. 11.1. A simplified description of the fiber spinning process along with an overview of the various flow instabilities that may occur in this process can be found in Chapter 1.

As discussed later, other morphological equations such as polymer crystallinity or orientation can also be added to the system. It is, however, important at this juncture to mention that while it is always possible to add as many equations to the model as we want, the hurdles are usually our level of understanding of the process and ability to formulate the equations. For example, certain optical characteristics of the fiber or film such as fiber luster or film clarity or any electrical, mechanical, aesthetical, or chemical character-istics of fiber or film for that matter can be adopted as additional variables on top of the usual fluid velocity, temperature, cross-sectional area, and stress variables. Of fundamental importance in the simulation is, however, how well we can formulate these equations to generate meaningful results for both theoretical and practical simulation purposes.

### 11.1.1.2. Filament Instability Modes in Spinning Processes

In this section, we briefly enumerate various filament instability modes encountered in melt spinning process. The first is the capillarity failure of filaments as described by Ziabicki (2) and Ide and White (3). This type of instability usually occurs because of the fluid surface tension inducing the breakup of filaments into drops or ligaments. The second is the ductile failure of filament described by Ide and White (4) and Chang and Lodge (5) due to the growth of a neck generated in the filament by high local stress. This type of filament instability is usually exhibited by extension thinning polymers (6,7). The third instability is the cohesive failure mode exemplified by instan-taneous reduction or break of a finite cross section exhibited by extension-thickening polymers (2–4,6,7). The above three modes of instability have been grouped to the convenient terminology of the "spinnability" of materials, i.e., the capability of materials to form fibers through spinning process, however loosely defined (2,3). The lack of this spinnability is usually manifested as filament or threadline breaks in fiber spinning, which constitutes one of the most important industrial problems directly connected to process productiv-ity and profitability.

As related to the second filament instability mode (the ductile failure) mentioned above, the phenomenon of the necklike deformation of spinline cross-sectional area occurring in high-speed spinning process has recently been studied along with the stress-induced spinline crystallization kinetics, as reviewed in Section 11.1.1.4.

The next instability called draw resonance is probably the most famous one, frequently occurring not only in melt spinning, but also in other

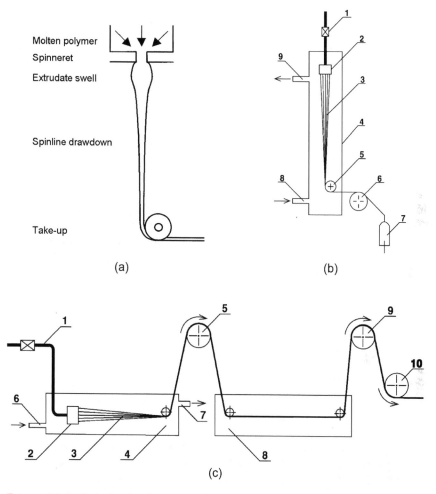

**FIGURE 11.1** Schematic diagrams of the three spinning processes. (a) Melt spinning. (b) Dry spinning: (1) metering pump, (2) spinneret, (3) spinning line, (4) drying tower, (5–7) take-up elements, (8 and 9) inlet and outlet of the drying gas. (c) Wet spinning: (1) inlet of the spinning drop, (2) spinneret, (3) spinning inlet, (4) spinning bath, (5) take-up godet, (6 and 7) inlet and outlet of the spinning bath, (8) plasticizing, (9 and 10) drawing elements. (From Ref. 2.)

extension deformation processes like film casting or film blowing where the drawing of fiber or film is involved in the environment of highly nonlinear process dynamics. This instability arises as the drawdown ratio [the ratio of velocities at the take-up position and the spinneret in Fig. 11.1(a)] is increased beyond its critical value and is manifested by sustained periodic variations in spinline variables such as the cross-sectional area and the tension. This is not a viscoelastic phenomenon because it can also appear in the process of Newtonian fluids. The draw resonance instability has aroused keen interest among and received extensive attention from researchers over the last four decades ever since it was first discovered in early 1960s by Christensen (8) and Miller (9) and named as such. The reason for this extraordinary popularity is many-fold. First, it is no doubt spectacular in its manifestation and easily observable in many processes. Second, it is seemingly inexplicable because of its oc-currence while process conditions are being maintained constant. Third, although it has been theoretically solved using linear stability analysis to be a Hopf bifurcation, its fundamental understanding has defied many efforts for so long. Finally, most importantly, it directly relates to the profitability of the industry as an essential process productivity problem in fiber spinning, film casting, and film blowing processes. It is only recently that the draw resonance criterion has been derived based on the traveling times of some spinline kinematic waves, which propagate through the spinline from the spinneret to the take-up position (10,11).

### 11.1.1.3. Historical Review of Experimental Findings

It is a formidable job to attempt to review the draw resonance literature in its entirety covering the last four decades because its enormity is simply overwhelming. Any review in this area accordingly cannot be complete in its scope or depth, and our review here also is no exception. We simply try to cite the references to the best of our knowledge, starting with experimental reports first, although many contain theoretical and experimental results together. The discovery of draw resonance instability was first made by Christensen (8) and Miller (9) on film casting and named as such, and then followed by Bergonzoni and DiCresce (12). Donnelly and Weinberger (13) and Weinberger et al. (14) first reported the onset of draw resonance on isothermal spinning of Newtonian silicone oil at the critical drawdown ratio near 20 and further determined the similar critical values for viscoelastic polymer materials as well, as shown in Fig. 11.2. Then many groups had contributed experimental findings subsequently. Most notable among them were Ishihara and Kase (15–17) in Japan, Bogue et al. (18–20) in Tennessee, White et al. (3,4,6,21–24) in both Tennessee and Akron, and Han et al. (25–27) at Polytechnic. Figure 11.3 shows one example of those results, illustrating the

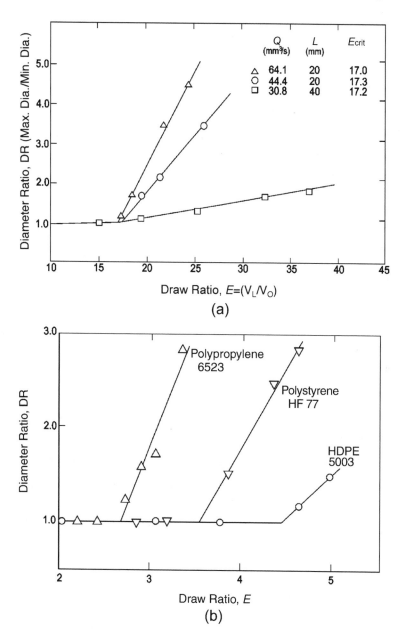

**FIGURE 11.2** Determination of critical drawdown ratio of (a) silicone oil and (b) PP, PS, and HDPE. (From Refs. 13 and 14.)

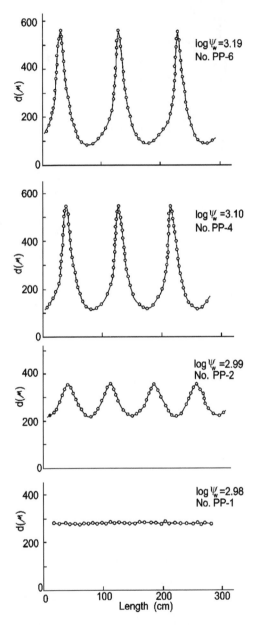

**FIGURE 11.3** Experimentally measured waveform of draw resonance of PP where d($\mu$) and $\psi_w$ denote spinline diameter and drawdown ratio, respectively. (From Ref. 17.)

periodic oscillations of draw resonance experimentally observed with different drawdown ratios (in this case, $\psi_w$).

### 11.1.1.4. Historical Review of Theoretical Results

The results of theoretical nature in the draw resonance literature are legion. Ever since the first historical effort to model the melt spinning process by Kase and Matsuo (28), there has been a continuous stream of research results coming out of various groups around the world. The first such successful results came from Pearson et al. (29–35) in the United Kingdom in the area of stability starting with isothermal Newtonian fluids and then extended to nonisothermal and viscoelastic cases. They put the governing equations to a rigorous mathematical analysis and obtained the first stability results. Gelder (36) first solved the stability eigenmatrix for a simple isothermal Newtonian spinning. Then came along Denn et al. (37–41) at Delaware who systematically tackled the stability problem of melt spinning and successfully obtained linear stability results. Figure 11.4 shows one example by Fisher and Denn (38), portraying the different stability regions determined by the Deborah number and the power-law index of the fluids. Kase et al. (42–44) in Japan also contributed significantly to the understanding of spinning stability by solving the hyperbolic equations of spinning rendered by the Lagrangian coordinate transformation. Beris at Delaware obtained many useful stability results for upper convective Maxwell (UCM) fluids, and they also dealt with the inertia effect and numerical aspects of the simulation (45,46).

As for the transient simulation of spinning, there have been several important results in the literature: Hyun et al. (10,11) at Korea University, Ishihara and Kase (15–17), Iyengar and Co (47) at Maine, and others (48,49). Sensitivity analysis of the various effects on spinning also has made progress (50–53). A rigorous perturbation method was pioneered and applied by Bechtel et al. (54–60) at Ohio State to the various spinning cases including nonisothermal, liquid crystal polymers to come up with many useful stability results. Interesting mathematical points of view on the spinning have also been presented by Renardy (61,62). The separate issue of coextrusion spinning stability was made much advance for the case of bicomponent spinning by Lee and Park (63) and Ji et al. (64).

The spinning with spinline crystallization and the related spinline necking phenomena has also received a lot of attention, resulting in many valuable research reports (65). Kikutani et al. (66,67) at Tokyo Institute of Technology elucidated the so-called necklike deformation observed on the high-speed spinline through a series of elaborate experiments by relating it to the spinline crystallization. As for the modeling of spinline crystallization, there have been notable achievements (68–76). Kulkarni and Beris (70) used an inhomogeneous structural model with crystallization kinetics to explain the mechanism of the necklike deformation in high-speed spinning. Of

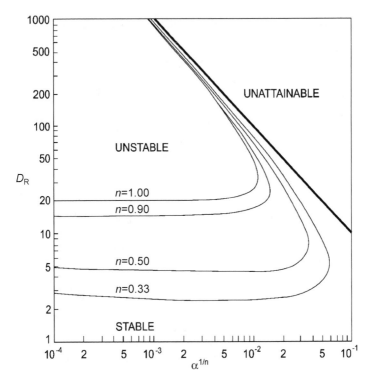

**FIGURE 11.4** Stability diagrams of White–Metzner model 1 with power-law viscosity where $\alpha^{1/n}$ and $D_R$ denote a dimensionless Deborah number and drawdown ratio, respectively. (From Ref. 38.)

particular importance is Doufas and McHugh (71–74) who, for the first time, demonstrated along with many valuable results the spinline necklike deformation using their newly devised, quite effective flow-induced crystallization model in their simulation. Lately, Joo et al. (75) successfully simulated nonisothermal spinline crystallization using a two-dimensional model.

The general review papers (34,77–83) and textbooks (84–92) on the instability subject of fiber spinning and other polymer processing have appeared in parallel with the above research progress following the first excellent review covering the whole gamut of the subject by Petrie and Denn (93).

### 11.1.2 Modeling of the Spinning Process

The modeling of melt spinning typically involves several balance equations formulated on the spinline, comprising continuity, motion, energy, and crystallization mechanisms, along with a suitable rheological constitutive equation for the material to be spun. These governing equations are expressed in a set of partial differential equations either in one- or two-dimensional spatial coordinates depending on whether or not the cross section of the filament is modeled into the system. Fixed boundary conditions are usually given in filament velocity, cross-sectional area, and temperature at the spinneret and in filament velocity at the take-up. The extrudate swell at the die might be included into the model if the distance coordinate begins at some point in the capillary upstream the spinneret, considering the prehistory of deformation happening inside the capillary. In this case, an integral constitutive equation such as K-BKZ model could be employed as described by Papanastasiou et al. (94). Otherwise, a differential constitutive equation is usually employed for modeling with the origin of the distance coordinate at the position of extrudate swell where the swell size becomes the initial cross-sectional area of the filament.

The dimensionless governing equations of melt spinning for the simplest case of isothermal Newtonian fluids spinning are as follows when all the secondary forces on the spinline such as inertia, gravity, surface tension, and air drag are neglected (10,15,29,36,37,41,65).

Continuity equation:

$$\frac{\partial a}{\partial t} + \frac{\partial (av)}{\partial x} = 0 \text{ where } t = \frac{\bar{t}\bar{v}_0}{L}, \quad x = \frac{\bar{x}}{L}, \quad a = \frac{\bar{a}}{\bar{a}_0}, \quad v = \frac{\bar{v}}{\bar{v}_0} \quad (11.1)$$

Equation of motion:

$$\frac{\partial (a(\tau_{xx} - \tau_{rr}))}{\partial x} = 0 \text{ where } \tau_{ij} = \frac{\bar{\tau}_{ij}L}{\eta_0 \bar{v}_0} \quad (11.2)$$

Constitutive equation (Newtonian model):

$$\tau_{xx} = 2\frac{\partial v}{\partial x}, \quad \tau_{rr} = -\frac{\partial v}{\partial x} \quad (11.3)$$

Boundary conditions:

$$a = 1, \quad v = 1, \quad \underline{\underline{\tau}} = \underline{\underline{\tau}}_0 \qquad \text{at } x = 0 \text{ for all } t \quad (11.4)$$

$$v = r \qquad \qquad \text{at } x = 1 \text{ for all } t \quad (11.5)$$

where $a$ denotes a dimensionless spinline cross-sectional area, $v$ is a dimensionless spinline velocity, $t$ is a dimensionless time, $x$ is a dimensionless spinline distance, $\tau_{xx}$ is a dimensionless axial stress, $\tau_{rr}$ is a dimensionless radial stress, $\eta_0$ is the zero shear viscosity, $L$ is the distance from spinneret to take-up, and $r$ is the drawdown ratio. Overbars denote dimensional variables. Subscript 0 denotes spinneret conditions.

For portraying more complex situations, we can include additional equations as required depending on the particular conditions to reckon with. For nonisothermal spinning, for example, the energy equation is added to account for various heat transfer modes arising in the system such as axial heat convection, radial cooling to the ambient air, radiation, and crystallization heat (59,60,65,68).

Energy equation:

$$
\begin{aligned}
\frac{\partial \theta}{\partial t} + v\frac{\partial \theta}{\partial x} = &-St\frac{v^{1/3}}{a^{5/6}}(\theta - \theta_a)\left(1 + \left(8\frac{v_y}{v}\right)^2\right)^{1/6} \\
&+ C_R\frac{1}{a^{1/2}}(\theta^4 - \theta_a^4) + C_H\left(\frac{\partial \phi}{\partial t} + v\frac{\partial \phi}{\partial x}\right)
\end{aligned}
\tag{11.6}
$$

$$
\text{where } St = \frac{1.67 \times 10^{-4}L}{\rho C_p \bar{a}_0^{5/6}\bar{v}_0^{2/3}}, \quad \theta = \frac{\bar{\theta}}{\bar{\theta}_0}, \quad C_R = \frac{\varepsilon_m \sigma_{SB}\bar{\theta}_0^3 L}{\rho C_p \bar{v}_0 \sqrt{\bar{a}_0}},
$$

$$
\phi = \frac{\bar{\phi}}{\bar{\phi}_\infty}, \quad C_H = \frac{\Delta H_f \bar{\phi}_\infty}{C_p \bar{\theta}_0}
$$

where $\theta$ denotes a dimensionless spinline temperature, $\theta_a$ is a dimensionless ambient temperature, $St$ is the Stanton number, $C_R$ is the dimensionless radiation coefficient, $v_y$ is a dimensionless cooling air velocity, $\rho$ is the fluid density, $C_p$ is the fluid heat capacity, $\Delta H_f$ is the crystallization heat, $C_H$ is the dimensionless crystallization heat, $\phi$ is a dimensionless crystallinity, $\bar{\phi}_\infty$ is the maximum crystallinity, $\varepsilon_m$ is the emissivity, and $\sigma_{SB}$ is the Stefan–Boltzman constant. Overbars denote dimensional variables and subscript 0 denotes spinneret conditions.

Next, the crystallinity equation taking care of nonisothermal, stress-induced crystallization occurring on the spinline can be added. An example can be found in Refs. 2, 65, 68, 70, and 75.

Crystallinity equation:

$$
\begin{aligned}
\frac{\partial \phi}{\partial t} + v\frac{\partial \phi}{\partial x} = &nK_m(1 - \phi)\left[\ln\left(\frac{1}{1 - \phi}\right)\right]^{(n-1)/n} \\
&\times \exp\left(-4\ln 2\left(\frac{\theta - \theta_{max}}{\theta_{hw}}\right)^2 + C_{cr}(C_{op}\tau)^2\right)
\end{aligned}
\tag{11.7}
$$

$$
\text{where } \theta_{max} = \frac{\bar{\theta}_{max}}{\bar{\theta}_0}, \quad K_m = \frac{K_{max}L}{\bar{v}_0}, \quad C_{op} = \frac{\bar{C}_{op}\eta_0\bar{v}_0}{\Delta_a L}
$$

where $\theta_{max}$ denotes the dimensionless temperature at maximum crystallization rate, $K_{max}$ is the maximum crystallization rate, $K_m$ is the dimensionless

$K_{\max}$, $\bar{C}_{op}$ is the stress optical coefficient, $C_{op}$ is the dimensionless $\bar{C}_{op}$, $\Delta_a$ is the intrinsic amorphous birefringence, $C_{cr}$ is the stress-induced crystallization parameter, $\theta_{hw}$ is the dimensionless crystallization temperature half width, and $n$ is Avrami index. Overbars denote dimensional variables.

The so-called secondary forces acting on the spinline on top of the rheological force can also be taken into consideration by adding each term to Eq. (11.2): inertia, gravity, air drag, and surface tension. The equation of motion fully loaded with these force terms is shown below (65).

Equation of motion (including the secondary forces):

$$C_{in}\left(\frac{\partial v}{\partial t} + v\frac{\partial v}{\partial x}\right) = \frac{1}{a}\frac{\partial(a(\tau_{xx} - \tau_{rr}))}{\partial x} + C_{gr} - C_{ad}v^{1.19}a^{-0.905}$$

$$+ C_{st}\frac{1}{a^{3/2}}\frac{\partial a}{\partial x} \tag{11.8}$$

where $C_{in} = \dfrac{\rho\bar{v}_0 L}{2\eta_0}$, $\quad C_{gr} = \dfrac{\rho g L^2}{2\eta_0\bar{v}_0}$, $\quad C_{ad} = \dfrac{3.122 \times 10^{-4}\bar{v}_0^{0.19}L^2}{2\eta_0\bar{a}_0^{0.905}}$,

$$C_{st} = \frac{\sqrt{\pi}HL}{2\sqrt{\bar{a}_0}\eta_0\bar{v}_0}$$

where $C_{in}$ indicates a dimensionless inertia coefficient, $C_{gr}$ is a dimensionless gravity coefficient, $C_{ad}$ is a dimensionless air drag coefficient, $C_{st}$ is a dimensionless surface tension coefficient, $g$ is the gravitational constant, and $H$ is the surface tension.

Finally, the rheological constitutive equation capable of adequately representing the viscoelasticity of materials in the model can take on many different forms since there is no unique constitutive equation to describe adequately the rheological behavior of polymeric liquids. Some of the typical examples are listed below (11,38,47,94–100).

White–Metzner model (11,38):

$$\text{Model 1}: \underline{\underline{\tau}} + \tilde{\lambda}_1\left(\frac{\partial\underline{\underline{\tau}}}{\partial t} + \underline{v}\cdot\underline{\nabla}\,\underline{\underline{\tau}} - \underline{\nabla}\,\underline{v}\cdot\underline{\underline{\tau}} - \underline{\underline{\tau}}\cdot(\underline{\nabla}\,\underline{v})^T\right) = 2\tilde{\eta}\underline{\underline{D}} \tag{11.9a}$$

where $\tilde{\lambda}_1 = 3^{(p-1)/2}\dfrac{K}{G}\left(\dfrac{\bar{v}_0}{L}\right)^p\left(\dfrac{\partial v}{\partial x}\right)^{p-1}$,

$\tilde{\eta} = 3^{(p-1)/2}\dfrac{K}{\eta_0}\left(\dfrac{\bar{v}_0}{L}\right)^{p-1}\left(\dfrac{\partial v}{\partial x}\right)^{p-1}$, $\quad 2\underline{\underline{D}} = (\underline{\nabla}\,\underline{v} + \underline{\nabla}\,\underline{v}^T)$

where $\underline{\underline{\tau}}$ denotes an extra stress tensor, $\underline{\underline{D}}$ is a strain rate tensor, $\eta_0$ is the zero-shear viscosity, $G$ is the modulus, $K$ is the coefficient in power-law viscosity,

$p$ is the power-law index, and $\tilde{\lambda}$ and $\tilde{\eta}$ are the power-law strain rate-dependent dimensionless relaxation time and viscosity, respectively.

$$\text{Model 2}: \underline{\underline{\tau}} + \tilde{\lambda}_2 \left( \frac{\partial \underline{\underline{\tau}}}{\partial \tau} + \underline{v} \cdot \nabla \underline{\underline{\tau}} - \nabla \underline{v} \cdot \underline{\underline{\tau}} - \underline{\underline{\tau}} \cdot (\nabla \underline{v})^T \right) = 2 \frac{\tilde{\lambda}_2}{De} \underline{\underline{D}} \qquad (11.9b)$$

$$\text{where } \tilde{\lambda}_2 = \frac{De}{1 + b\sqrt{3}De(\partial v / \partial x)}, \qquad De = \frac{\lambda_0 \bar{v}_0}{L}$$

where $\tilde{\lambda}_2$ denotes another strain rate-dependent dimensionless relaxation time, $\lambda_0$ is the relaxation time at zero strain rate, $De$ is the Deborah number, and $b$ is the parameter representing the strain rate dependency of relaxation times.

Phan-Thien–Tanner model (95–98):

$$K \underline{\underline{\tau}} + \lambda \left( \frac{\partial \underline{\underline{\tau}}}{\partial t} + \underline{v} \cdot \nabla \underline{\underline{\tau}} - \underline{\underline{L}} \cdot \underline{\underline{\tau}} - \underline{\underline{\tau}} \cdot \underline{\underline{L}}^T \right) = 2 \underline{\underline{D}} \qquad (11.10)$$

$$\text{where } K = \exp(\varepsilon De \operatorname{tr} \underline{\underline{\tau}}), \quad \underline{\underline{L}} = \nabla \underline{v} - \xi \underline{\underline{D}}$$

where $\underline{\underline{L}}$ denotes an effective strain rate tensor and $\varepsilon$ and $\xi$ denote PTT model parameters.

Giesekus model (47,99):

$$\underline{\underline{\tau}} + \gamma De \underline{\underline{\tau}}^2 + De \left( \frac{\partial \underline{\underline{\tau}}}{\partial t} + \underline{v} \cdot \nabla \underline{\underline{\tau}} - \nabla \underline{v} \cdot \underline{\underline{\tau}} - \underline{\underline{\tau}} \cdot (\nabla \underline{v})^T \right) = 2 \underline{\underline{D}} \qquad (11.11)$$

where $\gamma$ is a network mobility parameter.

K-BKZ model (94,100):

$$\underline{\underline{\tau}} = \int_{-\infty}^{t} \sum_k \frac{a_k}{\lambda_k} e^{-(t-t')/\lambda_k} \frac{\alpha}{(\alpha - 3) + \beta I_{\underline{\underline{B}}} + (1 - \beta) II_{\underline{\underline{B}}}} (\underline{\underline{B}}_t(t') - \underline{\underline{I}}) dt' \qquad (11.12)$$

where $\underline{\underline{B}}$ denotes the Finger deformation tensor, $a_k$ is the relaxation coefficient, $\alpha$ is the shear material parameter, $\beta$ is the extensional material coefficient, $\lambda_k$ is the relaxation time, and $I_{\underline{\underline{B}}}$ and $II_{\underline{\underline{B}}}$ are the first and second invariants of $\underline{\underline{B}}$, respectively. In nonisothermal cases, above constitutive models could include temperature-dependent relaxation times and viscosity functions.

Two comments are in order here. First, as with these constitutive equations, the filament cooling term in the energy equation and the crystallization term in both energy and crystallinity equations could take on many different expressions depending on which particular mechanisms are adopted. The other comment is that for portraying possible distributions of temperature, stress, velocity, and crystallinity across the spinline cross section, a two-dimensional model (both distance and radius being the spatial coordinates) could be adopted. Particularly, as related to the so-called skin-core structure inside the produced fiber, two-dimensional models are warranted for non-isothermal systems (59,60,74,75). While the detailed simulation results are different with all these different models, the stability question of the spinning process, the focus of this chapter, is fundamentally the same for all models.

### 11.1.2.1. Numerical Simulations

There are basically two categories in the numerical simulation of spinning processes: steady-state and transient solutions. In general, the steady-state solutions are rather easily available with conventional numerical schemes such as finite difference, finite element (75,94,101), or orthogonal collocation methods (102). Transient solutions need more involved numerical schemes (1,10,11,15–17,43,47–49,83,96,103).

### 11.1.3  Stability

### 11.1.3.1.  Linear Stability Analysis

Now we move into the realm of stability of spinning, the main theme of the chapter. The linear stability analysis adopting eigenvalue calculation procedures is dealt with first followed by the frequency response. As shown below, the linear stability analysis typically involves substituting the perturbation variables into the governing equations, followed by linearization to get an eigenmatrix system, and finally obtaining the critical values of (bifurcation) parameters that render the real part of the largest eigenvalue to vanish. These critical values will then demarcate the onset of the instability.

The governing equations are compactly represented in a vector residual form.

$$\underline{R}(\underline{q}, \dot{\underline{q}}) = \underline{0} \qquad (11.13)$$

where $\underline{q} = \underline{q}(\underline{a}, \underline{v}, \underline{\tau}_{xx}, \underline{\tau}_{rr}, \underline{\theta})$ is the state variable vector of spinline area, spinline velocity, axial stress, radial stress, and spinline temperature. The vector $\dot{\underline{q}}$ denotes the time derivative of $\underline{q}$.

At steady state, $\dot{\underline{q}} = 0$. The steady solution is $\underline{q} = \underline{q}_s$ from $\underline{R}(\underline{q}_s, \underline{0}) = \underline{0}$. The temporal perturbations are then introduced onto these steady profiles of state variables to get the following perturbation variables.

$$\underline{q}(t, x) = \underline{q}_s(x) + \underline{\delta q}(t, x) \qquad (11.14)$$

where $\underline{\delta q}$ represents the perturbed quantities of states variables.

On substituting these perturbation variables into the nondimensionalized governing equations, Eq. (11.13) becomes as follows after nonlinear terms neglected.

$$\underline{J}\,\delta q - \underline{\underline{M}}\,\delta \dot{q} = \underline{0} \tag{11.15}$$

where $\underline{J} \equiv \partial \underline{R}/\partial \underline{q}$ and $\underline{\underline{M}} \equiv -\,\partial \underline{R}/\partial \underline{\dot{q}}$ are the Jacobian and mass matrices, respectively.

Next, after replacing $\delta q(t, x)$ by $\underline{y}(x)e^{\Omega t}$, where $\Omega$ is a complex eigenvalue that accounts for the growth rate of the perturbation, and using a proper finite difference scheme, Eq. (11.15) finally becomes an eigenmatrix system.

$$\Omega \underline{\underline{M}}\,\underline{y} = \underline{J}\,\underline{y} \tag{11.16}$$

Now given the process conditions including the crucial take-up velocity, the same as the drawdown ratio, $r$, this eigenvalue problem is to be solved for $\Omega$. If the real part of the largest eigenvalue of $\Omega$ for a given $r$ is found positive, then the system is determined unstable because this means that any initial disturbance introduced in the perturbation variables of Eq. (11.14) grows unbounded with time. The eigenmatrix equation of Eq. (11.16) is, however, usually solved in a converse manner: solved for the critical $r_c$ which makes the real part of the largest eigenvalue vanish, signaling the onset of draw resonance. There have been other methods (37–41) to solve the differential equations directly without going through the eigenmatrix formulation. This method involves an assumed scalar value for $\Omega$ and the Runge–Kutta method with a shooting scheme, focusing on the largest eigenvalue instead of its entire vector, resulting in the same stability data.

Based on the results obtained through either method, various stability diagrams can be drawn for the spinning process as evidenced by many reports in the literature (20,32,33,37–41,46,53,63,64,83,96,104–106). Figure 11.5 displays an example of nonisothermal stability diagram (96,104).

### 11.1.3.2. Stability Analysis Using Frequency Response

The usual frequency response method puts sinusoidal perturbation input into the linearized system and then determines the onset of instability using stability conditions like the Nyquist criterion. The results by Kase and Araki (50) along this line were quite significant in that on top of the usual stability results, their linear transfer function analysis clearly showed the pivotal role played by the spinline tension in deciding the stability of spinning process. This point will be further illustrated later when we discuss the sensitivity analysis of the spinning. Useful results by the frequency response method have been reported recently (51,107). In Fig. 11.6, the amplitude of the take-up cross-sectional area is plotted against the perturbation frequency when different disturbances are introduced for the Phan-Thien–Tanner fluid spinning.

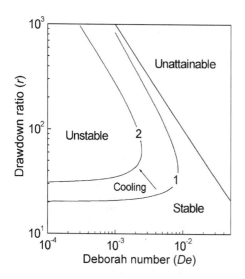

**FIGURE 11.5** Stability diagrams of Maxwell fluid. (1) Isothermal case, (2) non-isothermal case, $St = 0.1$, $\theta_a = 0.543$, $k = 5.8$. (From Ref. 104.)

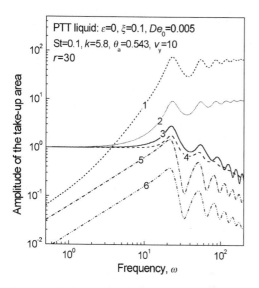

**FIGURE 11.6** Amplitudes of spinline cross-sectional area at take-up for non-isothermal PTT spinning with various disturbances introduced in extrusion temperature (1), spinneret area (2), take-up velocity (3), extrusion velocity (4), cooling air temperature (5), and cooling air velocity (6). (From Ref. 107.)

### 11.1.3.3. Stability Criterion Based on Traveling Kinematic Waves

In the foregoing sections, the spinning stability was discussed with focus on how the onset of the draw resonance instability could be precisely determined. Using the linear stability analysis method, either eigenmatrix or frequency response, the critical conditions triggering the onset can be rather easily obtained. Examining the stability diagrams in detail, we also can discern the way changes in particular process conditions influence the stability, i.e., stabilize or destabilize. However, despite the success of these stability analyses in determining the onset of draw resonance, there still remains the fundamental question: the physics behind the phenomenon as to why such instability arises is not clear. This stems from the fact that the linear stability analysis explained above only finds the unstable eigenmodes derived from the artificially meshed system, but does not answer what fundamental mechanism is involved to trigger the onset of the draw resonance.

In an effort to answer this perplexing question, Hyun et al. (10,11,105, 106,108–114) at Korea University has been approaching the problem from a different viewpoint. That is, focusing on the hyperbolic nature of the system, the behavior of the various kinematic waves traveling on the spinline from the spinneret to take-up is scrutinized to reveal whether there exist any relationships among the traveling times of these kinematic waves (10,11). It was found that traveling times from the spinneret to the take-up of the unity throughput waves (the waves where the product of the spinline cross-sectional area times the spinline velocity has the value of unity) and the maximum and minimum cross-sectional area waves (the spinline cross-sectional area waves which maintain the maximum and minimum values of the cross-sectional area at every point of the spinline) satisfy a certain relationship. (Further details of the traveling kinematic waves in the spinning process, a typical hyperbolic system, have been fully described in Ref. 10.) In other words, the comparison of the traveling times of these waves reveals that their relative magnitudes reverse the signs at the onset point of draw resonance, unequivocally meaning that there is an equation playing the role of the draw resonance criterion, as shown in Eq. (11.17).

$$(t_L)_1 + (t_L)_2 + \frac{T}{2} \gtreqless (\theta_L)_1 + (\theta_L)_2 \qquad \text{for } r \lesseqgtr r_C \qquad (11.17)$$

where $t_{L1}$ and $t_{L2}$ denote the traveling times of the unity-throughput waves on the descending and ascending curves of the throughput, respectively, $\theta_{L1}$ and $\theta_{L2}$ denote the traveling times of maximum (peak) and minimum (trough) cross-sectional area waves, respectively, $T$ is the period of oscillations, and $r_c$ is the drawdown ratio at onset point of draw resonance.

Here $t_L$ and $\theta_L$ are defined as

$$t_L = \int_0^1 \frac{dx}{U}, \quad \theta_L = \int_{x1}^1 \frac{dx}{W} \tag{11.18}$$

where $U$ and $W$ are the velocities of the unity throughput waves and the extremum (maximum or minimum) cross-sectional area waves, respectively, and $x_1$ denotes the first node in $x$-coordinate where the extremum waves first start. (At the spinneret, there exist no maximum or minimum cross-sectional area waves.)

Moreover, this draw resonance criterion of Eq. (11.17) has a theoretical meaning. If the drawdown ratio is less than the critical value, the left-hand side is larger than the right, meaning that the required traveling times for the two successive unity throughput waves plus one-half of the oscillation period is larger than the allowed times given by two successive cross-sectional area (peak and trough) waves, resulting in stability with any disturbances dying out with time. When the drawdown ratio takes on the critical value, the two sides become identical, meaning that the two unity throughput waves can travel from the spinneret to take-up while the two extremum cross-sectional area waves do the same. This illustrates that steady oscillations in draw resonance can occur at this critical onset point with its amplitude necessarily being infinitesimally small. If the drawdown ratio is larger than this critical value, the two unity throughput waves can always travel the spinning distance while one peak (maximum) and one trough (minimum) cross-sectional area waves do the same, resulting in persisting oscillation. These oscillations get progressively larger with sharp spikes as the drawdown ratio, $r$, gets larger, i.e., the deviation increasing from the onset point.

The reason why the unity throughput waves are chosen from the infinitely many throughput waves for the criterion equation is that they are the only waves that can travel the entire distance from the spinneret to take-up. Because of the unity boundary conditions for the throughput at the spinneret, only unity throughput waves start from the spinneret to travel toward the take-up, while all other nonunity throughput waves start somewhere down on the spinline where their nonunity values are first attained. Tracking down the unity throughput waves and the peak/trough cross-sectional area waves as in Figs. 11.7 and 11.8, the traveling times of these waves can be exactly computed to ascertain the criterion [details in Kim et al. (10) and Jung et al. (11)]. Figure 11.9 clearly shows the reversal of the sign of Eq. (11.17) at the onset point. This same draw resonance criterion equally applies to film blowing and film casting processes (1,108–111), confirming that the same physics and mechanism explain the draw resonance occurring in all these extensional deformational processes. It can be said that in the hyperbolic systems like these

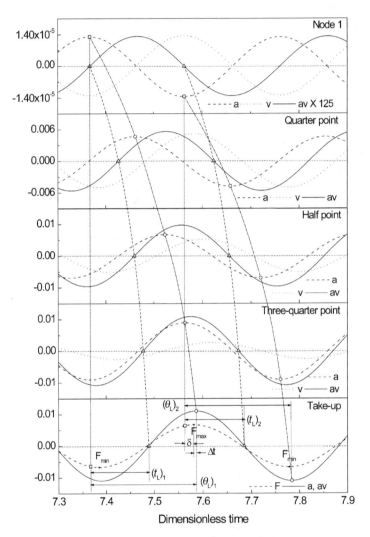

**FIGURE 11.7** Transient curves of spinline variables at five different spatial points of the spinline when $b = 0.4$, $De = 0.019$, $r = r_C = 27.97$ for White–Metzner model 2. (From Ref. 11.)

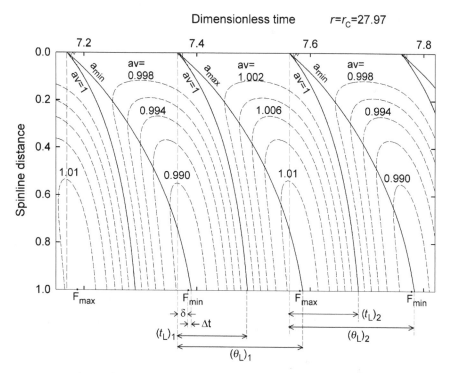

**FIGURE 11.8** Contours of constant $AV$ curves when $b = 0.4$, $De = 0.019$, $r = r_C = 27.97$ for White–Metzner model 2. (From Ref. 11.)

three, bifurcation occurs due to the interactions of the traveling kinematic waves when two successive throughput waves can travel the spinning distance while two successive cross-sectional area waves do the same, which is always possible if the drawdown ratio exceeds certain critical values.

The reason why one peak and one trough cross-sectional area wave come on each other's heels is because of the role spinline tension plays in transmitting the force from take-up to the spinneret. In other words, when a peak (maximum) wave starts at the spinneret, travels toward the take-up, and goes out the system through the take-up, the spinline tension becomes maximum causing a new trough (minimum) wave to appear at the spinneret. This wave then starts to travel toward the take-up and the spinline tension becomes minimum when this wave passes through the take-up, causing another peak (maximum) wave to appear at the spinneret. The whole cycle then repeats itself in a form of draw resonance. This mechanism of persistent draw resonance oscillations was explained in detail in Refs. 10 and 11.

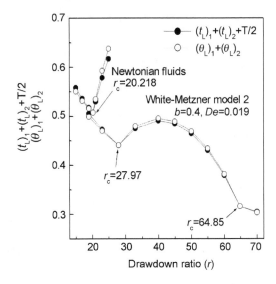

FIGURE **11.9** Required traveling time for two successive unity throughput waves, and allowed traveling time by $A_{max}$ and $A_{min}$ waves plotted against drawdown ratio $r$. (From Ref. 11.)

### 11.1.3.4.  Useful Approximate Method for Determining Draw Resonance

Finally, an interesting point as regards the criterion equation of Eq. (11.17) is the fact that it is always possible to derive its useful, approximate version (112,113). This is based on two things: one is that the traveling time of the unity throughput waves can be approximated by a simple equation involving the drawdown ratio, and the other is that the traveling times of two successive unity throughput waves is approximately equal to the fluid residence time or the traveling time of fluid elements from spinneret to take-up. Combining these facts, Eq. (11.17) can be replaced by its approximate version below (details in Ref. 113).

$$2t_L \approx \frac{2 \ln r}{r - 1} \approx \tau_L \text{ at } r = r_C \text{ (onset of draw resonance)} \qquad (11.19)$$

where the fluid residence time, $\tau_L$, is defined as

$$\tau_L = \int_0^1 \frac{dx}{v} \qquad (11.20)$$

It turns out that this deceivingly simple relation for approximately determining the onset of draw resonance yields fairly accurate results in fiber spinning, film casting, and film blowing. Figure 11.10 demonstrates one example in film casting process (113) where the approximate results by Eq. (11.19) and the exact ones by the linear stability analysis are contrasted. Figure 11.11 shows the same comparisons in fiber spinning and film blowing cases (83,114). Beyond the observations that the approximate results by Eq. (11.19) are quite close to the exact ones lies another important point. That is, the computations involved in Eq. (11.19) are extremely simple; we do not need to solve the partial differential equations of Eqs. (11.1)–(11.12) in their entirety nor any linear stability equations like Eq. (11.16) to approximately determine the onset of draw resonance, but instead, we only need the easily obtainable steady-state solutions and the fluid residence time. This fact provides a powerful conclusion regarding the utility of the approximate method of Eq. (11.19); for all practical purposes in industry and academia, the onset of draw resonance in fiber spinning, film casting, and film blowing can be quickly determined with reasonable accuracy usually within the errors of a few percent (113).

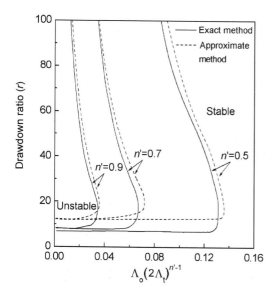

**Figure 11.10** Comparison of stability results between the exact and approximate methods for viscoelastic film casting where $\Lambda_0(2\Lambda_t)^{n'-1}$ denotes a dimensionless Deborah number. (From Ref. 113.)

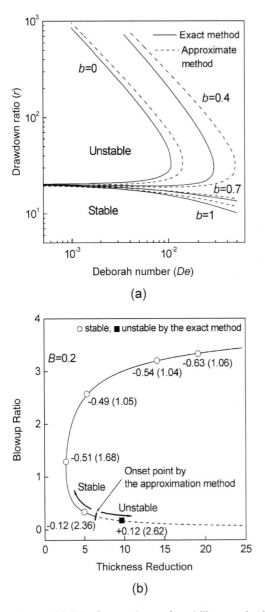

**FIGURE 11.11** Comparison of stability results between the exact and approximate methods for (a) White–Metzner spinning and (b) Newtonian film blowing. (From Ref. 83.)

### 11.1.3.5. Nonlinear Stability and Transient Simulation

The transient simulation of the spinning process can be carried out integrating the governing equations expressed by Eqs. (11.1)–(11.12) using one of appropriate numerical schemes mentioned before. Depending on whether the drawdown ratio, $r$, is smaller, equal to, or larger than the critical value, the system exhibits one of the three pictures, i.e., stable, the onset, or draw resonance limit cycle behavior, as shown in Fig. 11.12 where temporal pictures of the cross-sectional area at take-up are displayed. Figure 11.13 shows the same behavior in the phase space trajectories. Clearly, the draw resonance occur-

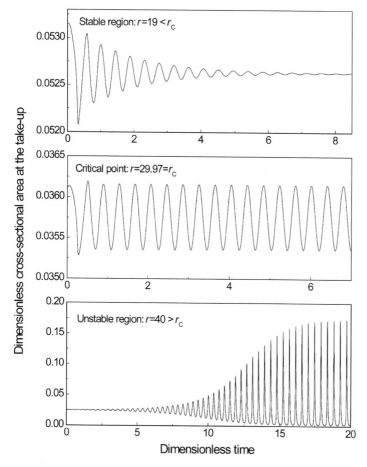

**FIGURE 11.12**  Transient response of the dimensionless cross-sectional area at take-up position. (From Ref. 96.)

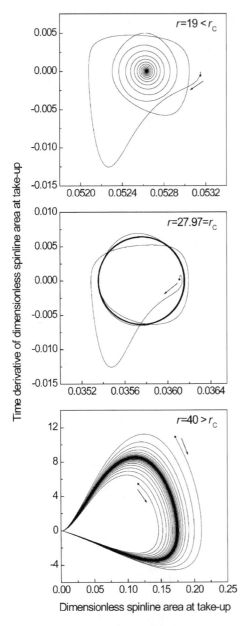

**FIGURE 11.13** Trajectories of the spinline area at take-up in the phase plane. (From Ref. 11.)

ring when the drawdown ratio is larger than a critical value is a supercritical Hopf bifurcation, where the limit cycle is stable on approaching from both within and outside.

### 11.1.3.6. Comparison of Transient Simulations with Experimental Results

The goodness of any simulation models is ultimately judged by how well the predictions match the real experimental results in terms of some characteristics of the system. In the draw resonance, the characteristics are usually three things, i.e., the shape, period, and severity of the oscillation. Whereas the shape and period are quite well predicted, the severity, measured as the ratio of the maximum and minimum cross-sectional areas (or diameters) of the product fiber, is reaching an ultimate value very slowly. This is because the period is well defined, well calculable from the model, while the exact magnitude of the spike, not easily estimated, steadily rises with increasing mesh number. Figure 11.14(a) shows a usual good matching between experiments and simulations in the period of draw resonance, while Fig. 11.14(b) displays an example where the severity of draw resonance numerically computed ever increases as the number of meshes increases.

### 11.1.4  Stabilization and Optimization of the Spinning Process

Now we move into the section dealing with the most important subject of any stability studies in engineering processing: how to utilize the theoretical results to improve the process productivity and product quality. In our case of draw resonance, this subject is then translated into how to eliminate or at least reduce the severity of draw resonance utilizing the theoretical simulation results.

### 11.1.4.1.  Sensitivity Analysis

First we examine the results of the sensitivity analysis, i.e., the effects of disturbances in various process conditions on the system performance, or specifically how cross-sectional area at take-up behaves with time when disturbances are introduced to the system. Two different approaches are possible in this sensitivity analysis: one is the transient simulation of the process behavior after disturbances introduced, and the other is the conventional frequency response analysis.

Table 11.1 reported by Jung et al. (52) shows the results of one particular sensitivity analysis where the cooling air velocity is shown having stabilizing effect in nonisothermal spinning of both Newtonian and Maxwell fluids. It can be seen here that increased cooling brings about higher spinline tension but lower tension sensitivity at the same time. However, the former is

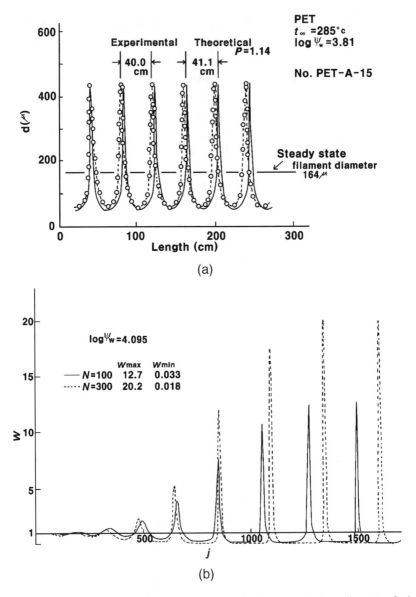

FIGURE 11.14 (a) Excellent agreement of theory and experiment of draw resonance instability where $d(\mu)$ denotes spinline diameter, $t_\infty$ is melt temperature, and $\psi_w$ is drawdown ratio. (b) Influence of the spatial mesh size on the severity of draw resonance where $j$ denotes dimensionless time, $w$ is spinline area, and $\psi_w$ is drawdown ratio. (From Ref. 17.)

TABLE 11.1  Sensitivities of Spinline Tension and Spinline Cross-Sectional Area at Take-Up to a Disturbance in the Take-Up Velocity ($St = 0.1$, $k = 5.8$, $\theta_a = 0.543$, $r = 50 \rightarrow 52.5$)

| $V_y$ | | $F$ (spinline tension) | $\|\Delta \ln F\|$ (first sensitivity) | $\|\Delta \ln a_L\|/\|\Delta \ln F\|$ (second sensitivity) | $\|\Delta \ln a_L\|$ (overall sensitivity) |
|---|---|---|---|---|---|
| *(a) Newtonian fluids* | | | | | |
| 0 | Unstable | 4.6071 | 0.08147 | 1.5742 | 0.1283 |
| 7.7 | Critical | 5.7269 | 0.06931 | 1.7037 | 0.1181 |
| 10 | Stable | 5.9222 | 0.06710 | 1.7255 | 0.1158 |
| 20 | Stable | 6.4928 | 0.06152 | 1.7841 | 0.1098 |
| *(b) A Maxwell fluid ($De_0 = 0.001$)* | | | | | |
| 0 | Unstable | 4.8975 | 0.07572 | 1.5170 | 0.1149 |
| 3.07 | Critical | 5.6064 | 0.06912 | 1.5800 | 0.1092 |
| 5 | Stable | 5.8753 | 0.06553 | 1.6081 | 0.1054 |
| 10 | Stable | 6.3687 | 0.05972 | 1.6597 | 0.0991 |
| 20 | Stable | 7.0254 | 0.05341 | 1.7264 | 0.0922 |

*Source*: Ref. 52.

quite understandable in the sense that lower spinline temperature increases the Maxwell fluid viscosity, which in turn increases the spinline tension. What is interesting here is the latter fact that this increased tension makes the tension sensitivity of the system to disturbances smaller. This reduced tension sensitivity is then obviously responsible for the increased stability. Figure 11.15 schematically illustrates the disturbance transmission mechanism as related to the sensitivity results of Table 11.1.

This finding of lowered tension sensitivity by increasing cooling is explained as follows. As the spinline tension increases and gets closer to its maximum, its ability to change, i.e., its sensitivity to disturbances, is bound to decrease, which in turn renders the system more stable. In fact, this is exactly

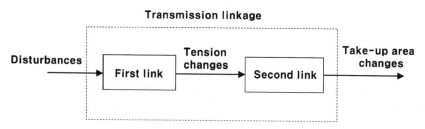

FIGURE 11.15  Schematic diagram illustrating the transmission linkage between disturbances and the spinline cross-sectional area at take-up. (From Ref. 52.)

the underlying mechanism of how the so-called draw resonance eliminator, an ingenious invention by Lucchesi et al. (115) at Union Carbide, works to dramatically increase the process productivity of film casting. By employing maximum cooling onto the cast film, the device was found to successfully increase the critical drawdown ratio more than double, i.e., eliminates the draw resonance at the existing production speeds. Here agreeing with the sensitivity results of Table 11.1, increased cooling resulted in decreased tension sensitivity while the tension itself increased. We can, as a matter of fact, perform similar sensitivity analysis on the effect of other process conditions. Figure 11.16 (53) shows the effect of Deborah number on spinning stability. Accordingly, as extension-thickening or -thinning fluids are used in the Phan-Thien–Tanner model, the opposite results are obtained: stabilizing in the former and destabilizing the latter.

One more important comment worth mentioning here as related to the tension sensitivity is in order. Is it ever possible to make the above tension sensitivity an absolute zero, or equivalently make the system absolutely stable? The answer to this question is yes. This is because in the theoretical sense, the tension sensitivity can always be made zero if we apply a constant force boundary condition at the take-up instead of the usual constant velocity as adopted in our studies. In other words, if the tension is made totally insensitive to any disturbances by staying constant, the system is necessarily stable. This point is illustrated in Fig. 11.17 by the results of the two different boundary conditions. As already explained before about how draw resonance perpetuates and repeats itself, the key component in the transmission linkage between process disturbances and fluctuations at the take-up is the spinline tension as shown by Fig. 11.15, which transmits process conditions from the take-up directly to the spinneret. The same point was evident in Table 11.1 data and also in Kase and Araki (50).

Finally, the frequency response method can produce similar sensitivity analysis results as demonstrated in Fig. 11.18 where the same opposite effects of the increasing Deborah number of Fig. 11.16 are well corroborated in the frequency domain.

### 11.1.4.2. Optimal Process Conditions for Stable Operation

Now following the above sensitivity analysis, we move to the next task of designing optimal process conditions for stable operations. Here we try to simplify and generalize as much as we can the principles and procedures for stabilizing and optimizing the spinning process by utilizing what we have learned from both the stability and sensitivity analyses. Although there exist many different factors affecting the spinning performance, there is a way that allows all those seemingly disparate things to be aligned in a theoretically consistent manner for the task of improving the stability.

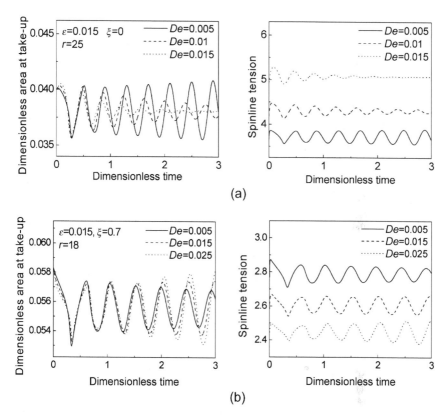

**FIGURE 11.16** Transient response of the cross-sectional area at take-up and spinline tension of the spinning of two different Phan-Thien and Tanner fluids: (a) an extension-thickening case and (b) an extension-thinning case. (From Ref. 53.)

The crux of the whole thing is, as already discussed in the last section, the fact that the spinline tension is to be increased in order to reduce the tension sensitivity and increase the stability. The draw resonance eliminator discussed before certainly obeys this strategy by increasing the tension through maximum cooling. Increasing the tension then dictates the directions of process factors to go. For example, as discussed in Table 11.1 and the draw resonance eliminator, to enhance the stability, cooling should always be increased. In addition, the fluid viscoelasticity or the Deborah number of the system should be increased for extension-thickening fluids and decreased for extension-thinning fluids as shown in Fig. 11.16. As for the secondary forces

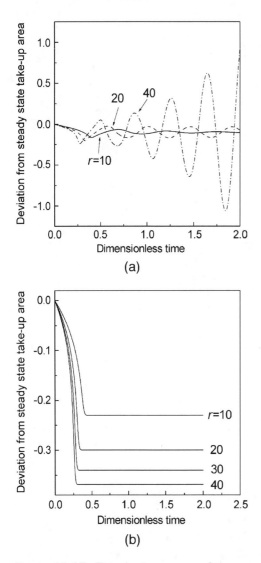

**FIGURE 11.17** Transient response of the cross-sectional area at take-up under (a) the constant take-up velocity boundary conditions and (b) the constant force boundary conditions. (From Ref. 104.)

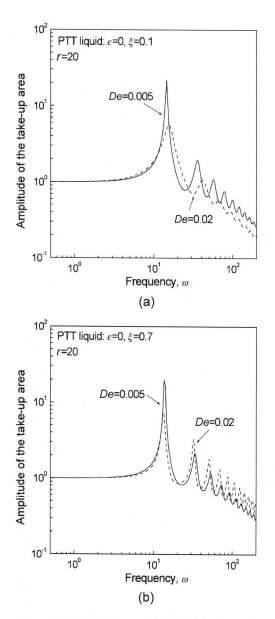

**FIGURE 11.18** Effect of fluid viscoelasticity on the sensitivity in frequency response analysis when a disturbance is introduced in take-up velocity. (a) An extension-thickening fluid and (b) an extension-thinning fluid. (From Ref. 107.)

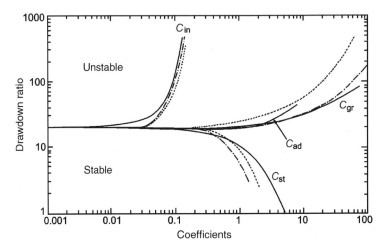

**FIGURE 11.19**　Effects of secondary forces on the stability in Newtonian spinning where $C_{in}$ denotes dimensionless inertia coefficient, $C_{gr}$ is dimensionless gravity coefficient, $C_{ad}$ is dimensionless air drag coefficient, and $C_{st}$ is dimensionless surface tension coefficient. (From Ref. 65.)

acting on the spinline, we have to look at the equation of motion, Eq. (11.8), to find out in which direction each force term should be changed to enhance the stability. The key in this case is the rheological force term, the first term in the right-hand side, whose increase should be resulted by changes in process factors for the enhanced stability. Using this measure, inertia, gravity, and air drag all have been found stabilizing while surface tension destabilizing, as amply displayed in Fig. 11.19.

Many detailed cases can arise in real processes, but having a generalized strategy proves important in industrial applications, and certainly this is the most valuable fruit of any theoretical pursuit of the system modeling, simulation, and analysis.

## 11.2 FILM BLOWING PROCESS

### 11.2.1 Introduction

Now we proceed to the film blowing process to examine its instability or, more specifically, the draw resonance instability, just as we did on the fiber spinning process in preceding sections. As already discussed before, the basic nature of film blowing process is similar to fiber spinning, i.e., extensional deformation, despite some distinct differences, i.e., biaxial extension and nonlinearity terms in the film blowing model. We thus move along the same path as in fiber spinning. We omit things if they are the same but include if new in film blowing. We focus on the biaxial nature of extensional deformation imbedded in the two force balance equations and also on the numerical schemes to bring about the transient solutions of the system (1).

### 11.2.1.1. Description of Film Blowing Process

The schematic diagram of film blowing is shown in Fig. 11.20. Although there have been other designs like double-bubble apparatus (116,117) for some polymers, in this chapter, we limit our discussion to this more prevalent design being used for most industrial operations. Unlike fibers where one-dimensional mechanical properties like tenacity are most important, the films

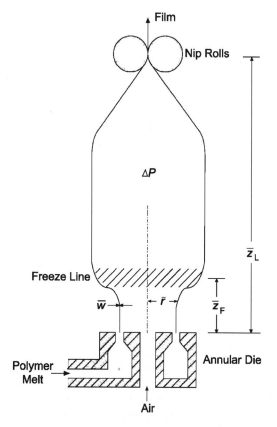

FIGURE 11.20  Schematic diagram of the tubular film blowing process.

coming out of the film blowing process should possess desirable two-dimensional mechanical properties like tear strength, for which obviously good control of biaxial extension of the film is essential.

Historically, film blowing is considered an engineering ingenuity in that a simple and robust design enables stretching of film in two directions simultaneously; the axial drawing of the film is provided by the pulling force of the nip rolls, whereas the circumferential drawing is accomplished by the air pressure inside the bubble. Through the separate control of these two drawings, i.e., the drawdown ratio (the ratio of film velocities at the nip rolls and the die exit in Fig. 11.20) and the blowup ratio (the ratio of the bubble radii at the freeze-line height and the die exit), desired two-dimensional film properties including the thickness reduction (the ratio of film thicknesses at the nip rolls and the die exit) are usually obtained the way we want.

### 11.2.1.2. Instability Modes in Film Blowing Process

As in fiber spinning, many different instabilities can be possible in film blowing. It has been customary to classify them into three different kinds: axisymmetric draw resonance, helical instability, and metastability as shown in Fig. 11.21. Among the many process variables, usually the following four are commonly considered conveying the essential information about the system dynamics: bubble radius, film thickness, freeze-line (or frost-line) height, and air pressure inside the bubble. During the oscillations, these variables exhibit periodically varying characteristics of the process. For example, all the above four variables manifest their temporal behavior with the same period in draw resonance, as seen in both time series and phase space trajectories.

Covering the important film surface irregularity problem caused by the melt fracture instability, the subject of melt fracture has been thoroughly investigated in Chapters 5–7 by Migler, Georgiou, and Dealy, respectively.

### 11.2.1.3. Historical Review of Film Blowing Literature

For the review of film blowing process, it is intended here to cover only those reports that are related to the subject of process instability in one way or another. Historically, film blowing had been evolved as one variant of tube extrusion with air blowing inside, and then refinements in equipment design and improved operations have steadily debuted on the industrial scenes. The theoretical analysis of the process in terms of its dynamic stability and the multiplicity of steady states has soon followed with many interesting results. The most notable milestone in the study of film blowing process was the first successful modeling of the process by Pearson and Petrie (118–122), which laid the foundation on the ensuing simulation and analysis efforts. The interest in this industrially important and theoretically intriguing process has

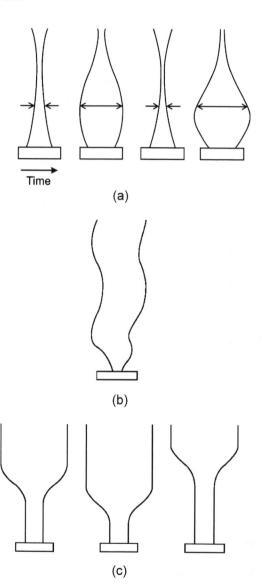

Time

(a)

(b)

(c)

FIGURE 11.21  Typical instabilities occurring in film blowing process. (a) Draw resonance, (b) helical instability, and (c) metastability.

continued during the past four decades with many valuable research findings and practical ideas.

Several research groups around the world have been noted for their significant contributions to the progress of film blowing process. White et al. (23,123–126) at Akron have compiled a systematic experimental data along with accompanying theoretical analysis, while Han et al. (127–131) at Polytechnic had made the first serious attempt to analyze the process. Campbell et al. (132–138) at Clarkson have investigated the process producing many good results using new models and new experimental design, Vlachopoulous et al. (139–142) at McMaster have developed efficient computer simulation programs with keen interests in the cooling mechanism, while Agassant et al. (143–146) in France have been conducting intensive investigation of the process as part of their overall research efforts on polymer processing in general. Lately, Carreau et al. (146–151) at Montreal have been actively pursuing a wide range of research interests in the process, reporting valuable results.

Since the seminal work by Pearson and Petrie (118–122) in the modeling of film blowing process, Cain and Denn (152,153) reported the most important research results about the stability of the process through a thorough modeling and analysis. Including Luo and Tanner (154), there also have been other notable results (155–164). Quite recently, Doufas and McHugh (165) applied their superb model of flow-induced crystallization to come up with valuable results. Also, Housiadas and Tsamopolous (166) and Hyun et al. (1,110,111) at Korea University have been reporting about modeling and stability in film blowing.

## 11.2.2 Modeling of Film Blowing Process

The governing equations of film blowing process comprise the same continuity, motion, energy, and constitutive equations as in fiber spinning, although there are three essential differences between the two. First, the biaxial extension performed by film blowing warrants two force balances in the model instead of one for fiber spinning. Second, a nonlinear term is involved in every equation due to the velocity being in the film direction not in the machine direction. Third, the nonisothermal cooling is directly coupled with the film deformation through the freezeline boundary conditions. These complexities are then responsible for the complications arising in the film blowing simulation, especially in obtaining transient solutions of the governing equations.

The modeling is usually done in one-dimensional formulation, i.e., the axial distance coordinate and time constitute the two independent variables of the system. The starting point of the modeling is, as mentioned earlier, the Pearson and Petrie model (118,119), which, despite some assumptions in the modeling, has demonstrated remarkable ability to portray the major dynamic features of the process. During the past 30 years, this model has played the dominant role while others have just included additional terms with minor modifications. The same trend is expected to continue. The only visible snag has been the inability on the part of theoreticians in coming up with transient solutions by solving governing equations. The reason for this apparent deficiency is considered twofold; steady-state solutions are often good enough for the purposes of the analysis and optimization, especially with the linear stability analysis always possible without transient solutions. The other is that it seems not easy to overcome severe numerical difficulties arising in solving the partial differential equations posed by the complex inherent nonlinearities imbedded in the process.

Very recently, this numerical problem in obtaining the transient solutions has been finally surmounted by Hyun et al. (1,110,111) in Korea through a couple of ideas incorporated into devising a new numerical scheme. Against the backdrop of all these points surrounding the modeling and simulation of the film blowing, we proceed to write down the governing equations of the system. Since the equations are available in the literature (85–92,118,119, 128,152–155,157,161,162), they are listed here without detailed explanations. We first start with the simplest isothermal dimensionless equations (85,118, 119,152,161).

Continuity equation:

$$\frac{\partial}{\partial}\left(rw\sqrt{1+(\partial r/\partial z)^2}\right)_z + \frac{\partial}{\partial z}(rwv) = 0 \tag{11.21}$$

where $t = \dfrac{\bar{t}\bar{v}_0}{\bar{r}_0}$, $z = \dfrac{\bar{z}}{\bar{r}_0}$, $r = \dfrac{\bar{r}}{\bar{r}_0}$, $v = \dfrac{\bar{v}}{\bar{v}_0}$, and $w = \dfrac{\bar{w}}{\bar{w}_0}$

Axial force balance:

$$2rw(\tau_{11} - \tau_{22})/\sqrt{1+(\partial r/\partial z)^2} + B(r_F^2 - r^2) = T_z \tag{11.22}$$

where $T_z = \dfrac{\bar{T}_z}{2\pi\eta_0\bar{w}_0\bar{v}_0}$, $B = \dfrac{\bar{r}_0^2\Delta P}{2\eta_0\bar{w}_0\bar{v}_0}$, $\Delta P = \dfrac{A}{\int_0^{\bar{z}_L}\pi\bar{r}^2 d\bar{z}} - P_a$, $\tau_{ij} = \dfrac{\bar{\tau}_{ij}\bar{r}_0}{2\eta_0\bar{v}_0}$

Circumferential force balance:

$$B = w\left(\frac{-(\tau_{11} - \tau_{22})(\partial^2 r/\partial z^2)}{\left(1+(\partial r/\partial z)^2\right)^{3/2}} + \frac{(\tau_{33} - \tau_{22})}{r\sqrt{1+(\partial r/\partial z)^2}}\right) \tag{11.23}$$

Constitutive equation (Newtonian model):

$$\tau_{ii} = D_{ii} \tag{11.24}$$

where

$$D_{11} = \frac{(\partial r/\partial z)}{1 + (\partial r/\partial z)^2} \frac{\partial(\partial r/\partial z)}{\partial t} + \frac{(\partial v/\partial z)}{\sqrt{1 + (\partial r/\partial z)^2}},$$

$$D_{22} = \frac{1}{w} \frac{\partial w}{\partial t} + \frac{(\partial w/\partial z)}{w} \frac{v}{\sqrt{1 + (\partial r/\partial z)^2}}$$

$$D_{33} = \frac{1}{r} \frac{\partial r}{\partial t} + \frac{(\partial r/\partial z)}{r} \frac{v}{\sqrt{1 + (\partial r/\partial z)^2}}$$

Boundary conditions:

$$v = 1, \quad w = 1, \quad r = 1, \quad \underline{\underline{\tau}} = \underline{\underline{\tau}}_0 \text{ at } z = 0 \tag{11.25a}$$

$$\frac{\partial r}{\partial t} + \frac{\partial r}{\partial z} \frac{v}{\sqrt{1 + (\partial r/\partial z)^2}} = 0, \frac{v}{\sqrt{1 + (\partial r/\partial z)^2}} = D_R \text{ at } z = z_F$$

$$\tag{11.25b}$$

where $r$ denotes a dimensionless bubble radius, $v$ is a dimensionless fluid velocity, $w$ is a dimensionless film thickness, $t$ is a dimensionless time, $z$ is a dimensionless axial distance, $\Delta P$ is the air pressure difference between inside and outside the bubble, $B$ is a dimensionless $\Delta P$, $A$ is the air amount inside bubble, $P_a$ is the atmospheric pressure, $T_z$ is a dimensionless axial tension, $\tau_{ii}$ are elements of dimensionless extra stress tensor, $z_L$ is the dimensionless distance between the die exit and the nip rolls, $z_F$ is the dimensionless freezeline height, $D_R$ is the drawdown ratio, $\eta_0$ is the zero-shear viscosity, and $D_{ij}$ are elements of dimensionless strain rate tensor. Overbars denote dimensional variables. Subscripts 0 and F denote conditions at die exit and freezeline height, respectively. Subscripts 1, 2, and 3 denote the flow direction, normal direction, and circumferential direction, respectively.

The approximations incorporated above are thin film approximation, uniform biaxial extension with shear deformation neglected, extrudate swell considered only as the initial conditions at the die exit, crystallization and secondary forces not included, and freezeline height fixed for isothermal cases. As with fiber spinning modeling in the previous sections, the additional equations are to be included as desired to portray more complicated situations. For nonisothermal case, energy equation is added with freezeline height in boundary conditions determined by the fluid solidification temperature (128,165).

Energy equation:

$$
\frac{\partial \theta}{\partial t} + \frac{U_c}{w}(\theta - \theta_a) + \frac{E_c}{w}(\theta^4 - \theta_a^4) + \frac{v}{\sqrt{1 + (\partial r/\partial z)^2}}\frac{\partial \theta}{\partial z}
$$

$$
+ C_H\left(\frac{\partial \phi}{\partial t} + \frac{v}{\sqrt{1 + (\partial r/\partial z)^2}}\frac{\partial \phi}{\partial z}\right) = 0 \tag{11.26}
$$

where $\theta = \dfrac{\bar{\theta}}{\bar{\theta}_0}$, $U_c = \dfrac{\bar{U}_c \bar{r}_0}{\rho C_P \bar{w}_0 \bar{v}_0}$, $E_c = \dfrac{\varepsilon_m \sigma_{SB} \bar{\theta}_0^3 \bar{r}_0}{\rho C_P \bar{w}_0 \bar{v}_0}$, $\phi = \dfrac{\bar{\phi}}{\phi_\infty}$, $C_H = \dfrac{\Delta H_f \bar{\phi}_\infty}{C_P \bar{\theta}_0}$

Boundary conditions:

$$
\theta = 1 \text{ at } z = 0 \text{ and } \theta = \theta_F \text{ at } z = z_F \tag{11.27}
$$

where $\theta$ denotes a dimensionless film temperature, $\theta_a$ is a dimensionless ambient temperature, $U_c$ is a dimensionless heat transfer coefficient, $E_c$ is a dimensionless radiation coefficient, $\rho$ is the fluid density, $C_p$ is the fluid heat capacity, $\Delta H_f$ is the crystallization heat, $C_H$ is the dimensionless crystallization heat, $\phi$ is a dimensionless crystallinity, $\bar{\phi}_\infty$ is the maximum crystallinity, $\varepsilon_m$ is the emissivity, and $\sigma_{SB}$ is the Stefan–Boltzmann constant.

For the case of viscoelastic film blowing, Phan-Thien–Tanner model (PTT) can be used as constitutive equations (97,98,111).

$$
K\underline{\underline{\tau}} + De\left(\frac{\partial \underline{\underline{\tau}}}{\partial t} + \underline{v} \cdot \underline{\nabla}\,\underline{\underline{\tau}} - \underline{\underline{L}} \cdot \underline{\underline{\tau}} - \underline{\underline{\tau}} \cdot \underline{\underline{L}}^T\right) = 2\frac{De}{De_0}\underline{\underline{D}} \tag{11.28}
$$

where $K = \exp\left(\varepsilon De\,\mathrm{tr}\,\underline{\underline{\tau}}\right)$, $\underline{\underline{L}} = \underline{\nabla}\,\underline{v} - \xi\underline{\underline{D}}$, $2\underline{\underline{D}} = \left(\underline{\nabla}\,\underline{v} + \underline{\nabla}\,\underline{v}^T\right)$,

$$
De = De_0\,\exp\left[k\left(\frac{1}{\theta} - 1\right)\right]
$$

where $k$ denotes the dimensionless activation energy, $De$ is the Deborah number, $De_0$ is the Deborah number at zero strain rate, and $\varepsilon$ and $\xi$ are PTT model parameters.

There can be, of course, many rheological models possible as was in fiber spinning depending on the modeler's preference. In order to take care of the on-line stress-induced crystallization, a crystallinity equation can be added with the crystallization heat term also added in the energy equation (68,71,75,165).

$$\frac{\partial \phi}{\partial t} + \frac{v}{\sqrt{1 + (\partial r/\partial z)^2}} \frac{\partial \phi}{\partial z} = nK_m(1 - \phi)\left[\ln\left(\frac{1}{1 - \phi}\right)\right]^{(n-1)/n}$$

$$\times \exp\left(-4\ln 2\left(\frac{\theta - \theta_{max}}{\theta_{hw}}\right)^2 + C_{cr}(C_{op}\tau)^2\right) \tag{11.29}$$

where $K_m = \dfrac{K_{max}\bar{r}_0}{\bar{v}_0}$, $C_{op} = \dfrac{2\bar{C}_{op}\eta_0\bar{v}_0}{\Delta_a\bar{r}_0}$

where $\theta_{max}$ denotes the dimensionless temperature at maximum crystallization rate, $K_{max}$ is the maximum crystallization rate, $K_m$ is the dimensionless $K_{max}$, $\bar{C}_{op}$ is the stress optical coefficient, $C_{op}$ is the dimensionless $\bar{C}_{op}$, $\Delta_a$ is the intrinsic amorphous birefringence, $C_{cr}$ is the stress-induced crystallization parameter, $\theta_{hw}$ is the dimensionless crystallization temperature half width, and $n$ is Avrami index.

The so-called secondary forces occurring on the film surface like gravity, inertia, air drag, and surface tension can also be reckoned with using corresponding terms to be added in the force balance equations. However, normally, only the gravity is significant enough to be added (89,90).

Axial force balance:

$$2rw(\tau_{11} - \tau_{22})/\sqrt{1 + (\partial r/\partial z)^2} + B(r_F^2 - r^2)$$

$$- 2\pi C_{gr} \int_0^z rw\sqrt{1 + (\partial r/\partial z)^2}\, dz = T_z \tag{11.30}$$

Circumferential force balance:

$$B = w\left(\frac{-(\tau_{11} - \tau_{22})(\partial^2 r/\partial z^2)}{\left(1(\partial r/\partial z)^2\right)^{3/2}} + \frac{(\tau_{33} - \tau_{22})}{r\sqrt{1 + (\partial r/\partial z)^2}}\right.$$

$$\left. - C_{gr}\frac{\partial r/\partial z}{\sqrt{1 + (\partial r/\partial z)^2}}\right) \tag{11.31}$$

where $C_{gr} = \rho g\bar{r}_0^2/2H_0\eta_0\bar{v}_0 \cdot C_{gr}$ denotes a dimensionless gravity coefficient.

## 11.2.2.1. Numerical Simulation

In general, the numerical simulation of film blowing is much more complicated than that of fiber spinning due to several factors including the biaxial extension, additional nonlinearity, and the multiplicity of steady states. Severe numerical difficulties are thus encountered in obtaining the transient solutions especially during instabilities like draw resonance. To surmount these hurdles, new numerical schemes have been continually tried in the past, and here we will follow the very recent one by Kim et al. (1,110,111) that have finally succeeded in obtaining the transient solutions from the governing equations of the film blowing. Before the details are discussed, the linear stability analysis is discussed next.

## 11.2.3 Stability

### 11.2.3.1. Linear Stability Analysis

Following the same procedure used for fiber spinning, the typical procedure of linear stability analysis of film blowing is described next. For simplicity, governing equations are compactly represented in a vectorial residual form.

$$\underline{R}(\underline{q}, \underline{\dot{q}}) = \underline{0} \tag{11.32}$$

where $\underline{q}$ is the state variable vector of bubble radius, film thickness, fluid velocity, axial distance, and stresses. The vector $\underline{\dot{q}}$ denotes the time derivative of $\underline{q}$. The reason why the axial distance is included above as a state variable is that as explained later, a coordinate transformation is adopted to make time and temperature as new independent variables instead of the original time and distance. This is required to handle a free-endpoint problem. Once time and temperature are chosen as new independent variables, the distance becomes a new dependent state variable.

At steady state, $\underline{\dot{q}} = 0$. The steady solution is $\underline{q} = \underline{q}_S$ from $\underline{R}(\underline{q}_S, \underline{0}) = \underline{0}$. The temporal perturbations are introduced onto the steady profiles of state variables to get perturbation variables.

$$\underline{q}(t, \theta) = \underline{q}_S(\theta) + \delta\underline{q}(t, \theta) \tag{11.33}$$

where $\delta q$ represents the perturbed quantities of state variables.

Insertion of Eq. (11.33) into the governing equations produces the following linearized vector equations.

$$\underline{J}\delta\underline{q} - \underline{M}\delta\underline{\dot{q}} = \underline{0} \tag{11.34}$$

where $\underline{J} \equiv \partial\underline{R}/\partial\underline{q}, \underline{M} \equiv -\partial\underline{R}/\partial\underline{\dot{q}}$ are the Jacobian and mass matrices, respectively.

Replacing $\delta q(t, \theta)$ by $\underline{y}(\theta)e^{\Omega t}$, where $\Omega$ is a complex eigenvalue that accounts for the growth rate of the perturbation, and rearranging Eq. (11.34) onto the orthogonal collocation nodes, the linear eigenmatrix equation such as Eq. (11.16) is obtained.

$$\Omega \underline{\underline{M}}\, y = \underline{\underline{J}}\, y \qquad (11.35)$$

Two typical stability diagrams are shown in Fig. 11.22(a) (152) and (b) (110,111) for Maxwell and Phan-Thien–Tanner fluids, respectively.

### 11.2.3.2.  Multiplicity of Steady States

The multiplicity of steady states is one of the most interesting characteristics of film blowing in contrast to fiber spinning having usually no multiplicity. Cain and Denn (152,153) thoroughly studied this subject with extensive documentation of various possible cases. Figure 11.23 shows the typical multiple steady-state maps for isothermal film blowing of Maxwell fluids (152) and Phan-Thien–Tanner fluids (110,111), respectively. One interesting point is with the UCM model, most constant $B$ (the dimensionless air pressure difference between the inside and outside bubble) contours terminate at a fixed value of thickness reduction near blowup ratio (BUR) = 1 point, while with the PTT model, the film thickness continues to cover the full range. This result is due to the unrealistically high-stress buildup by UCM model, and the PTT model is thus seen here performing better in portraying the film blowing process.

The finding of the multiplicities in film blowing process and their interesting transient behavior is very reminiscent of the chemical reactors where multiple steady states and their intricate transient behavior had been well studied and documented in 1960s and 1970s (167,168). Kim et al. (1,110,111) experimentally confirmed the existence of these multiple steady states and successfully simulated their transient behavior. Figure 11.24 shows such experimentally observed and numerically simulated three steady states in a nonisothermal film blowing (110,111).

### 11.2.3.3.  Stability Criterion Based on Traveling Kinematic Waves

The same stability criterion developed in fiber spinning and film casting exactly applies to film blowing case and so does the same approximate method. This point has already been explained in the earlier sections.

### 11.2.3.4.  Nonlinear Stability and Transient Simulation

As mentioned before, the transient behavior of film blowing has not been simulated due to the complexities of the system. It is only very recently that

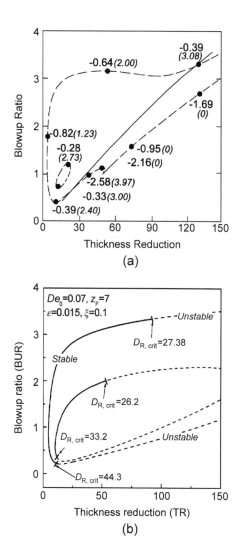

**FIGURE 11.22** Stability diagrams of (a) Maxwell fluids ($De_0 = 0.1$, $z_F = 5$) and (b) PTT fluids in isothermal case. (From Refs. 111 and 152.)

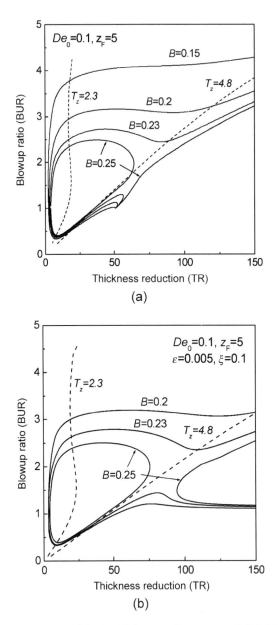

**FIGURE 11.23** Multiple steady states of (a) Maxwell and (b) PTT fluids in isothermal case. (From Refs. 111 and 152.)

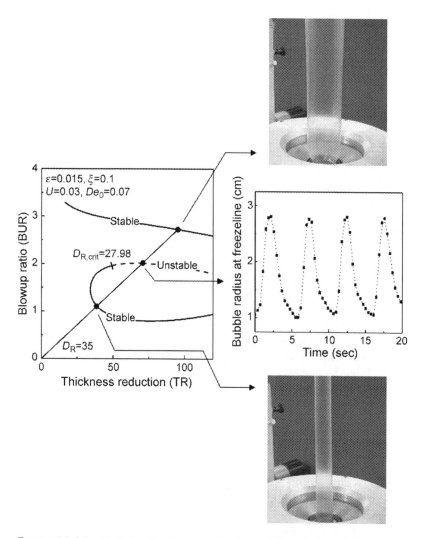

**FIGURE 11.24** Multiple steady states in the stability diagram of nonisothermal cases: theoretical results ($De_0 = 0.07$, $\varepsilon = 0.015$, $\xi = 0.1$, $U_C = 0.03$, $D_R = 35$, $\bar{\theta}_0 = 180\,^\circ$C) and experimental results ($D_R = 35$, $\bar{\theta}_0 = 180\,^\circ$C with LDPE resin from LG Chem). (From Ref. 1.)

those transient solutions have finally been obtained using a new numerical scheme by Kim et al. (1,110,111). In this section, their procedure is briefly explained.

The starting point of the effort is the simulation model of fiber spinning, which has been known for yielding the transient solutions readily. The same method based on successive iteration of each state variable simply does not work, however, for film blowing mainly because of the nonlinearity stemming from the velocity in film direction, not in machine direction. This forces the adoption of a Newton method combined with finite difference schemes in which full matrix equations are solved simultaneously for all variables. This is not feasible to pursue, either, in real numerical computations due to the immense computation time in order of weeks if computations are ever possible at all. Then comes the neat idea in reducing the dimensionality, i.e., introduction of orthogonal collocation method. Employing the minimum number of finite elements in $z$-coordinate with the minimum number of collocation points within each element, this method has successfully solved the transient problem of film blowing.

A couple of important new schemes were also incorporated in this new numerical method. First, the coordinate transformation has been made to convert the independent variables from $(t,z)$ to $(t,\theta)$ system. This was seen necessary to handle the moving freezeline because it transforms a free-endpoint problem into a fixed-endpoint problem, which is more amenable to numerical simulation. Second, the so-called cylindrical approximation in calculating the amount of the air pressure inside the bubble used by others (152,153,161,162) to enable the numerical computations at the expense of accurate moving boundary conditions has been dropped in favor of the correct tracing model of the moving and changing bubble shape. Combining all the above features into the model, Kim et al. (1,110,111) finally succeeded in obtaining heretofore unavailable transient solutions of film blowing, which then make possible thorough understanding of the dynamics and consequently the devising of strategies for stabilization and optimization of film blowing process.

Figure 11.25 demonstrates how remarkably the simulation results resemble the experimentally observed profiles, and Fig. 11.26 shows the transient behavior of the four dependent variables of film blowing in draw resonance in both time series and phase trajectories graphs.

### 11.2.4 Stabilization and Optimization of the Film Blowing Process

Developing strategies for stabilization and optimization of film blowing process is more challenging than for fiber spinning. This is due to the dynamically

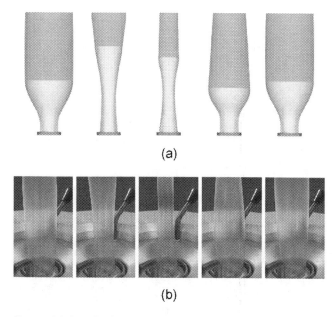

(a)

(b)

**FIGURE 11.25** Self-sustained periodic oscillations with time by (a) theory and (b) experiment when the system is in instability known as draw resonance (the same conditions as in Fig. 11.24).

complicated nature of the process engendered by its inherently intricate equipment design to perform biaxial extension and efficient surface cooling of the film. This, however, also provides additional degrees of controllability in operating the process as compared to fiber spinning. Other than the same process maneuverability as in fiber spinning, film blowing has additional manipulating variables like the air amount/pressure inside the bubble and the heat transfer modes of the outside cooling air. In addition, the fact that film blowing runs at much slower speeds than fiber spinning allows larger process sensitivities toward manipulating variables and consequently greater controllability of the process in general. The existence of multiple steady states in film blowing along with their complex stability behavior provides even more challenges and opportunities in operating the process. All these interesting points make the strategies for stabilization and optimization of film blowing a truly exciting subject for both industry and academia. Since the research efforts along this line are currently pursued and the details are still evolving, here, a brief discussion using an illustrative example is presented.

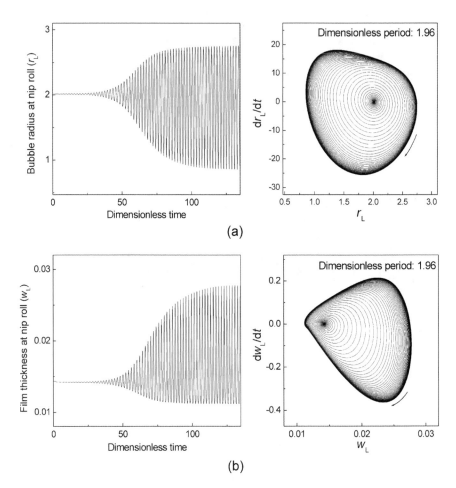

**FIGURE 11.26** Temporal pictures and phase space trajectories of the four state variables. (a) Bubble radius, (b) film thickness, (c) freezeline height, and (d) blowup air pressure.

### 11.2.4.1. Optimal Cooling Conditions in Film Blowing

Unlike the fiber spinning or film casting where cooling unequivocally stabilizes the system, i.e., the more cooling always brings about more stability, film blowing exhibits an optimal cooling condition to be the case, i.e., both too much or too little cooling can be detrimental. This is of course directly related to the multiplicity of steady states in film blowing. Figure 11.27 shows the stability diagram of particular process conditions where an initially unstable operating point can be moved to a stable region, but only

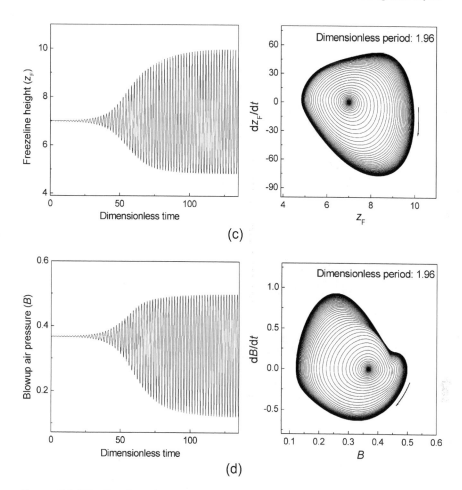

**FIGURE 11.26** Continued.

with optimal cooling, not too much or too little. In addition, in this optimization, to achieve the desired stable process conditions maintaining the same final film thickness, we have to adjust the air pressure/amount inside the bubble appropriately.

## 11.3   FURTHER R&D ISSUES IN FIBER SPINNING AND FILM BLOWING INSTABILITIES

In this final section, after the instabilities in fiber spinning and film blowing have been discussed using both theoretical models and experimental results,

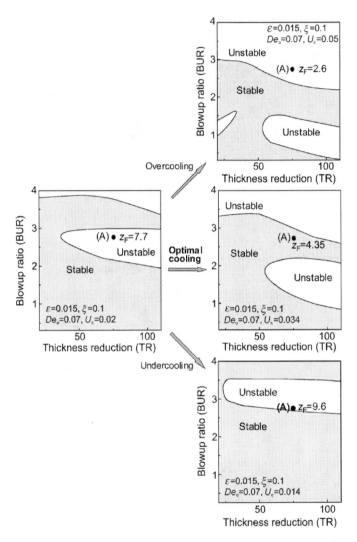

**FIGURE 11.27** Variations of stability windows when cooling conditions are changed.

it is intended to make brief comments as regards the future R&D efforts in this area.

First, all the models presented in this chapter have been one-dimensional; that is, the axial distance coordinate in the machine direction is the only spatial coordinate and thus every dependent variable was made a function of this variable and time. As many experimental data amply demonstrated, the one-dimensional models are sufficient in most cases for portraying the essential behavior of the system. But, of course, these one-dimensional models are not always good enough. Whenever the cross-sectional areas display inhomogeneous distributions of dependent variables like crystallinity, stress, velocity, temperature, etc. such as in skin-core structure in the product, two-dimensional models are warranted for both fiber spinning and film blowing.

In order to capture helical instability in film blowing, we need a three-dimensional model to simulate the process. As the dimension of the model increases, the immensity of modeling efforts jumps as in other physical processes. It suffices here to say that while the most essential points are being explained with one-dimensional models for fiber spinning and film blowing instabilities, for refined results, more two-dimensional and three-dimensional models should be considered in the future R&D.

Efforts to develop and refine the various models needed for the governing equations such as rheological and crystallization models should be continued. Orientation should be included in the future as a separate variable to describe the process and product. One of the immediate needs in the stability analysis is the study where crystallization kinetics is fully included in the governing equations. For the complete computer simulation of the whole process, we also need the prehistory and the spinline deformation combined. Another big task in simulating fiber spinning and film blowing is the inclusion of shear terms in modeling; that is, although the processes are of predominantly extensional deformation, the extent of influences of shear deformation should be closely looked into.

## ACKNOWLEDGMENTS

The authors are deeply indebted to Joo Sung Lee, a Ph.D. student for one of the authors (JC Hyun), for his dedicated efforts in conducting many detailed computations and drawings for fiber spinning and film blowing. The undertaking of this writing has been supported by research grants from the Korea Science and Engineering Foundation (KOSEF) through the Applied Rheology Center, an official KOSEF-created engineering research center at Korea University, Seoul, South Korea.

# REFERENCES

1. Hyun, J.C.; Kim, H.; Lee, J.S.; Song, H.-S.; Jung, H.W. Transient solutions of the dynamics in film blowing processes. J. Non-Newton. Fluid Mech., 2004. *in press*.

2. Ziabicki, A. *Fundamentals of Fibre Formation*; John Wiley & Sons: London, 1976.

3. Ide, Y.; White, J.L. The spinnability of polymer fluid filaments. J. Appl. Polym. Sci. 1976, *20*, 2511–2531.

4. Ide, Y.; White, J.L. Investigation of failure during elongational flow of polymer melts. J. Non-Newton. Fluid Mech. 1977, *2*, 281–298.

5. Chang, H.; Lodge, A.S. A possible mechanism for stabilizing elongational flow in certain polymeric liquids at constant temperature and composition. Rheol. Acta 1971, *10*, 448.

6. White, J.L.; Ide, Y. Instabilities and failure in elongational flow and melt spinning of fibers. J. Appl. Polym. Sci. 1978, *22*, 3057–3074.

7. Lee, S.; Kim, B.M.; Hyun, J.C. Dichotomous behavior of polymer melts in isothermal melt spinning. Korean J. Chem. Eng. 1995, *12*, 345–351.

8. Christensen, R.E. Extrusion coating of polypropylene. SPE J. 1962, *18*, 751.

9. Miller, J.C. Swelling behavior in extrusion. SPE Trans. 1963, *3*, 134.

10. Kim, B.M.; Hyun, J.C.; Oh, J.S.; Lee, S.J. Kinematic waves in the isothermal melt spinning of Newtonian fluids. AIChE J. 1996, *42*, 3164–3169.

11. Jung, H.W.; Song, H.-S.; Hyun, J.C. Draw resonance and kinematic waves in viscoelastic isothermal spinning. AIChE J. 2000, *46*, 2106–2111.

12. Bergonzoni, A.; DiCresce, A.J. The phenomenon of draw resonance in polymer melts. Polym. Eng. Sci. 1966, *6*, 45.

13. Donnelly, G.J.; Weinberger, C.B. Stability of isothermal spinning of a Newtonian fluid. Ind. Eng. Chem. Fundam. 1975, *14*, 334–337.

14. Weinberger, C.B.; Cruz-Saenz, G.F.; Donnelly, G.J. Onset of draw resonance during isothermal melt spinning: A comparison between measurements and predictions. AIChE J. 1976, *22*, 441–448.

15. Kase, S. Studies on melt spinning. IV. On the stability of melt spinning. J. Appl. Polym. Sci. 1974, *18*, 3279–3304.

16. Ishihara, H.; Kase, S. Studies on melt spinning. V. Draw resonance as a limit cycle. J. Appl. Polym. Sci. 1975, *19*, 557–565.

17. Ishihara, H.; Kase, S. Studies on melt spinning. VI. Simulation of draw resonance using Newtonian and power law viscosities. J. Appl. Polym. Sci. 1976, *20*, 169–191.

18. Matsumoto, T.; Bogue, D.C. Draw resonance involving rheological transitions. Polym. Eng. Sci. 1978, *18*, 564–571.

19. Nam, S.; Bogue, D.C. Dynamics of steady and unsteady melt spinning. Ind. Eng. Chem. Fundam. 1984, *23*, 1–8.

20. Tsou, J.-D.; Bogue, D.C. The effect of die flow on the dynamics of isothermal melt spinning. J. Non-Newton. Fluid Mech. 1985, *17*, 331–347.

21. Minoshima, W.; White, J.L. A comparative experimental study of the isothermal shear and uniaxial elongational rheological properties of low density, high

density and linear low density polyethylenes. J. Non-Newton. Fluid Mech. 1986, *19*, 251–274.

22. Minoshima, W.; White, J.L.; Spruiell, J.E. Experimental investigation of the influence of molecular weight distribution on melt spinning and extrudate swell characteristics of polypropylene. J. Appl. Polym. Sci. 1980, *25*, 287–306.

23. White, J.L.; Yamane, H. A collaborative study of the stability of extrusion, melt spinning and tubular film extrusion of some high-, low-, and linear-low density polyethylene samples. Pure Appl. Chem. 1987, *59*, 193–216.

24. Ide, Y.; White, J.L. Experimental study of elongational flow and failure of polymer melts. J. Appl. Polym. Sci. 1978, *22*, 1061–1079.

25. Han, C.D.; Lamonte, R.R.; Shah, Y.T. Studies on melt spinning III. Flow instabilities in melt spinning: melt fracture and draw resonance. J. Appl. Polym. Sci. 1972, *16*, 3307–3323.

26. Han, C.D.; Kim, Y.W. Studies on melt spinning. V. Elongational viscosity and spinnability of two-phase systems. J. Appl. Polym. Sci. 1974, *18*, 2589–2603.

27. Han, C.D.; Kim, Y.W. Studies on melt spinning. VI. The effect of deformation history on elongational viscosity, spinnability, and thread instability. J. Appl. Polym. Sci. 1976, *20*, 1555–1571.

28. Kase, S.; Matsuo, T. Studies on melt spinning. I. Fundamental equations on the dynamics of melt spinning. J. Polym. Sci. Part A 1965, *3*, 2541–2554.

29. Matovich, M.A.; Pearson, J.R.A. Spinning a molten threadline. Ind. Eng. Chem. Fundam. 1969, *8*, 512–520.

30. Pearson, J.R.A.; Matovich, M.A. Spinning a molten threadline: Stability. Ind. Eng. Chem. Fundam. 1969, *8*, 605–609.

31. Pearson, J.R.A.; Shah, Y.T. Stability analysis of the fiber spinning process. Trans. Soc. Rheol. 1972, *16*, 519–533.

32. Shah, Y.T.; Pearson, J.R.A. On the stability of nonisothermal fiber spinning— General case. Ind. Eng. Chem. Fundam. 1972, *11*, 150–153.

33. Pearson, J.R.A.; Shah, Y.T. On the stability of isothermal and nonisothermal fiber spinning of power-law fluids. Ind. Eng. Chem. Fundam. 1974, *13*, 134–138.

34. Pearson, J.R.A. Instability in non-Newtonian flow. Annu. Rev. Fluid Mech. 1976, *8*, 163.

35. Pearson, J.R.A.; Shah, Y.T.; Mhaskar, R.D. On the stability of fiber spinning of freezing fluids. Ind. Eng. Chem. Fundam. 1976, *15*, 31–37.

36. Gelder, D. The stability of fiber drawing processes. Ind. Eng. Chem. Fundam. 1971, *10*, 534–535.

37. Fisher, R.J.; Denn, M.M. Finite-amplitude stability and draw resonance in isothermal melt spinning. Chem. Eng. Sci. 1975, *30*, 1129–1134.

38. Fisher, R.J.; Denn, M.M. A theory of isothermal melt spinning and draw resonance. AIChE J. 1976, *22*, 236–246.

39. Fisher, R.J.; Denn, M.M. Mechanics of nonisothermal polymer melt spinning. AIChE J. 1977, *23*, 23–28.

40. Chang, J.-C.; Denn, M.M. Sensitivity of the stability of isothermal melt spinning to rheological constitutive assumptions. In *Rheology: Applications*;

Astarita, G., Marrucci, G., Nicolais, L., Eds.; Plenum Publishing: New York, 1980; Vol. 3. 9–13.

41. Chang, J.-C.; Denn, M.M.; Geyling, F.T. Effect of inertia, surface tension, gravity on the stability of isothermal drawing of Newtonian fluids. Ind. Eng. Chem. Fundam. 1981, *20*, 147–149.

42. Kase, S.; Kakajima, T. Growth of a dent on an isothermal fluid filament in uniaxial extension: A spinnability model in Lagrangian coordinate. Rheol. Acta 1980, *19*, 698–709.

43. Kase, S.; Katsui, J. Analysis of melt spinning transients in Lagrangian coordinates. Rheol. Acta 1985, *24*, 34–43.

44. Kase, S.; Nishimura, T. Classical wave equation in fluid filament stretching. Rheol. Acta 1988, *27*, 466–476.

45. Beris, A.N.; Liu, B. Time-dependent fiber spinning equations. 1. Analysis of the mathematical behavior. J. Non-Newton. Fluid Mech. 1988, *26*, 341–361.

46. Liu, B.; Beris, A.N. Time-dependent fiber spinning equations. 2. Analysis of the stability of numerical approximations. J. Non-Newton. Fluid Mech. 1988, *26*, 363–394.

47. Iyengar, V.R.; Co, A. Film casting of a modified Giesekus Fluid: Stability analysis. Chem. Eng. Sci. 1996, *51*, 1417–1430.

48. Yarin, A.L.; Gospodinov, P.; Roussinov, V.I. Stability loss and sensitivity in hollow fiber drawing. Phys. Fluids 1994, *6*, 1454–1463.

49. Gospodinov, P.; Yarin, A.L. Draw resonance of optical microcapillaries in non-isothermal drawing. Int. J. Multiph. Flow 1997, *23*, 967–976.

50. Kase, S.; Araki, M. Studies on melt spinning. VIII. Transfer function approach. J. Appl. Polym. Sci. 1982, *27*, 4439–4465.

51. Devereux, B.M.; Denn, M.M. Frequency response analysis of polymer melt spinning. Ind. Eng. Chem. Res. 1994, *33*, 2384–2390.

52. Jung, H.W.; Song, H.-S.; Hyun, J.C. Analysis of the stabilizing effect of spinline cooling in melt spinning. J. Non-Newton. Fluid Mech. 1999, *87*, 165–174.

53. Lee, J.S.; Jung, H.W.; Kim, S.H.; Hyun, J.C. Effect of fluid viscoelasticity on the draw resonance dynamics of melt spinning. J. Non-Newton. Fluid Mech. 2001, *99*, 159–166.

54. Bechtel, S.E.; Cao, J.Z.; Forest, M.G. Practical application of a higher order perturbation theory for slender viscoelastic jets and fibers. J. Non-Newton. Fluid Mech. 1992, *41*, 201–273.

55. Bechtel, S.E.; Cao, J.Z.; Forest, M.G. Illustration of an optimization procedure for fiber-spinning operation conditions: Maximum draw ratio under a thin-filament Maxwell model. J. Rheol. 1993, *37*, 237–287.

56. Bechtel, S.E.; Carlson, C.D.; Forest, M.G. Recovery of the Rayleigh capillary instability from slender 1-D inviscid and viscous models. Phys. Fluids 1995, *7*, 2956–2971.

57. Wang, Q.; Forest, M.G.; Bechtel, S.E. Modeling and computation of the onset of failure in polymeric liquid filaments. J Non-Newton. Fluid Mech. 1995, *58*, 97–129.

58. Forest, M.G.; Wang, Q.; Bechtel, S.E. 1-D models for thin filaments of liquid-

crystalline polymers: Coupling of orientation and flow in the stability of simple solutions. Physica. D. 1997, *99*, 527–554.

59. Henson, G.M.; Cao, D.; Bechtel, S.E.; Forest, M.G. A thin filament melt spinning model with radial resolution of temperature and stress. J. Rheol. 1998, *42*, 329–360.

60. Henson, G.M.; Bechtel, S.E. Radially dependent stress and modeling of solidification in filament melt spinning. Int. Polym. Proc. 2000, *15*, 386–397.

61. Renardy, M. Effect of upstream boundary conditions on stability of fiber spinning in the highly elastic limit. J. Rheol. 2002, *46*, 1023–1028.

62. Hagen, T.; Renardy, M. Eigenvalue asymptotic in non-isothermal elongational flow. J. Math Anal. Appl. 2000, *252*, 431–443.

63. Lee, W.-S.; Park, C.-W. Stability of a bicomponent fiber spinning flow. J. Appl. Mech. 1995, *62*, 511–516.

64. Ji, C.-C.; Yang, J.-C.; Lee, W.-S. Stability of Newtonian-PTT coextrusion fiber spinning. Polym. Eng. Sci. 1996, *36*, 2685–2693.

65. Ziabicki, A., Kawai, H., Eds. *High-Speed Fiber Spinning*; John Wiley & Sons: New York, 1985.

66. Kikutani, T.; Morinaga, H.; Takaku, A.; Shimizu, J. Effect of spinline quenching on structure development in high-speed melt spinning of PET. Int. Polym. Proc. 1990, *5*, 20–24.

67. Kikutani, T.; Radhakrishnan, J.; Sato, M.; Okui, N.; Takaku, A. High-speed melt spinning of PET. Int. Polym. Proc. 1996, *11*, 42–49.

68. Patel, R.M.; Bheda, J.H.; Spruiell, J.E. Dynamics and structure development during high-speed melt spinning of Nylon 6. II. Mathematical modeling. J. Appl. Polym. Sci. 1991, *42*, 1671–1682.

69. Schultz, J.M. Theory of crystallization in high-speed spinning. Polym. Eng. Sci. 1991, *31*, 661–666.

70. Kulkarni, J.A.; Beris, A.N. A model for the necking phenomenon in high-speed fiber spinning based on flow-induced crystallization. J. Rheol. 1998, *42*, 971–994.

71. Doufas, A.K.; McHugh, A.J.; Miller, C. Simulation of melt spinning including flow-induced crystallization. Part I. Model development and predictions. J. Non-Newton. Fluid Mech. 2000, *92*, 27–66.

72. Doufas, A.K.; McHugh, A.J. Simulation of melt spinning including flow-induced crystallization. Part II. Quantitative comparisons with industrial spinline data. J. Non-Newton. Fluid Mech. 2000, *92*, 81–103.

73. Doufas, A.K.; McHugh, A.J. Simulation of melt spinning including flow-induced crystallization. Part III. Quantitative comparisons with PET spinline data. J. Rheol. 2001, *45*, 403–420.

74. Doufas, A.K.; McHugh, A.J. Two-dimensional simulation of melt spinning with a microstructural model for flow-induced crystallization. J. Rheol. 2001, *45*, 855–879.

75. Joo, Y.L.; Sun, J.; Smith, M.D.; Armstrong, R.C.; Brown, R.A.; Ross, R.A. Two-dimensional numerical analysis of non-isothermal melt spinning with and without phase transition. J. Non-Newton. Fluid Mech. 2002, *102*, 37–70.

76. Ziabicki, A.; Tian, J. Necking in high-speed spinning revisited. J. Non-Newton. Fluid Mech. 1993, *47*, 57–75.
77. Denn, M.M. Continuous drawing of liquids to form fibers. Annu. Rev. Fluid Mech. 1980, *12*, 365–387.
78. White, J.L. Dynamics, heat transfer and rheological aspects of melt spinning: A critical review. Polym. Eng. Rev. 1981, *1*, 297.
79. Larson, R.G. Spinnability and viscoelasticity. J. Non-Newton. Fluid Mech. 1983, *12*, 303–315.
80. Petrie, C.J.S. Some remarks on the stability of extensional flows. Prog. Trends Rheol. 1988, *II*, 9–14.
81. Larson, R.G. Instabilities in viscoelastic flows. Rheol. Acta 1992, *31*, 213–263.
82. Malkin, A.Y.; Petrie, C.J.S. Some conditions for rupture of polymer liquids in extension. J. Rheol. 1997, *41*, 1–25.
83. Hyun, J.C. Draw resonance in polymer processing: A short chronology and a new approach. Kor.-Aust. Rheol. J. 1999, *11*, 279–285.
84. Denn, M.M. *Stability and Reaction and Transport Processes*; Prentice-Hall: Englewood Cliffs, 1975.
85. Middleman, S. *Fundamentals of Polymer Processing*; McGraw-Hill: New York, 1977.
86. Tadmor, Z.; Gogos, C.G. *Principles of Polymer Processing*; John Wiley & Sons: New York, 1979.
87. Petrie, C.J.S. *Elongational Flows*; Pitman: London, 1979.
88. Pearson, J.R.A.; Richardson, S.M. *Computational Analysis of Polymer Processing*; Applied Science Publishers: New York, 1983.
89. Pearson, J.R.A. *Mechanics of Polymer Processing*; Elsevier Applied Science: New York, 1985.
90. Agassant, J.-F.; Avenas, P.; Sergent, J.-Ph.; Carreau, P.J. *Polymer Processing*; Carl Hanser Verlag: New York, 1991.
91. Baird, D.G.; Collias, D.I. *Polymer Processing*; Butterworth-Heinemann: Newton, 1995.
92. Tanner, R.I. *Engineering Rheology*; Oxford University Press: New York, 2000.
93. Petrie, C.J.S.; Denn, M.M. Instabilities in polymer processing. AIChE J. 1976, *22*, 209–236.
94. Papanastasiou, T.C.; Macosko, C.W.; Scriven, L.E.; Chen, Z. Fiber spinning of viscoelastic liquid. AIChE J. 1987, *33*, 834–842.
95. Larson, R.G. *Constitutive Equations for Polymer Melts and Solutions*; Butterworth Publishers: Stoneham, 1988.
96. Jung, H.W. Process stability and property development in polymer extensional deformation processes. Ph.D. dissertation, Korea University, Seoul, Korea, 1999.
97. Phan-Thien, N.; Tanner, R.I. A new constitutive equation derived from network theory. J. Non-Newton. Fluid Mech. 1977, *2*, 353–365.
98. Phan-Thien, N. A nonlinear network viscoelastic model. J. Rheol. 1978, *22*, 259–283.

99. Giesekus, H. A simple constitutive equation for polymer fluids based on the concept of deformation-dependent tensorial mobility. J. Non-Newton. Fluid Mech. 1982, *11*, 69–109.

100. Papanastasiou, A.C.; Scriven, L.E.; Macosko, C.W. An integral constitutive equation for mixed flows: Viscoelastic characterization. J. Rheol. 1983, *27*, 387–410.

101. Keuning, R.; Crochet, M.J.; Denn, M.M. Profile development in continuous drawing of viscoelastic liquids. Ind. Eng. Chem. Fundam. 1983, *22*, 347–355.

102. Gupta, R.K.; Ballman, R.L. A study of spinline dynamics using orthogonal collocation. Chem. Eng. Commun. 1982, *14*, 23–33.

103. Schultz, W.W.; Zebib, A.; Davis, S.H.; Lee, Y. Nonlinear stability of Newtonian fibres. J. Fluid Mech. 1984, *149*, 455–475.

104. Hyun, J.C. Draw resonance in polymer processing: A short chronology of the research on its mechanism. Proceedings of the Korean Rheology Conference '99, The Korean Society of Rheology: Seoul, Korea, 1999; 75–83.

105. Jung, H.W.; Hyun, J.C. Stability of isothermal spinning of viscoelastic fluids. Korean J. Chem. Eng. 1999, *16*, 325–330.

106. Lee, J.S.; Jung, H.W.; Hyun, J.C. Melt spinning dynamics of Phan-Thien Tanner fluids. Kor.-Aust. Rheol. J. 2000, *12*, 119–124.

107. Jung, H.W.; Lee, J.S.; Hyun, J.C. Sensitivity analysis of melt spinning process by frequency response. Kor.-Aust. Rheol. J. 2002, *14*, 57–62.

108. Lee, J.S.; Hyun, J.C. Nonlinear dynamics and stability of film casting process. Kor.-Aust. Rheol. J. 2001, *13*, 179–187.

109. Lee, J.S.; Jung, H.W.; Song, H.-S.; Lee, K.-W.; Hyun, J.C. Kinematic waves and draw resonance in film casting process. J. Non-Newton. Fluid Mech. 2001, *101*, 43–54.

110. Kim, H.; Jung, H.W.; Hyun, J.C. Stability and nonlinear dynamics of film blowing. The 74th Annual Meeting of the Society of Rheology, Minneapolis, 2002.

111. Kim, H. Stability and nonlinear dynamics of tubular film blowing process. MS Dissertation, Korea University, Seoul, Korea, 2003.

112. Hyun, J.C. Theory of draw resonance: I. Newtonian fluids. AIChE J. 1978, *24*, 418–422; Part II. Power-law and Maxwell fluids 1978, *24*, 423–426.

113. Jung, H.W.; Choi, S.M.; Hyun, J.C. Approximate method for determining the stability of the film-casting process. AIChE J. 1999, *45*, 1157–1160.

114. Kim, B.M.; Choi, S.M.; Jung, H.W.; Hyun, J.C. An approximate method for determining the stability of film casting, film blowing, and fiber spinning. Australia–Korea Workshop on Polymer Melt and Polymer Solution Rheology; The Australian Society of Rheology: Parkville, Australia, 1996; 103–104.

115. Lucchesi, P.J.; Roberts, E.H.; Kurtz, S.J. Reducing draw resonance in LLDPE film resins. Plast. Eng. 1985, *41*, 87–90.

116. Kang, H.J.; White, J.L. Dynamics and stability of double bubble tubular film extrusion. Int. Polym. Proc. 1992, *7*, 38–43.

117. Song, K.; White, J.L. Single and double bubble tubular film extrusion of polybutylene terephthalate. Int. Polym. Proc. 2000, *15*, 157–165.

118. Pearson, J.R.A.; Petrie, C.J.S. The flow of a tubular film. Part I. Formal mathematical representation. J. Fluid Mech. 1970, 40, 1–19.
119. Pearson, J.R.A.; Petrie, C.J.S. The flow of a tubular film. Part 2. Interpretation of the model and discussion of solutions. J. Fluid Mech. 1970, 42, 609–625.
120. Petrie, C.J.S. A comparison of theoretical predictions with published experimental measurements on the blown film process. AIChE J. 1975, 21, 275–282.
121. Petrie, C.J.S. Memory effects in a non-uniform flow: A study of the behavior of tubular film of viscoelastic flow. Rheol. Acta 1973, 12, 92.
122. Petrie, C.J.S. Mathematical modeling of heat transfer in film blowing—A case study. Plast. Polym. 1974, 44, 259.
123. Minoshima, W.; White, J.L. Instability phenomena in tubular film, and melt spinning of rheologically characterized high density, low density and linear low density polyethylenes. J. Non-Newton. Fluid Mech. 1986, 19, 275–302.
124. Kanai, T.; White, J.L. Kinematics, dynamics and stability of the tubular film extrusion of various polyethylenes. Polym. Eng. Sci. 1984, 24, 1185–1201.
125. Kanai, T.; White, J.L. Dynamics, heat transfer and structure development in tubular film extrusion of polymer melts: A mathematical model and prediction. J. Polym. Eng. 1985, 5, 135.
126. White, J.L.; Cakmak, M. Orientation, crystallization, and haze development in tubular film extrusion. Adv. Polym. Technol. 1988, 8, 27–61.
127. Han, C.D.; Park, J.Y. Studies on blown film extrusion. I. Experimental determination of elongational viscosity. J. Appl. Polym. Sci. 1975, 19, 3257–3289.
128. Han, C.D.; Park, J.Y. Studies on blown film extrusion. II. Analysis of the deformation and heat transfer processes. J. Appl. Polym. Sci. 1975, 19, 3277–3290.
129. Han, C.D.; Park, J.Y. Studies on blown film extrusion. III. Bubble instability. J. Appl. Polym. Sci. 1975, 19, 3291–3297.
130. Han, C.D.; Shetty, R. Flow instability in tubular film blowing. 1. Experimental study. Ind. Eng. Chem. Fundam. 1977, 16, 49–56.
131. Han, C.D.; Kwack, T. Rheology–processing–property relationships in tubular blown film extrusion. I. High-pressure low-density polyethylene. J. Appl. Polym. Sci. 1983, 28, 3399–3418.
132. Ashok, B.K.; Campbell, G.A. Two-phase simulation of tubular film blowing of crystalline polymers. Int. Polym. Proc. 1992, 7, 240–247.
133. Campbell, G.A.; Cao, B. Modeling the blown-film process from die to frost line. Convert. Packag. June, 1987; 41–44.
134. Campbell, G.A.; Cao, B. The interaction of crystallinity, elastoplasticity, and a two-phase model on blown film bubble shape. J. Plast. Film Sheeting 1987, 3, 158.
135. Campbell, G.A.; Obot, N.T.; Cao, B. Aerodynamics in the blown film process. Polym. Eng. Sci. 1992, 32, 751–759.
136. Cao, B.; Campbell, G.A. Viscoplastic-elastic modeling of tubular blown film processing. AIChE J. 1990, 36, 420–430.
137. Cao, B.; Campbell, G.A. Air ring effect on blown film dynamics. Int. Polym. Proc. 1989, 4, 114–118.

138. Sweeney, P.A.; Campbell, G.A.; Feeney, F.A. Real time video techniques in the analysis of blown film instability. Int. Polym. Proc. 1992, 7, 229–239.
139. Vlachopoulous, J.; Sidiropoulos, V. The role of aerodynamics of cooling and polymer rheology in the film blowing process. Proceedings of the XIIIth International Congress on Rheology; The British Society of Rheology: Cambridge, UK, 2000; 3/403–3/405.
140. Sidiropoulos, V.; Vlachopoulos, J. Numerical study of internal bubble cooling (IBC) in film blowing. Int. Polym. Proc. 2001, 16, 48–53.
141. Sidiropoulos, V.; Tian, J.J.; Vlachopoulous, J. Computer simulation of film blowing. J. Plast. Film Sheeting 1996, 12, 107–129.
142. Sidiropoulos, V.; Vlachopoulos, J. The effects of dual-orifice air-ring design on blown film cooling. Polym. Eng. Sci. 2000, 40, 1611–1618.
143. Andre, J.M.; Demay, Y.; Agassant, J.F. Numerical modeling of the film blowing process. C.R. Acad. Sci. II B. 1997, 325, 621–629.
144. Andre, J.M.; Demay, Y.; Haudin, J.M.; Monasse, B.; Agassant, J.F. Numerical modeling of the polymer film blowing process. Int. J. Form. Process. 1998, 1, 187.
145. Debbaut, B.; Goublomme, A.; Homerin, O.; Koopmans, R.; Liebman, D.; Meissner, J.; Schroeter, B.; Reckmann, B.; Daponte, T.; Verschaeren, P.; Agassant, J.-F.; Vergnes, B.; Venet, C. Development of high quality LLDPE and optimised processing for film blowing. Int. Polym. Proc. 1998, 13, 262–270.
146. Laffargue, J.; Parent, L.; Lafleur, P.G.; Carreau, P.J.; Demay, Y.; Agassant, J.F. Investigation of bubble instabilities in film blowing process. Int. Polym. Proc. 2002, 17, 347–353.
147. Ghaneh-Fard, A.; Carreau, P.J.; Lafleur, P.G. On-line birefringence measurement in film blowing of a linear low density polyethylene. Int. Polym. Proc. 1997, 12, 136–146.
148. Ghaneh-Fard, A.; Carreau, P.J.; Lafleur, P.G. Study of instabilities in film blowing. AIChE J. 1996, 42, 1388–1396.
149. Ghaneh-Fard, A.; Carreau, P.J.; Lafleur, P.G. Study of kinematics and dynamics of film blowing of different polyethylenes. Polym. Eng. Sci. 1997, 37, 1148–1163.
150. Ghaneh-Fard, A.; Carreau, P.J.; Lafleur, P.G. Application of birefringence to film blowing. J. Plast. Film Sheeting 1996, 12, 68–86.
151. Lafleur, P.G.; Carreau, P.J.; Ghaneh-Fard, A. In-line birefringence measurements on film blowing of polyolefins. International Symposium on Orientation of Polymers; SPE: Boucherville, Canada, 1998; 270–280.
152. Cain, J.J.; Denn, M.M. Multiplicities and instabilities in film blowing. Polym. Eng. Sci. 1988, 28, 1527–1541.
153. Cain, J.J. A simulation of the film blowing process. Ph.D. dissertation, UC Berkeley, CA, 1987.
154. Luo, X.-L.; Tanner, R.I. A computer study of film blowing. Polym. Eng. Sci. 1985, 25, 620–629.
155. Gupta, R.K.; Metzner, A.B.; Wissbrun, K.F. Modeling of polymeric film-blowing processes. Polym. Eng. Sci. 1982, 22, 172–181.
156. Liu, C.-C.; Bogue, D.C.; Spruiell, J.E. Tubular film blowing. Part 2. Theoretical modeling. Int. Polym. Proc. 1995, 10, 230–236.

157. Yeow, Y.L. Stability of tubular film flow: A model of the film-blowing process. J. Fluid Mech. 1976, 75, 577–591.
158. Fleissner, M. Elongational flow of HDPE samples and bubble instability in film blowing. Int. Polym. Proc. 1988, 2, 229–233.
159. Micic, P.; Bhattacharya, S.N.; Field, G. Transient elongational viscosity of LLDPE/LDPE blends and its relevance to bubble stability in the film blowing process. Polym. Eng. Sci. 1998, 38, 1685–1693.
160. Field, G.J.; Micic, P.; Bhattacharya, S.N. Melt strength and film bubble instability of LLDPE/LDPE blends. Polym. Int. 1999, 48, 461–466.
161. Yoon, K.-S.; Park, C.-W. Stability of a blown film extrusion process. Int. Polym. Proc. 1999, 14, 342–349.
162. Yoon, K.-S.; Park, C.-W. Stability of a two-layer blown film coextrusion. J. Non-Newton. Fluid Mech. 2000, 89, 97–116.
163. Rutgers, R.P.G.; Clemeur, N.; Husny, J. The prediction of sharkskin instability observed during film blowing. Int. Polym. Proc. 2002, 17, 214–222.
164. Pontaza, J.P.; Reddy, J.N. Numerical simulation of tubular blown film processing. Numer. Heat Transf. A 2000, 37, 227–247.
165. Doufas, A.K.; McHugh, A.J. Simulation of film blowing including flow-induced crystallization. J. Rheol. 2001, 45, 1085–11004.
166. Housiadas, K.; Tsamopoulos, J. 3-Dimensional stability analysis of the film blowing process. Proceedings of the XIIIth International Congress on Rheology; The British Society of Rheology, Cambridge, UK, 2000; 3/152–3/154.
167. Aris, R. Elementary Chemical Reactor Analysis; Prentice-Hall: Englewood Cliffs, 1969.
168. Varma, A.; Morbidelli, M. Mathematical Methods in Chemical Engineering; Oxford University Press: New York, 1997.

# 12

## Coextrusion Instabilities

**Joseph Dooley**
The Dow Chemical Company, Midland, Michigan, U.S.A.

### 12.1 INTRODUCTION TO COEXTRUSION

Multilayer coextrusion is a process in which two or more polymers are extruded and joined together in a feedblock or die to form a single structure with multiple layers. Coextrusion avoids the costs and complexities of conventional multistep lamination and coating processes, where individual plies must be made separately, primed, coated, and laminated. Coextrusion readily allows manufacture of products with layers thinner than can be made and handled as an individual ply. Consequently, only the necessary thickness of a high-performance polymer is used to meet a particular specification of the product. In fact, coextrusion has been used commercially to manufacture unique films consisting of hundreds of layers, with individual layer thicknesses less than 100 nm (1).

Coextruded films are produced by a tubular-blown film process and a flat-die, chill-roll casting process, whereas coextruded sheet is produced with a flat die and a roll-stack process. Extruders used before the die and take-away equipment used afterward are standard equipment for manufacture of sheet or blown or cast film; however, the coextrusion dies are unique.

The choice of whether to use the blown or cast film process is normally dependent on the rate and final properties of the structure that are desired.

Cast film lines can typically run at a higher rate than a blown film line because the cooling efficiency of a chill roll is higher than using air to cool a bubble. However, the cast film process produces a product with uniaxial orientation rather than the biaxial orientation produced with the blown film process. In many cases, biaxial orientation is preferred to produce a film with more balanced physical properties.

Coextrusion die sizes vary and are dependent on the application. Blown film dies for the agricultural market can be as large as 2 m in diameter, which would produce a film approximately 6 m wide. Cast film dies for agricultural films have been made with widths as large as 10 m.

In addition to uses in bags and packaging, coextruded structures are also used in many other areas. Many coextruded sheets are made for use in thermoforming operations to form specific package or container shapes. Coextrusion is also used in the profile market. Pipes as well as window profiles have been made from coextruded structures.

## 12.2  METHODS TO PRODUCE COEXTRUDED STRUCTURES

### 12.2.1  Tubular Dies

Tubular coextrusion dies were the earliest dies used to make multilayer polymer film. Successful design requires formation of uniform concentric layers in the annular die land formed by the mandrel and adjustable or nonadjustable outer die ring. Early designs included center-fed dies that had the mandrel supported by a spider (2). Feedports arranged a concentric melt stream that was pierced by the mandrel as it flowed to the die exit, forming annular layers. Limitations of this early design were discontinuity and nonuniformity caused by spider-induced weld lines in the layers.

Another early design used stacks of distribution manifolds, so that as flow proceeded to the die exit, concentric layers were extruded on one another sequentially (3). The number of layers could be varied by changing the number of manifolds in the stack. The crosshead design of this die eliminated the spider support of the mandrel with its attendant weld-line problem.

The design most commonly used today is the multimanifold spiral mandrel tubular-blown film die. This die consists of several concentric manifolds, one within the other. The manifolds are supported and secured through the base of the die. Each manifold consists of a flow channel that spirals around the mandrel, allowing polymer to flow down the channel or leak across a land area to the next channel. This flow pattern smoothens out the flow of the polymer and minimizes any weld lines in the final film. Whereas early designs were limited to two or three layers, dies containing seven or more layers are now offered commercially.

These dies must achieve uniform concentric flow of all layers because it is impractical to provide circumferential thickness adjustment for each layer. Rheological data obtained at the intended extrusion temperature and shear rate are needed to size manifolds and channels for layer uniformity and minimum pressure drop. Frequently, spiral mandrel manifolds, common in single-layer dies, are used to improve circumferential distribution. A well-designed spiral mandrel manifold can be helpful, but streamlining is necessary to minimize stagnation, residence time, and purging. A manifold design is normally only optimum for a particular polymer. Employing a polymer with significantly different properties may require a different manifold insert in the die to obtain satisfactory layer distribution.

Most tubular-blown film lines are designed for oscillation of the die or winder to randomize film thickness variations at the windup and avoid buildup of gauge bands, which can cause problems with film flatness. More layers complicate bearing and sealing systems in an oscillating die, but designs have now been refined to employ new sealing materials that minimize polymer leakage. New designs incorporate temperature control of individual annular manifolds to permit coextrusion of thermally sensitive polymers.

Another style of tubular-blown film die is the stackable plate die. In this style of die, each layer is spread uniformly and formed into a tube in a single plate. Plates are then stacked on top of each other and the layers are added sequentially. This style of die is becoming popular for specific applications because the number of layers can be adjusted by simply changing the number of plates in the die. The major disadvantage for this style of die is that there is a large separating force between the plates and so many die bolts are required to hold the plates together. This means that the plates must be rather large in diameter in order to maintain structural integrity and this can produce longer flow paths and temperature differentials that can be detrimental to thermally sensitive polymers. Depending on the types of polymers to be used and the number of layers, the prices of the spiral mandrel and stacked plate dies can be comparable.

Flat dies, also called slit dies because the orifice is a wide rectangular opening, are used in chill roll, cast film, and sheet coextrusion. These dies are used almost exclusively for multilayer coextrusion with sheet thickness >250 $\mu$m, as well as in coextrusion coating processes (4), where a multilayer web is extrusion-coated onto a substrate such as paperboard, aluminum foil, polymer foam, or textile.

Another commercial application for flat-die coextrusion is biaxially oriented multilayer films (5) made with the tentering process to improve mechanical properties. Tentered film is biaxially oriented by stretching in the longitudinal and transverse directions, either sequentially or simultaneously, at uniform optimum temperature. In sequential stretching, the multilayer extrudate is cooled to a suitable orientation temperature on a first set of rolls, then stretched in the machine direction between a second set of rolls, which is driven faster than the first set. The uniaxially stretched film then enters a tentering frame, which has traveling clips that clamp the edge of the film. The clips are mounted on two tracks that diverge inside a temperature-controlled oven increasing film width to provide transverse stretch. The film is then heat-set and cooled. Simultaneous tentering frames, which feature accelerating clips that stretch the film longitudinally as they diverge transversely, are also used.

Two basic die types used in flat-die coextrusion systems are multimanifold dies and the feedblock/single-manifold die. A hybrid combines feedblocks with a multimanifold die.

### 12.2.2 Multimanifold Dies

Multimanifold dies have individual manifolds that extend the full width of the die. Each manifold is designed to distribute its polymer layer uniformly before combining with other layers outside the die (external combining) or inside the die before the final die land (internal combining). External-combining dies are typically limited to two-layer coextrusion because two slit orifices must be individually adjusted with die-lip adjusting bolts. The webs are combined at the roll nip. The vast majority of multimanifold dies are internal-combining rather than external-combining. This is due to the fact that better adhesion between the layers is normally found with internally combined structures because they are in thermal contact for a longer period of time and can form products with better interfacial adhesion.

In principle, internal-combining dies are similar to multimanifold tubular coextrusion dies except that the manifolds are flat. With these dies, it is possible to regulate flow across the width by profiling an adjustable restrictor bar in each manifold to help obtain uniform distribution. However, wide dies require numerous adjusting bolts on each layer manifold along with die-lip adjustment to control final thickness; this can make them difficult to operate. Multimanifold dies capable of coextruding five and six layers have been sold; they are expensive and require skilled operators. The principal advantage of multimanifold dies is the ability to coextrude polymers with very different viscosities because each layer is spread independently prior to combining.

A significant disadvantage of wide multimanifold dies is the difficulty in coextruding very thin layers, such as thin cap (surface) layers, or thin adhesive (tie) layers used to bond two dissimilar polymers. Frequently, these thin layers represent only 1% or 2% of the total structure thickness and therefore are extruded at a relatively low rate. With wide dies, it is difficult to obtain uniformity when extrusion rate per width is very low. Also, it is difficult to coextrude thermally sensitive polymers such as poly(vinyl chloride) (PVC) and poly(vinylidene chloride) copolymers (PVDCs) in wide dies because slow-moving material near the walls greatly increases residence time and thermal exposure.

### 12.2.3 Feedblock Method

The feedblock method of flat-die coextrusion, originally developed and patented by the Dow Chemical Company, uses a feedblock before a conventional single manifold die, as is shown in Fig. 12.1 (6,7).

A layered melt stream, which is prearranged ahead of the die inlet by the feedblock, is extended with the width of the die as it is reduced in thickness (Fig. 12.2). Polymer melts from each extruder can be subdivided into as many layers as desired in the final product. Feedports arrange metered layers in required sequence and thickness proportions. Modular feedblock design can be used to change the number, sequence, or thickness distribution of layers by changing a flow-programming module in the feedblock. Programming modules consist of machined flow channels designed to subdivide and direct the flow of each material to specific locations and proportions required by the product. This technique can also be used to minimize edge waste through tuning of the feedblock to produce the desired structure across the desired width of the film or sheet.

The shape of the multilayer melt stream entering the die inlet can be round, square, or rectangular, as long as the feedblock is properly designed to deliver the layers to the die with constant composition (8). Some feedblock

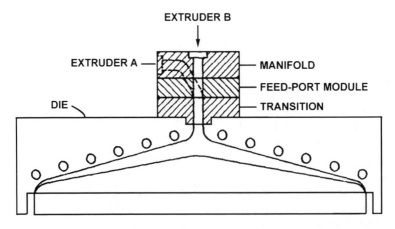

FIGURE 12.1 A feedblock-and-die combination.

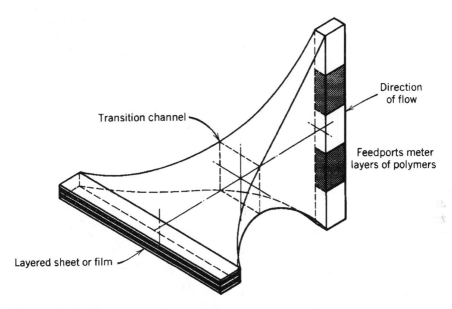

**FIGURE 12.2** The principle of the feedblock for coextruding multilayer film or sheet. The number of layers is equal to the number of feedports.

suppliers prefer round-die entry design for ease of machining or retrofitting to old dies. Others prefer square or rectangular die entries for ease of design and minimization of shape change as the layer interfaces are extended to the rectangular die orifice. A thermally sensitive polymer can be encapsulated by stable polymers so it does not contact the die walls, thus reducing residence time. The fact that the multilayer stream at the die inlet is narrow compared to die width makes it relatively easy to meter thin-surface or adhesive layers.

The versatility of the feedblock has made it the most popular flat-die coextrusion method. Large numbers of layers may be coextruded, layer structure may be readily altered with interchangeable modules, and thermally sensitive polymers may be protected by encapsulation.

One limitation of feedblocks is that polymer viscosities must be matched fairly closely because the combined melt stream must spread uniformly within the die. Severe viscosity mismatch results in nonuniform layers; the lower viscosity material tends to flow to the die edges. A crude rule of thumb is that polymer viscosities must be matched within a factor of three or four, which is a reasonably broad range for many commercially important coextrusions. Layer uniformity may be adjusted by varying melt temperature within limits dictated by heat transfer. Increasing temperature decreases viscosity, and material moves from the center to the edges; decreasing temperature has the opposite effect. Typically, the individual polymer melt temperatures differ by as much as 30–60°C. Beyond that, heat transfer tends to nullify further adjustment by temperature variation.

Often polymers are intentionally selected with a mismatch in viscosities in order to produce a product with specific physical properties. Layer nonuniformity expected with the mismatch is compensated by using shaped feedport geometry (i.e., the layers are introduced into the die nonuniformly, so that uneven flow within the die results in a satisfactorily uniform final distribution). Considerable art has been developed to extend the range of viscosity mismatch that can be accommodated in a feedblock system by using compensating feedport geometry (9,10). Some feedblocks are reportedly capable of coextruding polymers with viscosity mismatch of 100 or more (11–13).

## 12.3  RHEOLOGICAL CONSIDERATIONS FOR COEXTRUSION

Polymer rheology information is critical for designing coextrusion dies and feedblocks. The flow characteristics of the polymer must be considered when selecting materials for coextruded products. Viscosities of non-Newtonian polymers are dependent on extrusion temperature and shear rate, both of which may vary within the coextrusion die. The shear rate dependence is further complicated in that it is determined by the position and thickness of a polymer layer in the melt stream. A polymer used as a thin surface layer in a coextruded product experiences higher shear rate than it would if it were positioned as a central core layer.

The best-designed die or feedblock does not necessarily ensure a commercially acceptable product. Layered melt streams flowing through a coextrusion die can spread nonuniformly or can become unstable, leading to layer nonuniformities and even intermixing of layers under certain conditions.

## 12.4  INTERFACIAL DEFORMATION—VISCOUS
##        ENCAPSULATION

The importance of viscosity matching for layer uniformity was first studied in capillary flow of two polymers in bicomponent fiber production (14–17). Two polymers introduced side by side into a round tube experience interfacial distortion during flow if the viscosities are mismatched. The lower viscosity polymer tends to encapsulate the higher-viscosity polymer. It is possible for the low-viscosity polymer to encapsulate the higher-viscosity polymer totally. The degree of interfacial distortion due to viscosity mismatch depends primarily on the extent of the viscosity difference and the residence time.

Layer nonuniformities in feedblock-fed flat dies occur for the same reason when there is a large-enough viscosity mismatch. Low-viscosity polymer migrates to the die wall, producing encapsulation. This migration can start in the feedblock or in the die manifold as the layered stream spreads, resulting in increased layer thickness for low-viscosity polymer at the edges of the film or sheet. If low-viscosity polymer is a core layer, it not only becomes thicker at the edges, but may even wrap around higher-viscosity skin layers at the film edges.

Tubular-blown film dies are more tolerant of viscosity mismatch because the layers are arranged concentrically (i.e., there are no ends). However, a good die design is required to obtain concentric layers.

## 12.4.1 Experimental

Good layer uniformity in coextruded structures is usually a requirement for producing a structure with uniform properties. Many variables can affect layer uniformity. One of the most important variables is the viscosity of the polymer in each layer.

As was discussed earlier, when resins with significantly different viscosities are coextruded into a multilayer structure, a phenomenon known as "viscous encapsulation" can take place. This phenomenon is illustrated in Fig. 12.3, which shows a two-layer structure flowing down a tube. In this figure, the viscosity of layer "A" is less than layer "B." This figure shows how the less viscous layer (A) tends to move to the highest area of stress (near the tube walls) and thus encapsulate the layer with the higher viscosity (B).

Many studies have been done on the viscous encapsulation phenomenon. The earliest studies by Southern and Ballman (14) were on polymers with shear-thinning viscosities that varied in such a way as to crossover each other as a function of the system shear rate (i.e., the more viscous polymer at low shear rates became the less viscous polymer at higher shear rates). This

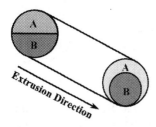

Viscosity of A < Viscosity of B

FIGURE 12.3   Viscous encapsulation in a tube.

allowed the researchers to cause the interface between the polymers to transform from convex to concave, depending on the shear rate at which the coextruded structure was processed and its relation to the viscosity crossover point. In these studies, the less viscous material always encapsulated the more viscous material. Similar studies have been done by Lee and White (18) and Han and Kim (19,20).

Several studies have been performed to look at the effect of viscosity and elasticity on the shape of the interface between coextruded layers. Studies by Southern and Ballman (21) and Khan and Han (22) led to the conclusion that the viscosity difference was more important in driving interface deformation than elasticity effects. This same result was found by White et al. (15) when they investigated the effect of normal stress differences on interface deformation. White et al.'s study showed that the more viscous material was always encapsulated by the less viscous material even when the first and second normal stress differences between the two polymers were considered. One important point to note, however, is that many of these studies were done using coextruded structures flowing through tube geometries. As will be discussed later, elastic effects become more pronounced when coextruded structures are processed through nonradially symmetric geometries.

Another study of viscous encapsulation was performed by Everage (23) in which he looked at the amount of encapsulation that took place as it related to the flow length in a tube. His results suggested that viscous encapsulation can be a relatively slow process because it required a very long die (L/D > 100) for complete encapsulation to occur for his system of materials. Results from other researchers (21,24) suggest that the residence time in the die is the more important factor influencing the degree of viscous encapsulation.

More recent studies by Dooley et al. (25) and Dooley and Rudolph (26) have shown that viscous encapsulation can be a relatively fast process, depending on the viscosity difference between the layers. This differs from the earlier results of Everage. In these studies, two-layer coextruded structures were made using different polystyrene resins in each layer with different colored pigments added to each layer to determine the location of the interface. A series of experiments was conducted, which showed that the addition of the pigments at the loadings used in these experiments did not affect the flow properties of the resins. These two-layer structures were extruded through a circular channel and the encapsulation velocity was measured experimentally.

Four high-impact polystyrene resins were used in the first study (25). These resins were referred to as polystyrene A, B, C, and D, respectively and their rheological properties are shown in Fig. 12.4. This figure shows that these resins have significantly different viscosities over the shear rate range tested. These particular polystyrene resins were chosen for this study in order

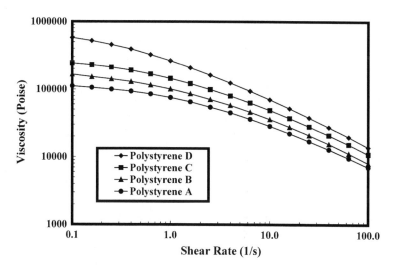

FIGURE 12.4   Viscosity comparison of polystyrene resins.

to produce as large a viscosity ratio between the layers in the coextruded structures as possible.

The primary coextrusion line used in this study consisted of a 31.75-mm-diameter, 24:1 length-to-diameter ratio (L/D) single-screw extruder for the substrate resin and a 19.05-mm-diameter, 24:1 L/D single-screw extruder for the cap resin. These extruders were attached to a feedblock that was designed to produce a two-layer structure consisting of a 20% cap layer and a 80% substrate layer. Attached to the exit of the feedblock was a die containing a circular cross section.

The circular die channel was designed to have a cross-sectional area of approximately 0.91 cm$^2$ based on a radius of 0.54 cm. The axial length of the die channel was 61 cm. This die channel was fabricated in two halves and bolted together, so that it could easily be split apart for removal of the polymer sample.

The coextrusion line was run for a minimum of 30 min to ensure that steady-state conditions had been reached. The normal extrusion rate was approximately 3.4 kg/hr for the polystyrene resins, which gave a wall shear rate of approximately 10 reciprocal seconds. The coextruded structures were extruded at 204°C. Variable-depth thermocouples were placed in the melt streams from each extruder just prior to the entry into the feedblock to ensure that the temperatures of the materials were the same prior to being joined together. When steady state was reached, the extruders were stopped simultaneously and the coextruded material was cooled while still in the die channel. After it had cooled to room temperature, the frozen polymer "heel" was removed from the die and examined. This procedure allowed the major deformations of the interface to be analyzed.

Of the four polystyrene resins studied, polystyrene D was the most viscous. Because polystyrene D resin was the most viscous, it was used as the substrate layer in a series of experiments in which each of the four polystyrene resins (A, B, C, and D) was coextruded in a two-layer structure as a 20% cap layer over a substrate of polystyrene D. At a shear rate of 10 reciprocal seconds, the viscosity ratios of the structures were 2.5, 2, 1.4, and 1 for cap layers of polystyrene resins A, B, C, and D, respectively. These viscosity ratios are defined as the viscosity of the substrate divided by the viscosity of the cap layer. In these experiments, the substrate layer of polystyrene D was always pigmented black, whereas the cap layer containing the different polystyrene resins was always pigmented white.

Figure 12.5 shows the solidified samples removed from the circular die channel for the experiments described above. This figure shows all of the samples viewed from the bottom of the channel, so that it is easier to observe the white cap layer flowing around and encapsulating the black substrate layer. The samples are labeled as the cap layer over the substrate layer (i.e., A/D represents a cap layer of polystyrene A resin coextruded over a substrate layer of polystyrene D resin).

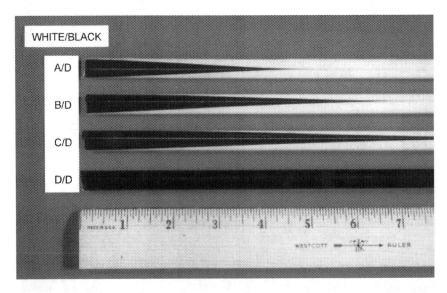

FIGURE 12.5 Samples removed from the circular die channel. Viscosity ratio decreases from 2.5 to 1 from top to bottom.

The first observation to note from Fig. 12.5 is that the D/D sample shows no sign of viscous encapsulation. This can be determined by the fact that only the black substrate layer is visible down the entire length of the die channel, and so no white cap layer has moved to encapsulate the substrate. This is to be expected because each layer is composed of the same resin processed at the same temperature.

The second important observation from Fig. 12.5 is that the higher is the viscosity difference between the cap layer and the substrate layer, the faster the viscous encapsulation occurs. The samples listed in order of the speed of encapsulation from fastest to slowest are A/D, B/D, C/D, and D/D, which have viscosity ratios of 2.5, 2, 1.4, and 1, respectively. Note that complete encapsulation of the A/D sample takes place in approximately 13 cm (~5 in.), whereas the C/D sample is not completely encapsulated even after 20 cm (~8 in.).

Figure 12.6 shows the A/D sample after it has been cut into 2.54-cm sections. This figure shows the sample cross sections beginning near the channel entry and then at distances of approximately 2.5, 5.1, 7.6, 10.2, and 15.2 cm from the channel entrance. Note how the white cap layer is flowing around and encapsulating the black substrate layer. This figure shows that the

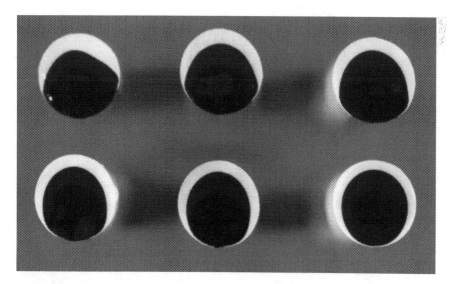

FIGURE 12.6 Cross-sectional images of the A/D sample beginning near the channel entry and then at distances of approximately 2.5, 5.1, 7.6, 10.2, and 15.2 cm from the channel entrance.

white cap layer completely encapsulates the black substrate somewhere between 10 and 15 cm downstream from the entry.

A similar study was done with polyethylene resins in place of polystyrene resins (26). The results of the polyethylene study were similar to the polystyrene study. Figure 12.7 shows cross-sectional cuts for combinations of polyethylene resins with similar and different viscosity ratios.

Note that when the viscosities of the polyethylene layers were similar, the interface location changed very slowly as the structure flowed down the channel. However, when the cap layer was significantly less viscous than the substrate layer, viscous encapsulation occurred fairly quickly.

### 12.4.2 Numerical

Although there have been many experimental studies of the viscous encapsulation phenomenon, there have been far fewer attempts to model this effect. Much of this is due to the difficulty in specifying boundary conditions at the various polymer and metal interfaces in a three-dimensional (3-D) simulation.

Studies have been done to determine the interface location for a coextruded structure in two dimensions (27,28). This is essentially a free surface flow in which the interface location between the layers must be determined. However, modeling three-dimensional coextrusion flows has

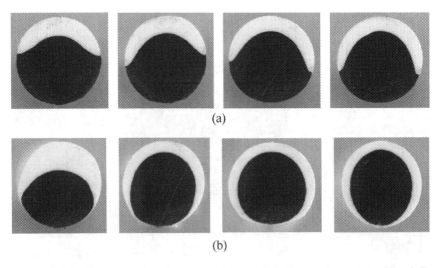

(a)

(b)

FIGURE 12.7 Cross-sectional images of the (a) A/B sample and (b) the A/D sample beginning near the channel entry and then at set distances down the channel to near the end of the channel.

proven to be much more difficult. Early efforts focussed on inelastic flows (29,30). However, unless viscoelastic flow effects are considered, the three-dimensional characteristics of the interface cannot be accurately modeled (31–33).

## 12.5 INTERFACE DEFORMATION – ELASTIC LAYER REARRANGEMENT

Although matching the viscosities of adjacent layers has proven to be very important, the effect of polymer viscoelasticity on layer thickness uniformity is also important (25,34–37).

It has been shown that polymers that are comparatively high in elasticity produce secondary flows normal to the primary flow direction in a die that can distort the layer interface. This effect becomes more pronounced as the width of a flat die increases. Appropriate shaping of the die channels can minimize the effect of layer interface distortion due to elastic effects.

Coextruding a structure that contains layers of polymers with low and high levels of elasticity can cause interface distortion due to the differences in elasticity between the layers in flat dies. The effect is typically not observed in tubular dies.

### 12.5.1 Experimental

Viscous encapsulation occurs when layers of different viscosities are coextruded to form a multilayered structure. Designing the structure with similar viscosities in the layers, or "viscosity matching," can minimize this effect. However, even with well-matched viscosities, coextruded structures have been processed in which layer deformation still occurs.

Dooley et al. (25) used a high-impact polystyrene resin, a low-density polyethylene resin, and a polycarbonate resin to evaluate the effect of elasticity on coextrusion. The rheological properties of these resins are shown in Figs. 12.8 and 12.9. These figures show that these resins have significantly different viscous and elastic flow properties. Figure 12.8 shows that the polystyrene and polyethylene resins are both shear-thinning, whereas the polycarbonate resin is more Newtonian in viscosity. Figure 12.9 shows the differences in elasticity between the resins based on their storage moduli. The polystyrene resin appears to be the most elastic (based on storage modulus values) followed by the polyethylene resin and then the polycarbonate resin.

Two-layer coextruded structures were made using the same polymer in each layer with different-colored pigments added to each layer to determine the location of the interface. The coextrusion line used in this study was

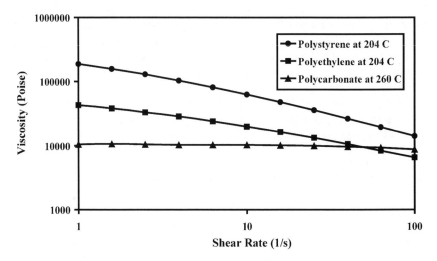

**FIGURE 12.8** Viscosity comparison of polystyrene, polyethylene, and polycarbonate resins.

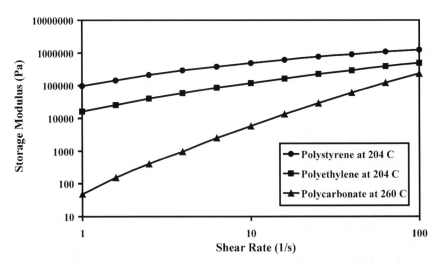

**FIGURE 12.9** Storage modulus comparison of polystyrene, polyethylene, and polycarbonate resins.

described earlier. The extruders were attached to a feedblock that was designed to produce a two-layer structure consisting of a 20% cap layer and a 80% substrate layer. Attached to the exit of the feedblock was a die containing one of the different channel geometries studied. Experiments were run on several different lines with different extruder sizes and feedblock designs, all producing similar results.

Several different die channel geometries were used in this study and three will be discussed here: a square channel, a circular channel, and a teardrop-shaped channel. The square, teardrop, and circular geometries were chosen because they are common shapes used in the design of feedblocks, dies, and transfer lines. The square channel had sides that were 0.95 cm long. The teardrop and circular channels were designed to have approximately the same cross-sectional area ($0.91 \text{ cm}^2$) as the square channel. The axial length of each of the die channels was 61 cm. These die channels were fabricated in two halves and bolted together, so that they could easily be split apart for removal of the polymer sample.

The coextrusion line was run for a minimum of 30 min to ensure that steady-state conditions had been reached. The normal extrusion rate was approximately 8.6 kg/hr for the polystyrene resin, which would give a wall shear rate in the range of 30–40 reciprocal seconds. Both the polystyrene and polyethylene resins were extruded at 204°C, whereas the polycarbonate resin was extruded at 260°C. Variable-depth thermocouples were placed in the melt streams from each extruder just prior to the entry into the feedblock to ensure that the temperatures of the materials were the same prior to being joined together. When steady state was reached, the extruders were stopped simultaneously and the coextruded material was cooled while still in the die channel. After it had cooled to room temperature, the frozen polymer "heel" was removed from the die and examined. This procedure allowed the major deformations of the interface to be analyzed.

An initial experiment consisted of running an 80/20 structure of polycarbonate resin into the square channel geometry die. The structure produced near the exit of the channel is shown in Fig. 12.10. This figure shows that the interface in the polycarbonate structure is fairly flat and very little deformation has occurred.

The resulting structure produced when a two-layered polystyrene structure was extruded through the square channel die is shown in Fig. 12.11. This structure is significantly different than that produced by the polycarbonate resin because there is extensive deformation of the interface. This structure shows that material has flowed up along the die walls to the corners and then turns and flows toward the center of the channel. Simultaneously, the material that was nearer the center of the channel is being pushed up toward the top of the channel.

FIGURE **12.10** Two-layer polycarbonate structure near the exit of the square channel.

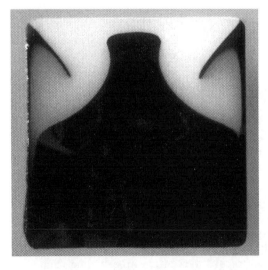

FIGURE **12.11** Two-layer polystyrene structure near the exit of the square channel.

This interface shape is obviously not due to viscous encapsulation because identical materials are present in each layer, so there can be no difference in viscosity to drive the viscous encapsulation. It is hypothesized that this interface shape is the result of elastic forces that produce secondary flows in the square geometry. These secondary flows would be present in a direction perpendicular to the main flow direction and be driven by second normal stress differences. These secondary flows have been discussed previously for polymer flows by White et al. (15) and Lee ad White (18). They noted that second normal stress differences could influence the interface shape along with the viscosity difference. They showed experimentally that in bicomponent tube flow, the less viscous layer always encapsulated the more viscous layer, regardless of the first and second normal stress differences between the two materials. It should be noted, however, that all of their studies were done in channels with circular cross sections.

However, because the secondary normal forces are small and difficult to measure, it has been assumed in the past that they could be ignored (38). Studies which show that these secondary flows are produced by differences in the normal forces (35–37) and are not due to viscous effects have been done. These differences in normal forces are produced when a viscoelastic material flows through a nonradially symmetric channel.

Figure 12.12 shows the predicted secondary flows in a square channel for an elastic material. The predicted flow patterns for a viscoelastic fluid flowing through a square channel have been shown previously to contain eight recirculation zones or vortices, two each per quadrant (35–37,39–43), as is shown in Fig. 12.12. These flow patterns appear to correspond well with the interface deformation shown for the polystyrene resin in Fig. 12.11. These flows would cause material to move up the walls and then turn back toward the center of the channel. These secondary flows would also cause the material near the center to be pushed upward toward the top of the channel. Numerical predictions of these secondary flow patterns will be covered in a subsequent section.

One interesting aspect of the elastic layer rearrangement shown in Fig. 12.11 is that it is a phenomenon that continues indefinitely as an elastic material flows down a square channel. This is different from viscous encapsulation because the driving force for viscous encapsulation tends to decrease as the materials flow down the channel as the less viscous material encapsulates the more viscous material and an energetically preferred state is reached. The constant layer rearrangement in an elastic material is shown in Fig. 12.13 for a two-layer polystyrene structure. This figure shows the progression of the elastic layer rearrangement as the structure flows down the channel by showing cuts at axial distances from the entry of 5, 20, 30, 40, 50, and 58 cm. The cut in the upper left-hand corner was taken from the sample near the

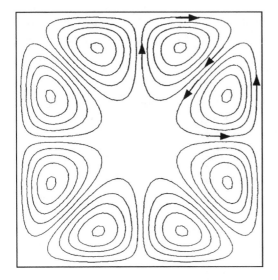

**FIGURE 12.12** Secondary flow patterns for an elastic material in a square channel.

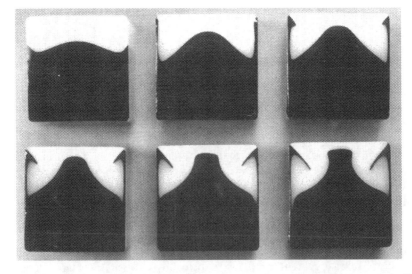

**FIGURE 12.13** Progression of a two-layer polystyrene structure as it flows down a square channel.

entry to the square channel, whereas the cut in the lower right-hand corner was taken from the sample near the exit of the square channel. This figure shows the steady deformation of the layer interface as it progresses down the channel.

Figure 12.14 shows a sample taken near the exit of the square channel for a two-layered structure composed of polyethylene. This figure shows results similar to those obtained for the polystyrene resin. However, the layer rearrangement for the polyethylene sample is not quite as extensive as the deformation in the polystyrene sample. This is consistent with the results of the measurements of the resin's storage moduli. If the storage modulus is used as an indication of elasticity of the resin, the polystyrene is the most elastic, the polyethylene is of intermediate elasticity, and the polycarbonate is the least elastic. This order of level of elasticity also corresponds to the amount of layer rearrangement observed in these samples, with polystyrene showing the most rearrangement and the polycarbonate the least amount of layer rearrangement.

As described previously, the layer rearrangements observed were hypothesized to be driven by secondary flows in the square channel caused by second normal stress differences. One way to test this hypothesis would be to extrude the materials through a channel with a radially symmetric geometry that would produce no secondary flows. This was done by extruding the three resins through a die with a circular cross section. Very little layer rearrange-

FIGURE 12.14   Two-layer polyethylene structure near the exit of the square channel.

ment was observed in each case as compared to the rearrangement seen in the square and teardrop channels. Figure 12.15 shows a cross-sectional cut of the polyethylene resin near the exit of the die with the circular cross section. The layer interface location in this sample is very similar to the interface location observed at the beginning of the channel, implying that the interface did not move substantially as the structure flowed down the channel. Similar behavior was also observed for the polycarbonate and polystyrene resins, even though they have substantially different viscoelastic properties compared to the polyethylene resin.

In experiments similar to those done with the square and circular channels, two-layered structures were also processed through a teardrop-shaped channel. The flow through a teardrop-shaped channel is very important industrially because many commercial film and sheet dies use teardrop-shaped channels in their distribution manifolds. Originally, many of the manifolds in film and sheet dies were circular in cross-sectional shape. However, the transition from a circular manifold to a rectangular-shaped land region caused some flow difficulties because of the abrupt change in geometry, especially in coextruded structures. This difficulty was overcome by using a tapered transition from the circular manifold to the rectangular land area, thus producing a teardrop-shaped channel. Because that cross-sectional shape is difficult to cut across the entire width of a large distribution manifold, many times the back of the manifold (in the circular section) is cut with a flat

FIGURE 12.15 Two-layer polyethylene structure near the exit of the circular channel.

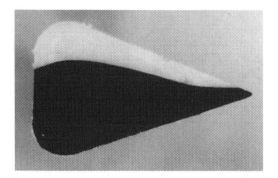

FIGURE 12.16   Two-layer polycarbonate structure near the exit of the teardrop channel.

section. This cross-sectional shape is the familiar teardrop shape that is used in many film and sheet die distribution manifolds.

Figure 12.16 shows the location of the interface near the exit of the teardrop channel for the polycarbonate resin. This sample shows a smooth interface with very little layer movement as the material flows down the channel. In contrast, a section cut from the two-layer polyethylene sample near the exit of the die is shown in Fig. 12.17, whereas a cut from the two-layer polystyrene sample is shown in Fig. 12.18.

The results obtained from the teardrop channel are consistent with those seen in the square channel in that the most layer movement was observed in the polystyrene sample, the least movement was observed in the polycarbonate sample, and the polyethylene sample was intermediate. The teardrop geometry, however, produced only six vortices or recirculation

FIGURE 12.17   Two-layer polyethylene structure near the exit of the teardrop channel.

FIGURE 12.18   Two-layer polystyrene structure near the exit of the teardrop channel.

zones (35,36) as compared to the eight observed in the square channel. This produced a flow pattern that is symmetric only about the centerline of the channel and is therefore different from the flow pattern seen in the square channel. These results are very significant from a commercial perspective because a large portion of the sheet and film dies produced in the industry contains teardrop-shaped distribution manifolds.

These results have many implications for the commercial coextrusion of viscoelastic polymers. Because the layer movements observed occurred without any differences in viscosity between the layers, this implies that layer rearrangement can occur in coextrusions in which the polymer viscosities are well matched. Also, because the layers continue to rearrange as the polymer flows down the channel, these results also imply that this effect will become more pronounced in sheet dies as they are scaled up to larger widths. The die used in these experiments would represent a die with a width of approximately 1.2 m. As discussed earlier, commercial cast film dies have been built with widths up to 10 m, or approximately eight times as wide as the die used in these experiments.

### 12.5.2  Numerical

The momentum and continuity equations for the steady-state flow of an incompressible viscoelastic fluid are given by:

$$-\nabla p + \nabla \cdot \mathbf{T} = 0 \qquad\qquad (12.1)$$
$$\nabla \cdot v = 0 \qquad\qquad (12.2)$$

where $p$ is the pressure field, $\mathbf{T}$ is the viscoelastic extra-stress tensor, and $v$ is the velocity field. Inertial and volume forces are assumed to be negligible.

The viscoelastic extra-stress tensor can be represented by $\mathbf{T}$. If a discrete spectrum of $N$ relaxation times is used, then $\mathbf{T}$ can be decomposed as follows:

$$\mathbf{T} = \sum_{i=1}^{N} \mathbf{T}i \tag{12.3}$$

where $\mathbf{T}_i$ is the contribution of the $i$th relaxation time to the viscoelastic stress tensor. For the extra stress contributions $\mathbf{T}_i$, a constitutive equation must be chosen. For example, the White–Metzner constitutive equation takes the form:

$$\mathbf{T}_i + \lambda_i \overset{\triangledown}{\mathbf{T}}_i = 2\eta_i \mathbf{D} \tag{12.4}$$

where $\lambda_i$ is the relaxation time, $\eta_i$ is the partial viscosity factor, $\mathbf{D}$ is the rate of deformation tensor, and the symbol $\triangledown$ stands for the upper-convected time derivative operator. This constitutive equation predicts a zero second normal stress difference.

The rate of deformation tensor $\mathbf{D}$ can be written as:

$$\mathbf{D} = \frac{1}{2}\left(\nabla v + \nabla v^{\mathrm{T}}\right) \tag{12.5}$$

A slightly more complex but more realistic viscoelastic equation is the Giesekus constitutive equation that has the form:

$$\mathbf{T}_i \left[\mathbf{I} + \frac{\alpha_i \lambda_i}{\eta_i} \mathbf{T}_i\right] + \lambda_i \overset{\triangledown}{\mathbf{T}}_i = 2\eta_i \mathbf{D} \tag{12.6}$$

where $\mathbf{I}$ is the unit tensor. In Eq. (12.6), $\alpha_i$ are additional material parameters of the model, which control the ratio of the second to the first normal stress difference. In particular, for low shear rates, $\alpha_1 = -2N_2/N_1$, where $\alpha_1$ is associated with the highest relaxation times $\lambda_i$.

Another differential viscoelastic constitutive equation that can be used is the Phan Thien Tanner model, one form of which is given by:

$$\exp\left[\frac{\varepsilon_i \lambda_i}{\eta_i} tr(\mathbf{T}_i)\right] \mathbf{T}_i + \lambda_i \left[\left(1 - \frac{\xi_i}{2}\right)\overset{\triangledown}{\mathbf{T}}_i + \frac{\xi_i}{2}\overset{\triangle}{\mathbf{T}}_i\right] = 2\eta_i \mathbf{D} \tag{12.7}$$

where again $\lambda_i$ and $\eta_i$ are the relaxation time and the partial viscosity factor, respectively, whereas the symbol $\triangledown$ stands for the upper-convected time derivative operator and $\triangle$ stands for the lower-convected time derivative operator. The parameters $\varepsilon_i$ and $\xi_i$ are material constants that relate to the extensional viscosity and second normal stress difference, respectively.

The finite element technique can be used to solve the set of equations described. Depending on the number of relaxation times used, the computational resources required to solve the problem can be extremely large. When more than a single relaxation time was required, the elastic viscous split stress (EVSS) algorithm (44) was used. This method divides the viscoelastic extra stress tensor into elastic and viscous parts as follows:

$$\mathbf{T}_i = \mathbf{S}_i + 2\eta_i \mathbf{D} \tag{12.8}$$

Using this algorithm, the equations expressed in terms of $\mathbf{S}_i$, $\mathbf{D}$, $v$, and $p$ can be solved. A postcalculation then gives the total viscoelastic extra-stress tensor $\mathbf{T}$. A quadratic interpolation is used for $v$, whereas a linear interpolation is used for $\mathbf{S}_i$, $\mathbf{D}$, and $p$.

Debbaut et al. (36) performed 2.5-D planar calculations on a cross section normal to the main flow direction for various channel geometries. In this technique, the pressure, the three velocity components, and the components of the extra-stress tensor $\mathbf{T}_i$ depend on two spatial variables only. This implies that the calculated fields remain constant in the main flow direction with the exception of the pressure.

Typical finite element meshes for the square, circular, and teardrop channel geometries are shown in Fig. 12.19. Note that in each geometry, the symmetry of the channel has been taken into consideration when developing the finite element mesh. For example, only one quadrant of the square and circular channels is meshed, whereas one half of the teardrop channel is tessellated.

The set of partial differential equations for the flows in these channels requires boundary conditions along the channel walls and at the inlet. The boundary conditions used in these initial simulations were zero fluid velocity along the external walls (no slip condition) and axes of symmetry along the other edges. A constant flow rate was imposed at the entry to the channel. For the initial simulations, the constant flow rate produced an entry velocity of approximately 1 cm/sec.

For the simulations, the viscometric properties of the three resins studied were used to develop the parameters for the viscoelastic constitutive

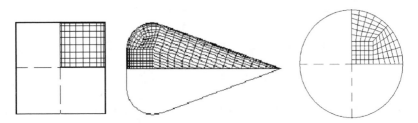

FIGURE 12.19  Representative two-dimensional finite element meshes used for the square, teardrop, and circular geometries.

equations. In the initial simulations, only a single relaxation time was used for each resin. These parameters were developed by fitting the viscometric data over a shear rate range of 10–100 $\text{sec}^{-1}$ because the average wall shear rate for a typical experiment was approximately 40 $\text{sec}^{-1}$.

The numerical results for the simulation of a polystyrene melt flowing through a square channel were shown in Fig. 12.12. These results show that there are two secondary flow vortices in each quadrant. These secondary flows circulate in a direction that would move material from the corners of the channel toward the center of the channel.

These results were obtained using the Giesekus model. Very similar results were found when using the Phan Thien Tanner model. However, when using the White–Metzner model, no secondary flow patterns were predicted. Because the White–Metzner model predicts a zero second normal stress difference, this illustrates the importance of the second normal stress difference in producing secondary flows.

The simulations for the polystyrene resin using the Giesekus model produced a diagonal secondary flow velocity that was approximately 1% of the main flow direction velocity. When a similar simulation was run for the polycarbonate resin, a secondary flow velocity of approximately 0.001% of the main flow velocity was calculated. This demonstrates why no secondary flows were observed experimentally for the polycarbonate resin. These data imply that the die channel would have to be approximately 1000 times as long in order to produce similar interface deformation in a polycarbonate structure compared to a polystyrene structure.

When a simulation was performed using the polystyrene resin in the circular channel with the White–Metzner constitutive model, no secondary flows were seen. This was expected because the White–Metzner model predicts a zero second normal stress difference. However, when similar simulations were performed using the Giesekus and Phan Thien Tanner models, again no secondary flows were predicted even though these models predict nonzero second normal stress differences and the polystyrene resin modeled in the simulations is highly elastic. These simulations show that the secondary flows are only generated in nonradially symmetric geometries.

Figure 12.20 shows the simulation results for the polystyrene resin in a teardrop channel using a Giesekus model. Recall that the teardrop shape channel approximates the geometry typically used in many die distribution manifolds. This figure shows that in the teardrop channel, the number of secondary flow vortices has been reduced to six instead of the eight, which were present in the square and rectangular channels. This figure also shows that the stronger secondary flows are along the longer channel dimension, whereas the weaker secondary flows are along the shorter channel side.

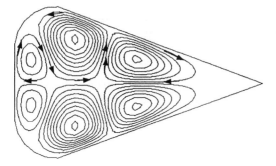

FIGURE 12.20 Secondary flow patterns for an elastic material in a teardrop channel.

Note that the direction of circulation of the secondary flows moves materials from the "corners" of the channel toward the center of the channel, and the material from the center is moved toward the flat walls. This is similar to what was observed with the square channel. Because this geometry is one of the most used in die manifold designs, it is imperative to understand the effects secondary flows can produce when using this type of manifold design to process multilayer coextruded structures.

### 12.5.3 Comparison of Experimental with Numerical Results

Figure 12.21 shows a comparison of numerical and experimental results for a polystyrene material in a square channel. This figure shows an excellent correlation between the predicted secondary flow patterns and the final

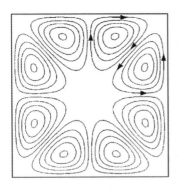 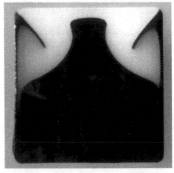

FIGURE 12.21 Comparison of numerical and experimental results for flow patterns of a polystyrene material in a square channel.

interface shape for the two sample from the end of the square channel. Comparing the upper right quadrants, it can be seen that material flows along the diagonal from the corner toward the center of the channel. It can also be seen that material from near the center of the channel flows outward toward the channel walls and then along the walls toward the corner. This flow pattern matches the secondary flow patterns predicted numerically.

Figure 12.22 shows a comparison between the numerical and experimental results for a polystyrene material in a teardrop channel. This comparison will also be done with the experimentally produced two-polystyrene structure. This experimental sample clearly shows the six secondary flow vortices that were predicted by the numerical simulation. Starting with the tip of the teardrop (on the right in the figure), it can be seen that material has flowed along the walls toward the tip and is then redirected and flows along the centerline toward the center of the channel. Near the center of the channel, the black layer has been elongated and stretched to very near the walls. On the left-hand side of the channel, the material flows along the centerline toward the wall where it splits and flows in opposite directions toward the corners. When the flow reaches the corners, it then turns and flows back toward the centerline. Note that the angle at which the material flows back toward the centerline from the corner is in good agreement between the numerical and experimental results.

In all of the previous simulations, only a single relaxation time was used in the simulations. At this point, it was decided to continue with the 2.5-D simulations but use an increased number of relaxation times to see if the accuracy of the simulations could be improved.

For this part of the study, only the low-density polyethylene resin was simulated using the Giesekus constitutive equation. The polyethylene was chosen rather than the polystyrene resin because the polystyrene experimental results had shown very thin layers of polymer flowing along the channel walls, whereas the polyethylene layers tended to be somewhat thicker. Trying to

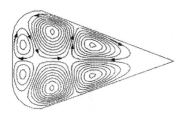
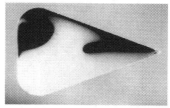

FIGURE 12.22 Comparison of numerical and experimental results for flow patterns of a polystyrene material in a teardrop channel.

duplicate very thin layer flow near a wall for the polystyrene resin would have required very fine finite element meshes and would have made the computations more difficult.

Based on the dynamic rheological properties of the polyethylene resin at a temperature of 204°C, a discrete spectrum of five relaxation times $\lambda_i$ ranging from $10^{-3}$ to 10 sec was chosen. The corresponding partial viscosities $\eta_i$ were fitted on the basis of the dynamic properties of the storage and loss moduli ($G'$ and $G''$, respectively), whereas the $\alpha_i$ parameters were selected based on the viscosity. The technique of identifying the material parameters is similar to that used by Laun (45) for an integral viscoelastic model.

The simulations performed were 2.5-D in nature (i.e., the calculation was performed in a cross section normal to the main flow direction and it was assumed that the calculated unknown fields did not vary along the main flow direction, except for the pressure gradient). In order to minimize the computational requirements, the EVSS algorithm was used. Once again as in the previous simulations, boundary conditions of no slip at the walls and symmetry planes were used.

For these simulations, it was desired to be able to track the interface between the layers of a two-layer polyethylene melt rather than to just predict the secondary flow patterns. A two-layer system was chosen in order to simplify the problem because only one interface would have to be predicted.

Once the velocity field has been calculated for a cross section, it is possible to track the fluid particles and determine the fluid configuration at several cross sections along the main flow direction. In order to do this, the particle pathlines must be calculated. This was done using a technique similar to the one developed by Goublomme et al. (46) for steady-state flows that has been used for the analysis of mixing. This technique uses a fourth-order Runge–Kutta method to integrate the pathline in an element.

Pathlines are calculated for a large number of material points randomly distributed throughout the computational domain in the $x$–$y$ plane at an initial downstream location of $z = 0$. For these simulations, 1500 points were used. Each point was labeled with a color (black or white) depending on its location in the two-layer structure at $z = 0$, and it maintains that color label as it follows the pathline. The pathlines of all of the particles are calculated until they reach a set downstream location $z = z'$. Once these calculations have been performed, the fluid particle locations can be examined at any downstream distance between $z = 0$ and $z = z'$.

When displaying the results of these simulations, only the white particles will be displayed against a black background. This will allow the interface between the white cap layer and the black substrate layer to be distinguished.

Figure 12.23 shows a comparison of experimental (top) and numerical (bottom) results for the flow of a two-layer polyethylene structure flowing

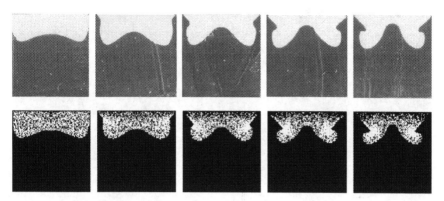

FIGURE 12.23   Comparison of experimental and numerical results of the flow of a two-layer polyethylene structure flowing down a square channel.

down a square channel. In this figure, results are shown at cuts taken at 5.1, 17.8, 30.5, 43.2, and 50.8 cm (from left to right, respectively) downstream from the entry of the channel. The simulations were performed using a total polymer flow rate of 8.6 kg/hr, with the white cap layer representing 20% of the structure and the black substrate being 80% of the structure.

Note that the first experimental sample in the top row does not have a completely flat interface between the white cap layer and the black substrate layer. This is because it was taken at 5.1 cm downstream from the entry, and so some interfacial distortion has already begun. In order to make a fair comparison between the experimental and numerical results, the location of the interface in the first simulation cut (on the left) was made to be identical to the experimental sample at 5.1 cm.

A comparison of the experimental and numerical results in Fig. 12.23 shows very good agreement in the location of the layer interface. Both the experimental and numerical results show the distinctive secondary flow patterns described in Section 12.5.2 for the flow of a viscoelastic material in a square channel. Both results show material flowing up the side walls of the channel and then turning back toward the center of the channel once they reach the corners of the channel. They also both show the characteristic upward flow of the material near the center of the interface. Considering the difficulty in producing these samples experimentally and the difficulty in numerically simulating this viscoelastic fluid, the agreement between the results is excellent.

Figure 12.24 shows a comparison of the experimental and numerical results of the flow of a two-layer polyethylene structure in a teardrop-shaped channel. Recall that the teardrop shape was chosen because of its industrial

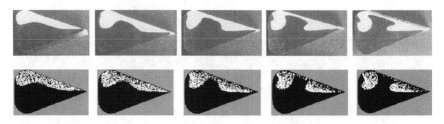

FIGURE 12.24 Comparison of experimental and numerical results of the flow of a two-layer polyethylene structure in a teardrop channel.

importance in the design of die manifolds. Once again, the cuts were taken at distances of 5.1, 17.8, 30.5, 43.2, and 50.8 cm from left to right. Also, the initial position of the interface in the numerical simulation was chosen to match the first experimental sample.

This figure shows that, experimentally, the interface undergoes significant deformation for this polyethylene resin. The black substrate material flows up along the left wall until it reaches the corner and then it turns back toward the centerline of the channel. The black substrate in the center of the channel is pushed upward toward the top wall where it is stretched in both directions along that wall. The white cap layer near the tip of the teardrop flows down to the centerline and then it flows along the centerline to the left. All of these layer deformations match the secondary flow patterns predicted earlier.

A good correlation between the numerical simulation results and the experimental results is shown in Fig. 12.24. This figure shows that the numerical simulations predict the flow of the substrate material into the corner and also up toward the top wall. Note also that the simulations predict the flow of the white cap material along the centerline in agreement with the experimental results. Overall, the agreement between the numerical and experimental results is very good.

In the previous simulations, a 2.5-D formulation was used to predict the flow patterns. At this point, it was decided to develop a 3-D simulation to see if the accuracy of the simulations could be improved even further.

For this part of the study, the low-density polyethylene resin was simulated using the Giesekus constitutive equation. Because a three-dimensional formulation was used, only four relaxation times were used (compared with the five relaxation times used in the 2.5-D formulation) in order to use reasonable computational resources. Based on the dynamic rheological properties of this resin at a temperature of $204^{\circ}$C, a discrete spectrum of four relaxation times $\lambda_i$ ranging from $10^{-3}$ to 1 sec was chosen. The corresponding partial viscosities $\eta_i$ were fitted on the basis of the dynamic properties of the storage and loss moduli ($G'$ and $G''$, respectively), whereas the $\alpha_i$ parameters were selected based on the viscosity.

The three-dimensional finite element mesh developed for these simulations is shown in Fig. 12.25. This mesh contains 1600 brick elements with 2106 vertices. Appropriate grading of the mesh is used near the channel walls to capture the flow gradients there.

Because this is now a three-dimensional simulation, the boundary conditions must be applied to planes rather than lines. In view of the symmetry of this geometry, only one quarter of the flow domain has been used in the computations. Boundary conditions of no slip at the walls along with suitable symmetry plane conditions were used. Fully developed viscoelastic extra stresses and velocity distributions were imposed at the inlet based on the specified flow rate. These were determined from the 2.5-D simulations performed earlier. A fully developed velocity distribution was imposed at the outlet.

In the 2.5-D simulations, the interface location was determined by tracking the path of a large number of arbitrarily selected fluid particles. For the three-dimensional simulations, a technique was used in which a pure advection equation was used for tracking the fluid motion. Let $c$ denote the advected variable that is constant along a given fluid trajectory. For a steady-state flow, the variable $c$ obeys the following equation:

$$v \cdot \nabla c = 0 \qquad (12.9)$$

From a qualitative point of view, $c$ could be considered as the fluid color: white in the top layer and black in the bottom layer. The finite element integration of a pure advection equation is not trivial. An upwinding technique has been used along with the addition of an anisotropic diffusion of $c$ along the flow trajectories and no crosswind diffusion (47,48). Knowledge

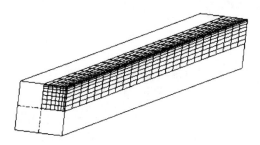

**FIGURE 12.25** Three-dimensional finite element mesh used for the square geometry.

of this variable allows tracking of any arbitrary material line or surface in the flow.

Figure 12.26 shows a comparison of experimental and numerical results of a two-layer polyethylene structure flowing down a square channel. The experimental results are shown in the upper half of the figure, whereas the corresponding numerical results are shown in the lower half of the figure. The experimental results are similar to those shown previously in Fig. 12.23, but the experiments were run at a slightly lower flowrate (7.9 vs. 8.6 kg/hr). The numerical simulations are once again started with an interface location that matches the first experimental sample.

Comparison of the experimental and numerical results in Fig. 12.26 shows that the numerical simulations predict the interface deformation trends well, but there are significant differences. The simulations appear to predict the interface deformation in the early part of the channel fairly well, but the differences between the two become more pronounced further down the channel. Near the exit of the channel, the simulations predict a greater thickness of material flowing along the vertical walls than is observed experimentally. Also the simulation underpredicts how far upward the material near the center of the interface moves compared to the experimental results.

One of the strengths of the three-dimensional simulation is the ability to better view the actual deformation of the interface in three dimensions. This is shown graphically in Fig. 12.27. This figure shows the deformation that the plane (or interface) between the top and bottom layers actually undergoes as the material flows down the channel.

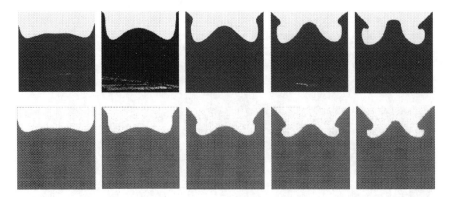

FIGURE 12.26  Comparison of experimental and numerical results of the flow of a two-layer polyethylene structure flowing down a square channel at a rate of 7.9 kg/hr.

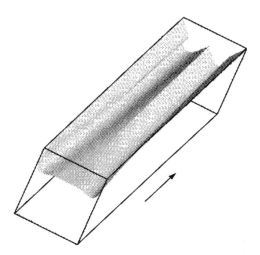

**FIGURE 12.27** Numerical results showing the layer interface deformation of a two-layer polyethylene structure flowing down a square channel.

It is interesting to note that the three-dimensional simulations appear to be less accurate in predicting the experimental results than the 2.5-D simulations. This may be due to the fact that the number of relaxation times was reduced from five to four in order to make the computer simulation requirements more reasonable. It appears that using a more appropriate relaxation spectrum is more important than using a fully three-dimensional simulation in order to better predict experimental results for these particular simulations.

## 12.6  INTERFACE DEFORMATION—INTERLAYER INSTABILITY

Interfacial instability is an unsteady-state process in which the interface location between layers varies locally in a transient manner. Interface distortion due to flow instability can cause thickness nonuniformities in the individual layers while still maintaining a constant thickness product. These instabilities result in irregular interfaces and even layer intermixing in severe cases.

At very low flow rates, the interface is smooth as is shown in Fig. 12.28a. At moderate output rates, low amplitude waviness of the interface is observed (see Fig. 12.28b), which is barely noticeable to the eye and may not interfere with the functionality of the multilayer film. At higher output rates, the layer distortion becomes more severe (Fig. 12.28c). If a large amplitude waveform

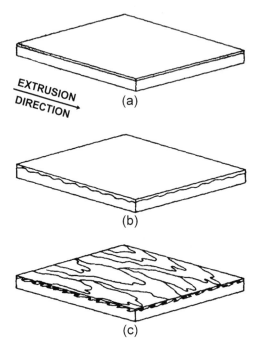

FIGURE 12.28 Interlayer instability examples at (a) low flow rates with a smooth interface, (b) moderate flow rates with a wavy interface, and (c) high flow rates with a very distorted interface.

develops in the flowing multilayer stream within the die, the velocity gradient can carry the crest forward and convert it into a fold. Multiple folding results in an extremely jumbled, intermixed interface. This type of instability, commonly called zig-zag instability, has been observed in tubular-blown film dies, multimanifold dies, and feedblock/single manifold dies.

This instability has been theorized to develop in the die land, and its onset has been correlated with a critical interfacial shear stress for a particular polymer system (49–55). Several authors dispute this relationship (56–58), but it is still a correlation that is commonly used.

The zig-zag type of interfacial instability can be reduced or eliminated by increasing skin layer thickness, increasing die gap, reducing total rate, or decreasing skin polymer viscosity. These methods may be used singly or in combination. These remedies reduce interfacial shear stress, and stable flow results when it is below the critical stress for the polymer system being coextruded. Most often, skin layer polymer viscosity is decreased. In feedblock coextrusion, the resultant viscosity mismatch imposed by this remedy can cause variations in layer thickness as discussed earlier. Shaped skin layer feedslots are then typically used to compensate and produce a uniform product.

Interfacial instability in a number of coextruded polymer systems has been studied to examine the effects of polymer melt properties on the instability. Much of the early work concentrated on viscosity differences between the layers (59–63) and how that could produce instabilities. Other studies have looked at surface/interfacial tension (64–67), shear-thinning effects (68–70), and viscosity and elasticity effects (71–79).

A unique set of experiments was conducted (80–84) in which a periodic flow rate change was introduced into a two-layer structure flowing through a channel. The experimental data collected were used to construct stability diagrams showing stable and unstable flow regimes.

Although interfacial instability is usually an undesirable condition, some studies have shown some that it can be beneficial in some instances. If the instability is large enough to cause mechanical mixing of the layers at the interface, the interfacial strength can be improved (85,86).

The majority of the studies on interfacial instabilities have been done on flows typical of cast film or sheet. However, the zig-zag type of instability is also seen in blown film systems (87,88).

Another type of instability has been observed in feedblock coextrusion of axisymmetric sheet (89). A wavy interface is also characteristic of this instability, but the wave pattern is more regular when viewed from the surface. This instability, commonly called wave instability, is hypothesized to originate in the die well ahead of the die land, and internal die geometry influences both the severity and pattern. For a given die geometry, the severity of instability increases with structure asymmetry and some polymers are more susceptible to unstable flow than others. It has been suggested that this type of instability may be related to the extensional viscosity properties of the polymers used in the coextruded structure (90–95). Examples of both zig-zag and wave instabilities are shown in Fig. 12.29.

Care should be taken when designing the joining geometry in a feedblock or die. In order to minimize instabilities, the layers should have similar velocities at the merging point. The joining of the layers should occur in a geometry that is as parallel as is realistically possible rather than joining in a perpendicular manner. The layers should also merge into a channel that is of an appropriate height, which does not force one layer to flow into the other.

Because wave instabilities appear to be related to the extensional viscosities of the individual layers, this implies that all of the previously mentioned design criteria for layer joining are important for this type of instability as well. In addition, the spreading of the layers in a film or sheet die is also important. Because this type of instability is related to extensional

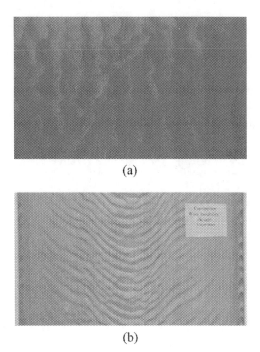

(a)

(b)

FIGURE 12.29    Interlayer instability patterns: (a) zig-zag; (b) wave.

viscosity, the rate at which the layers are stretched in the die will affect the forces in each layer. In structures containing materials with high extensional viscosities, the die should be designed to spread the layers slowly and at a uniform rate. This will help minimize wave pattern instabilities.

Modeling of interfacial instabilities is difficult because of the internal free surfaces that are present between the layers. Much of the literature devoted to modeling interlayer instability uses linear stability analysis to try to predict regions of stable and unstable flows (96–103). Other methods have been attempted with varying degrees of success (104–108). One of the more recent hypotheses that has been developed for wave-type instabilities is the effect of normal forces at the interface (109–113) and how they can be correlated to interfacial instabilities.

## 12.7  CONCLUSIONS

The coextrusion process has many advantages over other processes such as lamination because the joining of the layers and the formation of the final product are all done in one step. However, there are limitations. The two main difficulties that can arise when using the coextrusion process are interlayer instability and layer thickness nonuniformity.

Interlayer instability is a transient local layer thickness change that can be manifest in at least two ways: zig-zag and wave instabilities. Zig-zag instabilities are high-frequency, random-thickness changes that are typically correlated with a critical interfacial shear stress and occur in the die lip region. Wave instabilities are characterized by parabolic waves that normally extend across the width of the film or sheet. It is believed to occur in the die prior to the lips and has been associated with polymer extensional viscosity properties.

Layer thickness nonuniformity generally refers to a more global process in which there is a general layer thickness gradient across the film or sheet. There are at least two processes that can cause layer thickness nonuniformity that have been identified: viscous encapsulation and elastic layer rearrangement. The first of the two processes, viscous encapsulation, occurs when a less viscous polymer encapsulates a more viscous polymer in a coextruded structure, thereby producing a structure with a different layer thickness across the structure. The second process, elastic layer rearrangement, occurs when an elastic polymer flows through a nonradially symmetric channel, producing secondary flows. These secondary flows will drive polymer flow in a prescribed motion across the flow channel that can lead to layer thickness nonuniformity in a coextruded structure.

## REFERENCES

1. Schrenk, W.J.; Pinsky, J. Coextruded iridescent film. TAPPI Paper Synthetics Proceedings, 1976; 141–145.
2. Raley, G.E. Melt Extrusion of Multi-Wall Plastic Tubing. US Patent 3,223, 761, December 14, 1965.
3. Schrenk, W.J. Die. US Patent 3,308,508, March 14, 1967.
4. Wooddell, G.L. Characteristics of multilayer structures prepared by coextrusion extrusion coating. TAPPI Paper Synthetics Proceedings, 1980; 119 pp.
5. McCaul, J.P.; Hohaman, J.J. Biaxially oriented barrier coextrusions. TAPPI/ PLC Conference Proceedings, 1984; 633 pp.
6. Chisholm, D.; Schrenk, W.J. Method of Extruding Laminates. US Patent 3,557,265, January 19, 1971.
7. Schrenk, W.J. Multilayer film from a single die. Plast. Eng. March 1974, 30, 65.
8. Thomka, L.M.; Schrenk, W.J. Tooling for flat-die coextrusion. Mod. Plast. April 1972, 49, 62.
9. Cloeren, P. An overview of the latest developments in flat die coextrusion. SPE-RETEC Advances in Extrusion Technology 1991.
10. Dooley, J.; Stout, B. An experimental study of the factors affecting the layer

thickness uniformity of coextruded structures. SPE-ANTEC Tech. Pap. 1991, *37*, 62.

11. Cloeren Company. Coextrusion takes a giant step into the future. Mod. Plast. August 1983, *60* (28), 22.

12. Cloeren, P. Variable Thickness Extrusion Die. US Patent 4,197,069, April 8, 1980.

13. Cloeren, P. Method for Forming Multi-Layer Laminates. US Patent 4,152,387, May 1, 1979.

14. Southern, J.H.; Ballman, R.L. Stratified bicomponent flow of polymer melts in a tube. Appl. Polym. Sci. 1973, *20*, 175.

15. White, J.L.; Ufford, R.C.; Dharod, K.R.; Price, R.L. Experimental and theoretical study of the extrusion of two-phase molten polymer systems. J. Appl. Polym. Sci. 1972, *16*, 1313.

16. Han, C.D. A study of bicomponent coextrusion of molten polymers. J. Appl. Polym. Sci. 1973, *17*, 1289.

17. Everage, A.E., Jr. Theory of stratified bicomponent flow of polymer melts. I. Equilibrium Newtonian tube flow. Trans. Soc. Rheol. 1973, *17*, 629.

18. Lee, B.L.; White, J.L. An experimental study of rheological properties of polymer melts in laminar shear flow and of interface deformation and its mechanisms in two-phase stratified flow. Trans. Soc. Rheol. 1974, *18*, 467.

19. Han, C.D.; Kim, Y.W. Further observations of the interface shape of conjugate fibers. J. Appl. Polym. Sci. 1976, *20*, 2609.

20. Han, C.D.; Kim, Y.W. Use of the coextrusion technique for producing flame-retardant and antistatic fibers. J. Appl. Polym. Sci. 1976, *20*, 2913.

21. Southern, J.H.; Ballman, R.L. Additional observations on stratified bicomponent flow of polymer melts in a tube. J. Polym. Sci. 1975, *13*, 863.

22. Khan, A.A.; Han, C.D. On the interface deformation in the stratified two-phase flow of viscoelastic fluids. Trans. Soc. Rheol. 1976, *20*, 595.

23. Everage, A.E. Theory of stratified bicomponent flow of polymer melts: II. Interface motion in transient flow. Trans. Soc. Rheol. 1975, *19*, 509.

24. Lee, B.L.; White, J.L. An experimental study of rheological properties of polymer melts in laminar shear flow and of interface deformation and its mechanisms in two-phase stratified flow. Trans. Soc. Rheol. 1974, *18*, 467.

25. Dooley, J.; Hyun, K.S.; Hughes, K.R. An experimental study on the effect of polymer viscoelasticity on layer rearrangement in coextruded structures. Polym. Eng. Sci. July 1998, *38* (7), 1060.

26. Dooley, J.; Rudolph, L. Viscous and Elastic Effects in Polymer Coextrusion. TAPPI, Polymers, Laminations, Adhesives, Coatings, and Extrusions Conference, Paper 12-5, 2002.

27. Musara, S.; Keunings, R. Co-current axisymmetric flow in complex geometries: numerical simulation. J. Non-Newton. Fluid Mech. April 1989, *32*, 253.

28. Mitsoulis, E. Multilayer sheet coextrusion: analysis and design. Adv. Polym. Technol. 1988, *8*, 225.

29. Karagiannis, A.; Hrymak, A.N.; Vlachopoulos, J. Three dimensional studies on bicomponent extrusion. Rheol. Acta 1990, *29*, 71.

30. Gifford, W.A. A three dimensional analysis of coextrusion. Polym. Eng. Sci. 1997, *37*, 315.

31. Torres, A.; Hrymak, A.N.; Vlachopoulos, J.; Dooley, J.; Hilton, B.T. Boundary conditions for contact lines in coextrusion flows. Rheol. Acta 1993, *32*, 513.

32. Puissant, S.; Vergnes, B.; Agassant, J.F.; Demay, Y.; Labaig, J.J. Two-dimensional multilayer coextrusion flow in a flat coathanger die. Part 2. Experiments and theoretical validation. Polym. Eng. Sci. 1996, *36*, 936.

33. Chiou, J.Y.; Wu, P.Y.; Tsai, C.C.; Liu, T.J. An integrated analysis for a coextrusion process. Polym. Eng. Sci. 1998, *38*, 49.

34. Dooley, J.; Hilton, B.T. Layer rearrangement in coextrusion. Plast. Eng. February 1994, *50* (2), 25.

35. Dooley, J.; Dietsche, L. Numerical simulation of viscoelastic polymer flow-effects of secondary flows on multilayer coextrusion. Plast. Eng. April 1996, *52* (4), 37.

36. Debbaut, B.; Avalosse, T.; Dooley, J.; Hughes, K. On the development of secondary motions in straight channels induced by the second normal stress difference: experiments and simulations. J. Non-Newton. Fluid Mech. April 1997, *69* (2–3), 255.

37. Debbaut, B.; Dooley, J. Secondary motions in straight and tapered channels: experiments and three-dimensional finite element simulation with a multimode differential viscoelastic model. J. Rheol. November/December 1999, *43* (6), 1525.

38. Han, C.D. *Rheology in Polymer Processing*; Academic Press: New York, 1976.

39. Green, A.E.; Rivlin, R.S. Steady flow of non-Newtonian fluids through tubes. Q. Appl. Math. 1956, *14* (3), 299.

40. Townsend, P.; Walters, K.; Waterhouse, J. Secondary flows in pipes of square cross-section and the measurement of the second normal stress difference. J. Non-Newton. Fluid Mech. 1976, *1* (2), 107.

41. Dodson, A.G.; Townsend, P.; Walters, K. Non-Newtonian flow in pipes of non-circular cross section. Comput. Fluids 1987, *2* (3–4), 121.

42. Thangham, S.; Speziale, C.G. Non-Newtonian secondary flows in ducts of rectangular cross-section. Acta Mech. 1987, *68* (3–4), 121.

43. Wheeler, J.A.; Whissler Steady flow of non-Newtonian fluids in a square duct. Trans. Soc. Rheol. 1966, *10* (1), 353.

44. Rajagopalan, D.; Armstrong, R.C.; Brown, R.A. Finite element methods for calculation of steady viscoelastic flows using constitutive equations with a Newtonian viscosity. J. Non-Newton. Fluid Mech. 1990, *36*, 159.

45. Laun, J.M. Description of the non-linear shear behavior of a low density polyethylene melt by means of an experimentally determined strain dependent memory function. Rheol. Acta 1978, *17*, 1.

46. Goublomme, A.; Draily, B.; Crochet, M.J. Numerical prediction of extrudate swell of a high density polyethylene. J. Non-Newton. Fluid Mech. 1992, *44*, 171.

47. Hughes, T.J.R.; Brooks, A. In *Finite Elements in Fluids*; Gallager, R.H., Norrie, D.H., Oden, J.T., Zienkiewicz, O.C., Eds.; Wiley: New York, 1982; Vol. 4. Chapter 3.

48. Brooks, A.; Hughes, T.J.R. Streamline-upwind/Petrov–Galerkin formulation for convection dominated flows with particular emphasis on the incompressible Navier–Stokes equations. Comput. Methods Appl. Mech. Eng. 1986, *32*, 199.

49. Schrenk, W.J.; Bradley, N.L.; Alfrey, T., Jr.; Maack, H. Interfacial flow instability in multilayer coextrusion. Polym. Eng. Sci. 1978, *18*, 620.

50. Schrenk, W.J.; Alfrey, T. Coextruded multilayer polymer films and sheets (Chapter 15). In *Polymer Blends*; Paul, D.R., Newman, S., Eds.; Academic Press: New York, 1978.

51. Han, C.D.; Shetty, R. Studies on multilayer film coextrusion. II. Interfacial instability in flat film coextrusion. Polym. Eng. Sci. 1978, *18* (3), 180.

52. Han, C.D.; Rao, D.A. Studies on wire coating extrusion. II. The rheology of wire coating coextrusion. Polym. Eng. Sci. 1980, *20*, 128.

53. Schrenk, W.J. Multilayer polymer flow in flat die coextrusion. SPE-RETEC Conference Proceedings. PACTEC 5 1980; Vol. 1, 291.

54. Dooley, J.; Tung, H. Coextrusion. *Encyclopedia of Polymer Science and Technology*; John Wiley and Sons, Inc.: New York, 2002.

55. Butler, T. Effects of flow instability in coextruded films. TAPPI J. 1992, *75* (9), 205.

56. Kim, Y.J.; Chin, H.B.; Han, C.D. Interfacial instability in multilayer film. SPE-ANTEC Tech. Pap. 1982, *28*, 487.

57. Han, C.D.; Kim, Y.J.; Chin, H.B. Rheological investigation of interfacial instability in two-layer flat-film coextrusion. Polym. Eng. Rev. July 1984, *4* (3), 177.

58. Chen, J.H.; Ren, H.L. Research on unstable interface mechanism in coextrusion. SPE-ANTEC Tech. Pap. 1989, *35*, 206.

59. Yih, C.S. Instability due to viscosity stratification. J. Fluid Mech. 1967, *27*, 337.

60. Hickox, C.E. Instability due to viscosity stratification in axisymmetric pipe flow. Phys. Fluids 1971, *14*, 251.

61. Anturkar, N.R.; Papanastasiou, T.C.; Wilkes, J.O. Stability of multilayer extrusion of viscoelastic liquids. AIChE J. 1990, *36* (5), 710.

62. Anturkar, N.R.; Wilkes, J.O.; Papanastasiou, T.C. Estimation of critical stability parameters by asymptotic analysis in multilayer extrusion. Polym. Eng. Sci. 1993, *33* (23), 1532.

63. Waters, N.D. The stability of two stratified power–law liquids in Couette flow. J. Non-Newton. Fluid Mech. 1983, *12*, 85.

64. Hooper, A.P.; Boyd, W.G. Shear-flow instability at the interface between two viscous fluids. J. Fluid Mech. 1983, *128*, 507.

65. Chen, K.P.; Joseph, D.D. Elastic short wave instability in extrusion flows of viscoelastic liquids. J. Non-Newton. Fluid Mech. 1992, *42* (1–2), 189.

66. Renardy, Y. Weakly nonlinear behavior of periodic disturbances in two-layer plane channel flow of upper-convected Maxwell liquids. J. Non-Newton. Fluid Mech. 1995, *56*, 101.

67. Barger, M.; Ramanathan, R. Observations with miscible polymer pairs during microlayer coextrusion. SPE-ANTEC Tech. Pap. 1995, *41* (2), 1699.

68. Waters, N.D.; Keeley, A.M. The stability of two stratified non-Newtonian liquids in Couette flow. J. Non-Newton. Fluid Mech. 1987, *24*, 161.

69. Khomami, B. Interfacial stability and deformation of two stratified power law fluids in plane poiseuille flow. 1. Stability analysis. J. Non-Newton. Fluid Mech. 1990, *36*, 289.

70. Khomami, B. Interfacial stability and deformation of two stratified power law fluids in plane Poiseuille flow. 2. Interface deformation. J. Non-Newton. Fluid Mech. 1990, *37*, 19.

71. Khan, A.A.; Han, C.D. A study on the interfacial instability in the stratified flow of two viscoelastic fluids through a rectangular duct. Trans. Soc. Rheol. 1977, *21*, 101.

72. Renardy, Y. Stability of the interface in two-layer Couette flow of upper convected Maxwell liquids. J. Non-Newton. Fluid Mech. 1988, *28*, 99.

73. Chen, K.P. Interfacial instability due to elastic stratification in concentric co-extrusion of two viscoelastic fluids. J. Non-Newton. Fluid Mech. 1991, *40* (2), 155.

74. Chen, K.P. Elastic instability of the interface in Couette flow of viscoelastic liquids. J. Non-Newton. Fluid Mech. 1991, *40* (2), 261.

75. Su, Y.Y.; Khomami, B. Interfacial stability of multilayer viscoelastic fluids in slit and converging channel die geometries. J. Rheol. 1992, *36*, 357.

76. Su, Y.Y.; Khomami, B. Purely elastic interfacial instabilities in superposed flow of polymeric fluids. Rheol. Acta 1992, *31*, 413.

77. Hinch, E.J.; Harris, O.J.; Rallison, J.M. The instability mechanism for two elastic liquids being coextruded. J. Non-Newton. Fluid Mech. 1992, *43* (2–3), 311.

78. Chen, K.P.; Zhang, Y. Stability of the interface in coextrusion flow of two viscoelastic fluids through a pipe. J. Fluid Mech. 1993, *247*, 489.

79. Wilson, H.J.; Rallison, J.M. Short wave instability of coextruded elastic liquids with matched viscosities. J. Non-Newton. Fluid Mech. 1997, *72* (2–3), 237.

80. Wilson, G.M.; Khomami, B. An experimental investigation of interfacial instabilities in multilayer flow of viscoelastic fluids. 1. Incompatible polymer systems. J. Non-Newton. Fluid Mech. 1992, *45*, 355.

81. Wilson, G.M.; Khomami, B. An experimental investigation of interfacial instabilities in multilayer flow of viscoelastic fluids. II. Elastic and nonlinear effects in incompatible polymer systems. J. Rheol. 1993, *37*, 315.

82. Wilson, G.M.; Khomami, B. An experimental investigation of interfacial instabilities in multilayer flow of viscoelastic fluids. III. Compatible polymer systems. J. Rheol. 1993, *37*, 341.

83. Khomami, B.; Wilson, G.M. An experimental investigation of interfacial instability in superposed flow of viscoelastic fluids in a converging/diverging channel geometry. J. Non-Newton. Fluid Mech. 1995, *58*, 47.

84. Khomami, B.; Ranjbaran, M.M. Experimental studies of interfacial instabilities in multilayer pressure-driven flow of polymeric melts. Rheol. Acta 1997, *36*, 345.

85. Ranjbaran, M.M.; Khomami, B. The effect of instabilities on the strength of the interface in two-layer plastic structures. Polym. Eng. Sci. 1996, *36*, 1875.

86. Wang, L.; Shogren, R.L.; Carriere, C. Preparation and properties of thermoplastic starch–polyester laminate sheets by coextrusion. Polym. Eng. Sci. 2000, 40 (2), 499.
87. Vlcek, J.; Vlachopoulos, J.; Perdikoulias, J. Investigating interface instabilities in coextrusion. SPE-ANTEC Tech. Pap. 1993, 39 (3), 3365.
88. Perdikoulias, J.; Tzoganakis, C. Interfacial instability in blown-film coextrusion of polyethylenes. Plast. Eng. 1996, 52 (4), 41.
89. Schrenk, W.J.; Marcus, S.A. New developments in coextruded high barrier plastic food packaging. TAPPI/PLC Conference Proceedings, 1991; 23.
90. Ramanathan, R.; Schrenk, W.J. Flow visualization: a contemporary experimental tool to decipher stability of multilayer coextrusion flows. SPE-RETEC Conference Proceedings. Coextrusion VI—The Real World of Coextrusion 1991, 23.
91. Shanker, R.; Ramanathan, R. Effect of die geometry on flow kinematics in extrusion dies. SPE-ANTEC Tech. Pap. 1995, 41, 65.
92. Ramanathan, R.; Shanker, R.; Rehg, T.; Jons, S.; Headley, D.L.; Schrenk, W.J. Wave pattern instability in multilayer coextrusion—an experimental investigation. SPE-ANTEC Tech. Pap. 1996, 42, 224.
93. Perdikoulias, J.; Tzoganakis, C. Interfacial instabilities during coextrusion. SPE-ANTEC Tech. Pap. 1997, 43, 351.
94. Tzoganakis, C.; Perdikoulias, J. Interfacial instabilities in coextrusion flows of low-density polyethylenes: experimental studies. Polym. Eng. Sci. 2000, 40 (5), 1056.
95. Martyn, M.T.; Gough, T.; Spares, R.; Coates, P.D. Visualization of melt interface in a coextrusion geometry. SPE-ANTEC Tech. Pap. 2001, 47 (2), 1925.
96. Anturkar, N.R.; Papanastasiou, T.C.; Wilkes, J.O. Stability of coextrusion through converging dies. J. Non-Newton. Fluid Mech. 1991, 41 (1–2), 1.
97. Pinarbasi, A.; Liakopoulos, A. Stability of two-layer flow of Carreau–Yasuda and Bingham-like fluids. J. Non-Newton. Fluid Mech. 1995, 57, 227.
98. Laure, P.; Le Meur, H.; Demay, Y.; Saut, J.; Scotto, S. Linear stability of multilayer plane poiseuille flows of Oldroyd B fluids. J. Non-Newton. Fluid Mech. 1997, 71, 1.
99. Ganpule, H.K.; Khomami, B. A theoretical investigation of interfacial instabilities in the three layer superposed channel flow of viscoelastic fluids. J. Non-Newton. Fluid Mech. 1998, 79, 315.
100. Ganpule, H.K.; Khomami, B. The effect of transient viscoelastic properties on interfacial instabilities in superposed pressure driven channel flows. J. Non-Newton. Fluid Mech. 1999, 80, 217.
101. Ganpule, H.K.; Khomami, B. An investigation of interfacial instabilities in the superposed channel flow of viscoelastic fluids. J. Non-Newton. Fluid Mech. 1999, 81, 27.
102. Khomami, B.; Su, K.C. An experimental/theoretical investigation of interfacial instabilities in superposed pressure-driven channel flow of Newtonian and well characterized viscoelastic fluids. Part I. Linear stability and encapsulation effects. J. Non-Newton. Fluid Mech. 2000, 91, 59.

103. Khomami, B.; Renardy, Y.; Su, K.C.; Clarke, M.A. An experimental/ theoretical investigation of interfacial instabilities in superposed pressure-driven channel flow of Newtonian and well characterized viscoelastic fluids. Part II. Nonlinear stability. J. Non-Newton. Fluid Mech. 2000, *91*, 85.

104. Hrymak, A.N.; Vlachopoulos, J. Simulation of coextrusion flows of polymer melts. Seikei Kako 1998, *10* (9), 695.

105. Mavridis, H.; Shroff, R.N. Multilayer extrusion: experiments and computer simulation. Polym. Eng. Sci. 1994, *34* (7), 559.

106. Rincon, A.; Hrymak, A.; Vlachopoulos, J.; Dooley, J. Transient simulation of coextrusion flows in coat-hanger dies. SPE-ANTEC Tech. Pap. 1997, *43*, 335.

107. Matsunaga, K.; Kajiwara, T.; Funatsu, K. Numerical simulation of multi-layer flow for polymer melts. A study of the effect of viscoelasticity on interface shape of polymers within dies. Polym. Eng. Sci. 1998, *38* (7), 1099.

108. Valette, R.; Laure, P.; Demay, Y.; Fortin, A. Convective instabilities in the coextrusion process. Int. Polym. Proc. 2001, *16* (2), 192.

109. Perdikoulias, J.; Zatloukal, M.; Tzoganakis, C. Predicting the onset of inter-facial instabilities in coextrusion flows. TAPPI/PLC Conf. Proc. 1999, *2*, 1003.

110. Tzoganakis, C.; Zatloukal, M.; Perdikoulias, J.; Saha, P. Viscoelastic stress calculation in multilayer coextrusion dies. SPE-ANTEC Tech. Pap. 2000, *46* (1), 46.

111. Zatloukal, M.; Tzoganakis, C.; Vlcek, J.; Saha, P. Numerical simulation of polymer coextrusion flows. A criterion for detection of "wave" interfacial in-stability. Int. Polym. Proc. 2001, *16* (2), 198.

112. Zatloukal, M.; Vlcek, J.; Tzoganakis, C.; Saha, P. The effect of layer stretching on the onset of "wave" interfacial instabilities in coextrusion flows. SPE-ANTEC Tech. Pap. 2001, *47* (1), 44.

113. Zatloukal, M.; Vlcek, J.; Tzoganakis, C.; Saha, P. Viscoelastic stress calculation in multilayer coextrusion dies: die design and extensional viscosity effects on the onset of "wave" interfacial instabilities. Polym. Eng. Sci. 2002, *42* (7), 1520.

# 13

## Tiger Stripes: Instabilities in Injection Molding

**A.C.B. Bogaerds**
Eindhoven University of Technology, Eindhoven,
and DSM Research, Geleen, The Netherlands

**G.W.M. Peters and F.P.T. Baaijens**
Eindhoven University of Technology, Eindhoven, The Netherlands

### 13.1 INTRODUCTION

A numerical approach is presented that allows one to investigate the linear viscoelastic stability behavior of simplified injection molding flows. Important questions that can be addressed with this approach are: "What are critical flow conditions for the onset of an instability?" and "How does this relate to the rheology of the polymer?" An even more intriguing question that can be studied is "What is the physical mechanism that drives the secondary flow?" Important aspects of such a study are the numerical algorithm and the choice of the rheological model. The numerical algorithm is crucial for a correct prediction of the stability behavior of complex flow and, therefore, much attention is paid to the numerical aspects. Most numerical schemes produce approximate solutions that are not real solutions of the original problem. Moreover, the outcome of the stability analysis will depend on the capability of the nonlinear viscoelastic constitutive model to describe real polymer melts.

### 13.1.1 Background

Industrial processing of polymeric materials is often limited by the occurrence of flow instabilities. This can have severe implications on production rates as well as final product properties. Some of the most notorious examples include polymer extrusion where increasing flow rates can lead to the well-known shortwave sharkskin instability (this volume, Chapter 5 by Migler) or, when production speeds are increased further, gross melt fracture (this volume, Chapter 7 by Dealy). Another example can be found in the extrusion of multilayer films where a jump of the normal stresses and viscosity mismatch allows for spatial growing of oscillations of the fluid/fluid interface (this volume, Chapter 12 by Dooley). The focus of this chapter will be on viscoelastic flow instability during the filling stage of the injection molding process. A direct result of this instability is a specific surface defect characterized by shiny and dull bands roughly perpendicular to the flow direction, which alternate on the upper and lower surfaces of the mold as shown in Fig. 13.1. These defects, which are referred to as flow marks, tiger stripes, or ice lines, have been observed in a variety of polymer systems including polypropylene (17), acrylonitrile–styrene–acrylate (ASA) (20), ethylene–propylene block copolymers (65) and polycarbonate (PC)/acrylonitrile–butadiene–styrene (ABS) blends (38,41). The occurrence of these defects limits the use of injection-molded parts especially in unpainted applications such as car bumpers.

The nature of the alternating bands depends on the polymer material. With polypropylene and ASA injection molding, flow marks appear as dull, rough bands on the normally smooth and shiny surface (16,17,20). Scanning electron micrographs revealed that the flow marks have a striated surface topology showing hills and valleys oriented in the flow direction (20). For polymer blend systems, Hamada and Tsunasawa (38) suggest that the differ-

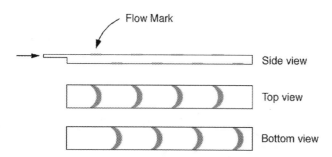

**FIGURE 13.1** Characteristic pattern for flow mark surface defects.

ences in reflectivity can be associated with differences in blend composition at the flow marks. During steady injection molding of PC/ABS blends, these authors found that the polycarbonate phase seems to preferentially coat the mold wall, leaving a shiny surface. In contrast, the flow mark bands were found to contain a higher concentration of ABS and were cloudy. By selectively etching the ABS component, approximate streamline patterns could be observed in cross sections of the products. When the smooth PC-rich surface was being deposited, the blend morphology showed a symmetric, smooth flow pattern approaching the free surface. However, when the flow front passed through the region where flow marks were being deposited, the steady flow pattern near the free surface had been disrupted and was no longer symmetric.

Other recent experimental findings have also concluded that the surface defects are the result of unstable flows near the free surface, similar to the flow pattern that is shown in Fig. 13.2 (17,20,38,41,65). The two most common mechanisms that have been proposed for the unstable flow are slip at the wall (20,41,65) and an instability at the stagnation point (17,65). Because of the limited availability of rheological data, there is no clear understanding of the rheological dependence of the instability, although Chang (20) found that materials with a higher recoverable shear strain had less severe flow mark surface defects.

A similar unstable flow mechanism was postulated by Rielly and Price (74) to explain the transfer of pigments during injection molding of high-density polyethylene. If a small amount of red pigment or crayon was placed on one mold surface, a transfer mark would be present on the *opposite* wall downstream of the original mark. The transfer was attributed to an "end-over-end" flow pattern that was found to depend on injection speed and mold thickness. The type of polymer (i.e., the rheological behavior) was also important because transfer marks were not observed for a cellulose acetate or a polystyrene polymer (74). Wall slip has been proposed as a possible mechanism for the transfer marks by Denn (22), but it may also have been caused by the same flow instability that causes flow mark surface defects (Wissbrun, K.F. Plastic flow in injection molds, private communication, 2001).

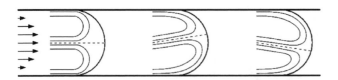

FIGURE 13.2 Unstable flow may cause surface defects.

Because of the complexity of the injection molding process (three-dimensional; nonisothermal flow; fully elastic material rheology with many time scales; crystallization; morphology changes in blends, fiber, or particulate reinforcement) it is not possible to address every aspect fully (43). A large body of work has focused on one of the many different aspects of the injection molding process. For example, the kinematics of injection molding of viscous shear thinning materials is fairly well understood (43). Whereas no simulations have been performed specifically investigating flow mark surface defects, the fountain flow near the advancing free surface (where the stagnation point instability has been postulated) has been investigated, initially by Rose in 1961. As fluid elements move toward the advancing interface, they "spill over towards the wall region being vacated by the advancing interface" (75) as illustrated in Fig. 13.3 (left).

The effect of the fountain flow on quenched stresses in injection-molded products was examined in detail by Tadmor (88) and more recently by Mavridis et al. (62). The deformation history experienced by the fluid elements in the fountain flow can have a significant impact on molecular orientation and trapped stresses in the injection-molded product. This is especially true in the surface layer since material that is deposited on the mold surface, with the polymer molecules in a stretched state, will be rapidly cooled, creating a "skin layer" with high residual stress. Material closer to the core region cools more slowly so the molecular stretch and orientation can relax (62,88). Because it is the skin layer that determines surface reflectivity, the uniformity of the elongational flow at the stagnation point will have a direct impact on surface quality.

There are significant difficulties with incorporating elasticity into simulations of the free surface flow because of the geometric "stick–slip" singularity that exists at the contact point where the free surface intersects the mold wall as summarized by Shen (80). Elastic constitutive equations are known to make geometric singularities more severe (36,39). To make elastic injection molding simulations tractable, many researchers have incorporated

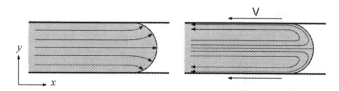

**FIGURE 13.3** Kinematics of fountain flow region: Reference frame of mold (left) and reference frame of the moving interface (right).

slip along the wall near the singularity (62,77). Various formulations for the slip condition do not seem to have a strong effect on the steady-state kinematics in the fountain flow region, but all seem to ease the difficulties associated with numerical calculations, especially for elastic constitutive equations (61,62,80).

Perhaps due to these difficulties associated with the geometric singularity, there have been few fully elastic simulations of injection molding flow (i.e., coupled velocity and stress calculations with an elastic constitutive relation). The few studies that have used realistic constitutive equations for polymer melts, such as the Leonov model (62), White–Metzner model (45,46), and Oldroyd-B model (77), were mostly focused on modeling the deformation of tracer particles by the fountain flow or predicting quenched elastic stresses in the final product and, unfortunately, did not investigate the stresses in the fountain flow. As for other complex flows, such as flow around a cylinder, there have been numerous studies using various numerical methods and viscoelastic constitutive equations as summarized in a review by Baaijens (2).

### 13.1.2 Experimental Motivation

At the DSM research laboratories, a series of injection molding experiments were carried out on several commercial, impact-modified polypropylene compounds (16,17). The tests were performed on a standard bar-shaped ruler mold with a length of 300 mm, width of 30 mm, and thickness of 3 mm (Fig. 13.4). The frequency and severity of the flow mark surface defects were recorded as a function of several molding parameters including mold and melt temperature and mold design as well as geometric factors such as mold width, injection screw diameter and buffer size. From the results, several potential mechanisms that had been proposed to explain the occurrence of flow marks were discarded. Because the defects did not depend on buffer size or screw and

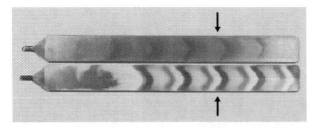

FIGURE 13.4  Two-color injection molding experiment compared with traditional injection-molded sample. (From Ref. 17.)

nozzle geometry, the possibility of an upstream instability in the nozzle or gate was ruled out. The mold surface was modified by coating the mold with a very thin layer of silicone oil or coating one side of the mold with a fluoropolymer. No effect on the frequency of the surface defect was found, so slip at the wall was discarded as the cause of the flow marks. This leaves the possibility of instability during mold filling.

To further investigate this possible mechanism, a new two-color injection molding technique was developed. The ruler mold was filled with polymer where the bottom 47% of the polymer had been dyed black. If the flow is stable, the white material should flow along the symmetry line in the center toward the free surface where it will be split by the stagnation point, leaving a thin coating on the top and bottom surfaces of the bar. Instead, both surfaces of the bar displayed alternating black and white strips that corresponded both in location and in frequency to the surface defects in the original experiments (Fig. 13.4). This technique allows the investigation of the causes of the surface defects, independent of the behavior of the polymer on the cold mold walls.

Short-shot experiments were also performed using the two-color injection molding technique. Fittings were placed in the mold allowing the ruler mold to be only partially filled. These experiments were carried out using a block of white polymer with a thin strip of black polymer along the centerline. The results for a series of tests where the mold was filled to different volume fractions are shown in Fig. 13.5. In a stable flow, the black material should coat both mold surfaces. However, instead of the symmetric fountain flow pattern expected at the interface, the black strip is first swept to the bottom

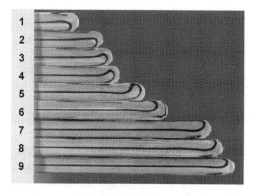

FIGURE 13.5   Short shots with two color injection molding of filled polypropylene compound. (From Ref. 17.)

side and then flipped around to the top side. The alternating colors of the surface coating exactly match the black and white striped pattern observed when the mold is completely filled. These results clearly strengthen the argument that the surface defects are caused by an instability in the fountain flow. The effects of the flow instability are apparent only in the fountain flow region and in the thin "skin layer" on the surface of the finished product. Additionally, an important observation of these experiments is that the channel flow far from the free surface remains stable.

These two-color injection molding experiments were used to reexamine the dependence of the instability on various parameters. One surprising result is that the instability does not depend on the mold temperature. However, the visibility of the surface defects in traditional injection molding experiments is strongly dependent on the mold temperature. For sufficiently high mold temperatures, the surface defect disappears because the polymers are able to relax before they solidify, but the two-color injection molding shows that the flow instability is not affected.

These experimental results lead to several simplifying assumptions when designing the model injection molding problem for the numerical simulations. This chapter will focus on two-dimensional injection molding flows. Since the instability does not depend on the temperature of the mold wall, isothermal calculations will be performed, neglecting temperature effects.

### 13.1.3 Analysis

Prediction of the stability of complex flows requires the response of the flow to arbitrary disturbances. In general, a flow is said to be stable when the perturbed flow returns to its initial or unperturbed state. Because all physical systems interact with their environment in one way or the other, perturbations of real-life flows cannot be avoided.

Quite naturally, some questions arise with regard to the analysis of the stability of general complex flows. For example, the magnitude of the initial disturbance at which the flow becomes unstable may be of significant importance. Consider Fig. 13.6, which gives two possible response diagrams to general perturbations. The horizontal axis defines the dimensionless flow parameter that characterizes the flow strength, whereas the vertical axis denotes a scalar measure for the perturbation of the flow. Note that due to the relatively slow motion and high viscosity of the polymer melt, inertia effects can be neglected and the only critical parameter that drives the flow is the dimensionless Weissenberg number (We). Although Fig. 13.6 serves only illustrative purposes and cannot possibly describe all real-life viscoelastic flows, it does show how the stability of the flow can depend on finite ampli-tude perturbations. For the branch that is referred to as a supercritical insta-

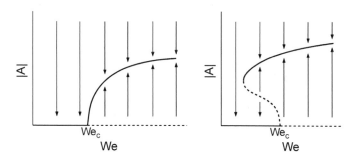

**FIGURE 13.6** Typical examples of branching diagrams for a supercritical (left) and a subcritical bifurcation (right). The horizontal axis denotes the branching parameter, which is usually the dimensionless Weissenberg number (We) for polymer melt flows. The vertical axis denotes a scalar norm of the perturbed flow. For both bifurcations there exists a critical parameter (We$_c$) for which the flow loses stability when the imposed perturbation is small (i.e. $|A| \to 0$).

bility, the critical Weissenberg number at which instability sets in does not depend on the magnitude of the initial perturbation, whereas a subcritical (nonlinear) instability can be initiated by finite perturbations for We < We$_c$ (23). Hence, the limit of stability for the supercritical case can be obtained from a local analysis in which only infinitesimal disturbances are considered, whereas the subcritical instability requires a global (nonlinear) analysis on top of the local analysis.

The full stability analysis of the viscoelastic flow problem obviously includes a (weakly) nonlinear analysis to determine, for instance, the character of the bifurcation point We$_c$ (if We$_c$ exists). However, from a numerical point of view, the full nonlinear analysis of the viscoelastic stability problem of a complex flow is not feasible and may not even be desirable yet. From an experimental point of view, the branching point (We$_c$) of a finite amplitude or subcritical instability is not well defined because one should exclude all finite perturbations from the experiments. Very often, the steady flow breaks up into unstructured flow patterns depending on the magnitude of the initial perturbation, and reproducible experiments are consequently difficult to obtain. The instability found in the experiments described above results in a recurring flow pattern, which might indicate that the instability sets in for infinitesimal disturbances. In any case, the linear analysis provides an upper bound for which the flow is guaranteed to lose stability. The main concern of this work is to determine the critical Weissenberg number (We$_c$) from a local analysis. If this point exists, it can be obtained from the analysis of infinitesimal disturbances using the linearized governing set of equations.

As suggested by the experiments in the previous section, the steady flow breaks down and a periodic motion is superposed on the steady flow. If this periodic secondary flow is dictated by the small amplitude (linear) dynamics, $We_c$ may be regarded as a so-called Hopf bifurcation point. Mathematically, this means that the linearized set of equations has a pair of complex eigenvalues crossing the axis of neutral stability (23,78). However, it cannot be excluded that the periodic motion observed in the injection molding experiments is the result of nonlinear interactions between the perturbed variables. Nevertheless, if $We_c$ exists and irrespective of the nature of this branching point, the first task for any stability analysis of complex flows such as these is to find this critical bifurcation point.

From a processing perspective, the most interesting questions are related to the position of the branching point. Because processing conditions directly translate into the Weissenberg number, numerical analyses of the stability problem of polymer melt flows might provide the necessary insight in the limiting conditions of the forming process. Next to this, another very important question that should be posed is how the stability of the flow is influenced by the (nonlinear) material behavior of the polymer melt. One of the major advantages of numerical modeling of the stability problem is the fact that it provides an indication on how to change processing parameters or even the molecular structure of the polymer. To relate this molecular structure to the stability characteristics of complex flows, application of realistic rheological models is required. Hence, the development of constitutive models that are able to predict realistic dynamics of polymer melts is of critical importance for the stability analysis of these types of flows.

The primary goal of this research is to analyze the stability problem of viscoelastic processing flows. To be able to predict the stability characteristics of these complex flows (which are complex regarding both kinematics as well as material properties), some important issues should be addressed thoroughly. First, an appropriate constitutive model for the stress should be selected. Apart from providing accurate predictions for the steady base flows, such a model should also be able to capture the essential dynamics of the polymer melt. Second, the apparent loss of temporal stability of many numerical schemes for viscoelastic flow computations should be taken into account. Especially for moderate to high Weissenberg number flows this can become a severe problem when spurious (i.e., purely numerical) instabilities are computed while the flow itself is physically stable.

The influence of the chosen rheological model on the stability dynamics is investigated and an efficient numerical algorithm is presented that is both numerically stable and accurate. Special attention is paid to deformations of the computational domain that result from perturbations of fluid/gas or fluid/

fluid interfaces. Finally, preliminary analyses of the fountain flow stability problem are presented and discussed.


## 13.2 GOVERNING EQUATIONS

Only incompressible, isothermal, and inertialess flows are considered. In the absence of body forces, these flows can be described by a reduced equation for conservation of momentum [Eq. (13.1)] and conservation of mass [Eq. (13.2)]:

$$\nabla \cdot \sigma = \vec{0} \tag{13.1}$$

$$\nabla \cdot \vec{u} = 0 \tag{13.2}$$

with $\nabla$ the gradient operator and $\vec{u}$ the velocity field. The Cauchy stress tensor $\sigma$ can be written as

$$\sigma = -p\mathbf{I} + \tau \tag{13.3}$$

with an isotropic pressure $p$ and the extra stress tensor $\tau$. The set of equations is supplemented with the kinematic conditions that describe the temporal evolution of the free surface:

$$\vec{n} \cdot \frac{\partial \vec{x}}{\partial t} = \vec{u} \cdot \vec{n} \tag{13.4}$$

where $\vec{x}$ denotes the local position vector, which describes the free surface, and $\vec{n}$ the associated outward normal vector.

In order to obtain a complete set of equations, the extra stress should be related to the kinematics of the flow. The choice of this constitutive relation will have a major impact on the results of the stability analysis (34). Motivated by the excellent quantitative agreement of the Pom-Pom constitutive predictions and dynamical experimental data (33,42,91), we use the approximate differential form of the eXtended Pom-Pom (XPP) model to capture the rheological behavior of the fluid. As is customary for most polymeric fluids, the relaxation spectrum is discretized by a discrete set of $M$ viscoelastic modes:

$$\tau = \sum_{i=1}^{M} \tau_i \tag{13.5}$$

For a (branched) polymer melt, this multimode approach introduces a set of equivalent Pom-Poms each consisting of a backbone and a number of dangling arms. The original differential form of the Pom-Pom model for a single viscoelastic mode as it was introduced by McLeish and Larson (64) is formulated as a decoupled set of equations for the evolution of orientation of the backbone tube **S**:

$$\overset{\triangledown}{\mathbf{A}} + \frac{1}{\lambda_b}\left[\mathbf{A} - \frac{\mathbf{I}}{3}\right] = 0 \qquad \mathbf{S} = \frac{\mathbf{A}}{\mathrm{tr}(\mathbf{A})} \tag{13.6}$$

and evolution of the backbone tube stretch ($\Lambda$):

$$\frac{\partial \Lambda}{\partial t} = \Lambda[\mathbf{D} : \mathbf{S}] - \frac{e^v(\Lambda - 1)}{\lambda_s}[\Lambda - 1] \qquad \text{for } \Lambda \le q \tag{13.7}$$

which relate to the polymeric stress by

$$\tau = G(3\Lambda^2\mathbf{S} - \mathbf{I}) \tag{13.8}$$

The upper convected derivative of **A**, a tensor that acts as an auxiliary variable to define the orientation of the backbone tube, is defined by

$$\overset{\triangledown}{\mathbf{A}} = \frac{\partial \mathbf{A}}{\partial t} + \vec{u} \cdot \vec{\nabla}\mathbf{A} - \mathbf{L} \cdot \mathbf{A} - \mathbf{A} \cdot \mathbf{L}^T \tag{13.9}$$

The characteristic time scale of the relaxation of the backbone orientation is defined by $\lambda_b$, whereas relaxation of the tube stretch is controlled by $\lambda_s$. The parameter $v$ in Eq. (13.7) is taken, based on the ideas of Blackwell et al. (6), as $v = 2/q$, with $q$ the number of dangling arms at both ends of the backbone. The plateau modulus is represented by $G$ whereas the kinematics of the flow are governed by fluid velocity $\vec{u}$, velocity gradient $\mathbf{L} = \vec{\nabla}\vec{u}^T$, and rate of deformation $\mathbf{D} = (\mathbf{L} + \mathbf{L}^T)/2$.

Recently, Verbeeten et al. (91) have modified the Pom-Pom model and effectively combined the set of Eqs. (13.6), (13.7), and (13.8) into a single relation for the extra stress. Furthermore, they were able to extend the model with a second normal stress difference, which is absent in the original approximate Pom-Pom formulation. The XPP model is defined by

$$\overset{\triangledown}{\tau} + \left\{\frac{1}{\lambda_b}\left[\frac{\alpha}{G}\tau \cdot \tau + \mathcal{F}\tau + G(\mathcal{F} - 1)\mathbf{I}\right]\right\} - 2G\mathbf{D} = 0 \tag{13.10}$$

with an auxiliary scalar valued function $\mathcal{F}$:

$$\mathcal{F} = 2re^v(\Lambda - 1)\left[1 - \frac{1}{\Lambda}\right] + \frac{1}{\Lambda^2}\left[1 - \frac{\alpha\mathrm{tr}(\tau \cdot \tau)}{3G^2}\right] \tag{13.11}$$

and tube stretch:

$$\Lambda = \sqrt{1 + \frac{\mathrm{tr}(\tau)}{3G}} \tag{13.12}$$

The ratio of both relaxation times is defined as $r = \lambda_b/\lambda_s$ and again, $v$ is taken as $2/q$. The second normal stress difference ($N_2$) is controlled with the additional parameter $\alpha(N_2 \neq 0$ for $\alpha \neq 0)$, which amounts to anisotropic relaxation of the backbone orientation.

## 13.3 FINITE ELEMENT ANALYSIS

In this section we present the numerical model that is used to determine the linear stability characteristics of the injection molding process. In general, linear stability analysis requires an expansion of the governing equations on the computational domain in which only first-order terms of the perturbation variables are retained. Hence, neglecting higher order terms, we may express the physical variables as the sum of the steady-state and perturbed values. For instance, we can write for the polymeric stress:

$$\tau(\vec{x}, t) = \tilde{\tau}(\vec{x}) + \tau'\,(\vec{x}, t) \tag{13.13}$$

where the tilde above $\tau$ denotes the steady-state value and $\tau'$ denotes the perturbation of the extra stress. Once the steady-state values of the unknowns are obtained, the resulting evolution equations for the perturbation variables are solved as a function of time with initially random perturbations of the extra stress variables. The transient calculations are continued until exponential growth (or decay) is obtained for the $L_2$ norm of the perturbation variables or until the norm of the perturbation has dropped below a threshold value. To simplify the governing linear stability formulations in the sequel of this chapter, the accent on the perturbation variables will be omitted since these are the primary degrees of freedom of the analysis.

The computational domain on which we perform our analysis is presented in Fig. 13.7. The governing equations for both the steady state and the stability computations are solved using the stabilized DEVSS-G/ SUPG method (14,37,87). The spatial discretization based on continuous

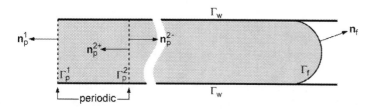

**FIGURE 13.7** Computational domain for fountain flow. In addition to the solid walls ($\Gamma_w$) there is a free surface ($\Gamma_f$) with outward normal $n_f$. The upstream boundary conditions are imposed on a periodic domain ($\Gamma_p^1, \Gamma_p^2$).

interpolation for all variables has been shown to produce accurate estimates of the stability problem on several occasions using a fully implicit temporal integration scheme (10,15,34). Here, we employ an efficient operator splitting θ scheme for the temporal evolution of the perturbed variables.

Due to the presence of a free surface ($\Gamma_f$, Fig. 13.7) during flow, the computational domain does not remain constant during evolution of an arbitrary disturbance. Hence, the next section describes a temporal integration scheme of the weak formulation of the perturbation equations, which is able to take linearized deformations of the free surface into account.

One of the major concerns when the linear stability behavior of viscoelastic flows is investigated by means of temporal integration of the governing equations using a finite element method (FEM) is the appearance of "spurious" solutions that result from the spatial discretization of the differential operator. For one-dimensional generalized eigenvalue problems, these artifacts of the numerical solution methods are fairly well documented (11,32,71). However, on complex flows, for which FEM has its practical relevance, the amount of literature is far less extensive (13,15,83,87).

We will first describe our FEM on stationary domains before moving on to domains that do not need to remain stationary due to the presence of free surfaces or fluid/fluid interfaces. Hence, if the governing set of linearized equations is considered:

$$\lambda \left( \frac{\partial \boldsymbol{\tau}}{\partial t} + \tilde{\vec{u}} \cdot \nabla \boldsymbol{\tau} + \vec{u} \cdot \nabla \tilde{\boldsymbol{\tau}} - \tilde{\bar{\mathbf{G}}} \cdot \boldsymbol{\tau} - \boldsymbol{\tau} \cdot \tilde{\bar{\mathbf{G}}}^T - \overline{\mathbf{G}} \cdot \tilde{\boldsymbol{\tau}} - \tilde{\boldsymbol{\tau}} \cdot \overline{\mathbf{G}}^T \right)$$
$$+ \ \boldsymbol{\tau} - \eta(\overline{\mathbf{G}} + \overline{\mathbf{G}}^T) = 0 \tag{13.14}$$

$$-\nabla \cdot \left( \boldsymbol{\tau} + \alpha(\nabla \vec{u} - \overline{\mathbf{G}}^T) \right) + \nabla p = \vec{0} \tag{13.15}$$

$$\nabla \cdot \vec{u} = 0 \tag{13.16}$$

$$\overline{\mathbf{G}}^T - \nabla \vec{u} = 0 \tag{13.17}$$

with $\overline{\mathbf{G}}$ the auxiliary velocity gradient tensor and $\alpha$ the stabilizing parameter as defined for the DEVSS method (37). Again, the tildes denote the base flow variables. In order to obtain an efficient time marching scheme for Eqs. (13.14)–(13.17), we consider an operator splitting method to perform the temporal integration. The major advantage of operator splitting methods is the decoupling of the viscoelastic operator into parts that are "simpler" and can be solved more easily than the full problem. Due to the stiffness of the viscoelastic stability problem as defined above, it is worthwhile to consider a $\theta$ scheme to perform the temporal integration (28,29). Hence, if we write Eqs. (13.14)–(13.17) as

$$\frac{\partial x}{\partial t} = \mathcal{A}(x) = \mathcal{A}_1(x) + \mathcal{A}_2(x) \tag{13.18}$$

the $\theta$ scheme is defined following (29)

$$\frac{x^{n+\Theta} - x^n}{\Theta \Delta t} = \mathcal{A}_1\left(x^{n+\Theta}\right) + \mathcal{A}_2(x^n) \tag{13.19}$$

$$\frac{x^{n+1-\Theta} - x^{n+\Theta}}{(1 - 2\Theta)\Delta t} = \mathcal{A}_1(x^{n+\Theta}) + \mathcal{A}_2(x^{n+1-\Theta}) \tag{13.20}$$

$$\frac{x^{n+1} - x^{n+1-\Theta}}{\Theta \Delta t} = \mathcal{A}_1(x^{n+1}) + \mathcal{A}_2(x^{n+1-\Theta}) \tag{13.21}$$

with time step $\Delta t$ and $\theta = 1/\sqrt{2}$ in order to retain second-order accuracy. Formally, only the constitutive equation and the perturbation equation for the interface contain the temporal derivatives, which implies that the first term of Eq. (13.18) should be multiplied by a diagonal operator with the only nonzero entrees being the ones corresponding to these equations. The remaining problem is to define the separate operators $\mathcal{A}_1$ and $\mathcal{A}_2$. In essence, we like to choose $\mathcal{A}_1$ and $\mathcal{A}_2$ in such a way that solving Eqs. (13.19)–(13.21) requires far less computational effort as compared to solving the implicit problem while the stability envelope of the time integrator remains sufficiently large. If the simplified problem $\mathcal{A}_1 = \beta \mathcal{A}$ and $\mathcal{A}_2 = (1 - \beta)\mathcal{A}$ is considered, the stability envelopes are plotted in Fig. 13.8 for different values of $\beta$. The arrows point towards the region of the complex plane for stable time integration. Obviously, setting $\beta = 0$ yields only a small portion of the complex plane, whereas $0.5 \le \beta \le 1.0$ results in a scheme that is unconditionally stable. Based on the argument that we split the viscoelastic operator into a kinematic problem and a transport problem for the advection of polymer stress, we can define $\mathcal{A}_1$ and $\mathcal{A}_2$ from the approximate location of the governing eigenvalues. For instance, the viscous (Stokes) problem has eigenvalues that are essentially real and negative. The absolute value has the tendency to grow very fast with mesh refinement (for a one-dimensional diffusion problem using a low-order FEM, $\max(|\mu|) = O(N^2)$, with $N$ the number of grid points) and it is convenient to define $\mathcal{A}_1$ as the kinematic problem for given polymer stress. On the other hand, the eigenspectrum of the remaining advection operator in the constitutive equation is located close to the imaginary axis (however, not on the imaginary axis due to the introduction of Petrov–Galerkin weighting

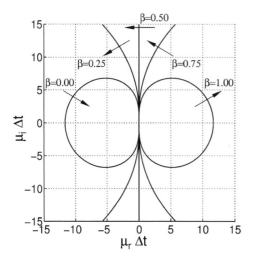

**FIGURE 13.8** Stability envelope of the $\theta$ scheme for $\mathcal{A}_1 = \beta\mathcal{A}$ and $\mathcal{A}_2 = (1 - \beta)\mathcal{A}$ with $\mu$ the spectrum of $\mathcal{A}$. Regions for stable time integration are indicated by the arrows for different values of $\beta$ and it can be seen that for $0.5 \leq \beta \leq 1.0$ this $\theta$ scheme is unconditionally stable for this special choice of $\mathcal{A}_1$ and $\mathcal{A}_2$.

functions later on) and we define $\mathcal{A}_2$ as the transport of extra stress. We can then define our $\theta$ scheme ($\mathcal{A}_1$ and $\mathcal{A}_2$) for the temporal evolution of polymer stress perturbations as

$$
\mathcal{A}_1 = - \begin{bmatrix} \lambda(\vec{u} \cdot \nabla \tilde{\tau} - \overline{\mathbf{G}} \cdot \tilde{\tau} - \tilde{\tau} \cdot \overline{\mathbf{G}}^T) + \tau - \eta(\overline{\mathbf{G}} + \overline{\mathbf{G}}^T) \\ -\nabla \cdot \left( \tau + \alpha(\nabla \vec{u} - \overline{\mathbf{G}}^T) \right) + \nabla p \\ \nabla \cdot \vec{u} \\ \overline{\mathbf{G}}^T - \nabla \vec{u} \end{bmatrix}
\tag{13.22}
$$

and:

$$
\mathcal{A}_2 = - \begin{bmatrix} \lambda(\tilde{\vec{u}} \cdot \nabla \tau - \tilde{\overline{\mathbf{G}}} \cdot \tau - \tau \cdot \tilde{\overline{\mathbf{G}}}^T) \\ 0 \\ 0 \\ 0 \end{bmatrix}
\tag{13.23}
$$

Based on the above definitions for $\mathcal{A}_1$ and $\mathcal{A}_2$, the kinematic (elliptic saddle point) problem for $\vec{u}$, $p$ and $\overline{\mathbf{G}}$, $\tau$ is updated implicitly in the first [Eq. (13.19)] and last [Eq. (13.21)] step of the $\theta$ scheme, whereas the transport of polymeric stress is updated explicitly. Conceptually, this operator splitting is very similar to the $\theta$ scheme that was developed by (76) and later applied to study the linear dynamics of complex viscoelastic flows by (85) and (81). The difference being the fact that the term

$$\lambda\left(\vec{u}\cdot\nabla\tilde{\tau} - \overline{\mathbf{G}}\cdot\tilde{\tau} - \tilde{\tau}\cdot\overline{\mathbf{G}}^T\right) - \eta\left(\overline{\mathbf{G}} + \overline{\mathbf{G}}^T\right),\tag{13.24}$$

is now contained within $\mathcal{A}_1$ rather than $\mathcal{A}_2$. The redefinition of $\mathcal{A}_1$ and $\mathcal{A}_2$ is motivated by the fact that the earlier $\theta$ schemes can only be applied to assess the linear stability dynamics of viscoelastic flows with a sufficiently large purely Newtonian contribution to the extra stress. For real polymer melts this generally is an unacceptable modification of the material description.

Notice that based on Eq. (13.22) there is hardly any gain in computational efficiency since the constitutive relation cannot be decoupled from the remaining equations in a weighted residual formulation. However, the updated polymeric stress at $t^{n+\theta}$ can be written as

$$\tau(t^{n+\theta}) = \mathbf{P}^{-1}\cdot\left[\frac{\lambda}{\Theta\Delta t}\tau(t^n) - F_1\left(\vec{u}(t^{n+\theta}), \overline{\mathbf{G}}(t^{n+\theta})\right) - F_2\left(\tau(t^n)\right)\right]\tag{13.25}$$

with

$$\mathbf{P} = \frac{\lambda + \Theta\Delta t}{\Theta\Delta t}\mathbf{I}\tag{13.26}$$

for the UCM model. The tensor functionals $F_1$ and $F_2$ depend only on the kinematics of the flow at $t^{n+\theta}$ and the polymeric stress at $t^n$:

$$F_1\left(\vec{u}(t^{n+\theta}), \overline{\mathbf{G}}(t^{n+\theta})\right) = \lambda\left(\vec{u}\cdot\nabla\tilde{\tau} - \overline{\mathbf{G}}\cdot\tilde{\tau} - \tilde{\tau}\cdot\overline{\mathbf{G}}^T\right) - \eta\left(\overline{\mathbf{G}} + \overline{\mathbf{G}}^T\right)\tag{13.27}$$

and

$$F_2\left(\tau(t^n)\right) = \lambda\left(\tilde{\vec{u}}\cdot\nabla\tau - \overline{\overline{\mathbf{G}}}\cdot\tau - \tau\cdot\overline{\overline{\mathbf{G}}}^T\right)\tag{13.28}$$

Substitution of Eq. (13.25) into (13.22) yields a modified momentum equation in which the only degrees of freedom are the kinematics of the flow at $t^{n+\theta}$. The kinematics of the flow and the constitutive relation can then be solved separately. If the appropriate finite element approximation spaces for $\vec{u}$, $\tau$, $p$, and $\overline{\mathbf{G}}$ are defined, the complete weak formulation of our $\theta$ scheme can be written as follows:

Problem $\theta$—FEM step 1a: Given the base flow variables and the perturbation stress at $t^n$, find $\vec{u}$, $p$, and $\overline{\mathbf{G}}$ at $t^{n+\theta}$ such that for all admissible test functions:

$$\left( \nabla \vec{\boldsymbol{\Phi}}_u^T, \mathbf{P}^{-1} \cdot \left[ \frac{\lambda}{\Theta \Delta t} \hat{\boldsymbol{\tau}} - \boldsymbol{F}_1(\vec{u}, \overline{\mathbf{G}}) - \boldsymbol{F}_2(\hat{\boldsymbol{\tau}}) \right] + \alpha \left( \nabla \vec{u} - \overline{\mathbf{G}}^T \right) \right)$$

$$-\left( \nabla \cdot \vec{\boldsymbol{\Phi}}_u, p \right) = 0 \tag{13.29}$$

$$\left( \Phi_p, \nabla \cdot \vec{u} \right) = 0 \tag{13.30}$$

$$\left( \Phi_\gamma, \overline{\mathbf{G}}^T - \nabla \vec{u} \right) = 0 \tag{13.31}$$

with $(\ ,\ )$ the usual $L_2$ inner product on the (stationary) domain $\Omega$, $\Phi_i$ the apropriate test functions and the hat above the perturbation variable, the perturbation variable evaluated at the previous fractional time step. The kinematics of the flows are solved separately from the constitutive relation and an update for the polymeric stress is now readily obtained from:

Problem $\theta$—FEM step 1b: Given the base flow variables and the kinematic perturbation variables at $t^{n+\theta}$, find $\tau$ at $t^{n+\theta}$ such that for all admissible test functions:

$$\left( \Phi_\tau + \frac{h^e}{|\tilde{\vec{u}}|} \tilde{\vec{u}} \cdot \nabla \Phi_\tau, \lambda \left( \frac{\boldsymbol{\tau} - \hat{\boldsymbol{\tau}}}{\Theta \Delta t} + \vec{u} \cdot \nabla \tilde{\boldsymbol{\tau}} + \tilde{\vec{u}} \cdot \nabla \hat{\boldsymbol{\tau}} - \overline{\overline{\mathbf{G}}} \cdot \hat{\boldsymbol{\tau}} - \hat{\boldsymbol{\tau}} \cdot \overline{\overline{\mathbf{G}}}^T \right. \right.$$

$$\left. \left. -\overline{\mathbf{G}} \cdot \tilde{\boldsymbol{\tau}} - \tilde{\boldsymbol{\tau}} \cdot \overline{\mathbf{G}}^T \right) + \boldsymbol{\tau} - \eta \left( \overline{\mathbf{G}} + \overline{\mathbf{G}}^T \right) \right) = 0 \tag{13.32}$$

The second step involves the solution of the transport problem:

Problem $\theta$—FEM step 2: Given the base flow variables and the perturbation variables at $t^{n+\theta}$, find $\tau$ at $t^{n+1-\theta}$ such that for all admissible test functions:

$$\left( \Phi_\tau + \frac{h^e}{|\tilde{\vec{u}}|} \tilde{\vec{u}} \cdot \nabla \Phi_\tau, \lambda \left( \frac{\boldsymbol{\tau} - \hat{\boldsymbol{\tau}}}{(1 - 2\Theta)\Delta t} + \hat{\vec{u}} \cdot \nabla \tilde{\boldsymbol{\tau}} + \tilde{\vec{u}} \cdot \nabla \boldsymbol{\tau} - \overline{\overline{\mathbf{G}}} \cdot \boldsymbol{\tau} \right. \right.$$

$$\left. \left. -\boldsymbol{\tau} \cdot \overline{\overline{\mathbf{G}}}^T - \overline{\hat{\mathbf{G}}} \cdot \tilde{\boldsymbol{\tau}} - \tilde{\boldsymbol{\tau}} \cdot \overline{\hat{\mathbf{G}}}^T \right) + \hat{\boldsymbol{\tau}} - \eta \left( \overline{\hat{\mathbf{G}}} + \overline{\hat{\mathbf{G}}}^T \right) \right) = 0. \tag{13.33}$$

The third fractional step [Eq. (13.21)] corresponds to symmetrization of the $\theta$ scheme and is similar to step 1. In the above weak formulation, addi-

tional stabilization of the hyperbolic constitutive equation is obtained by inclusion of SUPG weighting functions (14) using some characteristic grid size
$h^e$.

The above decoupling of the constitutive relation from the remaining
equations provides an efficient time integration technique that is second-order
accurate for linear stability problems. The efficiency becomes even more
evident when real viscoelastic fluids are modeled for which the spectrum of
relaxation times is approximated by a discrete number of viscoelastic modes.
For simplicity, the procedure is described for the UCM model. However, if
nonlinear models such as the PTT, Giesekus, or the XPP model are considered, as is common for polymer melts, a generalization of the $\theta$ scheme is
readily obtained since this requires only a redefinition of $P$ in Eq. (13.26) and
$P^{-1}$ can be evaluated either analytically or numerically.

A choice remains to be made for the approximation spaces of the
different perturbation variables. As is known from solving Stokes flow
problems, velocity and pressure interpolations cannot be chosen independently and need to satisfy the LBB condition. Likewise, interpolation of
velocity and extra stress has to satisfy a similar compatibility condition. We
report calculations using low-order finite elements using similar spatial
discretizations as were defined in Refs. (15) and (87) (continuous bilinear
interpolation for viscoelastic stress, pressure, and $\overline{G}$ and continuous biqua-
dratic interpolation for velocity).

Evidently, the set of equations should be supplemented with suitable
boundary conditions on parts of the boundary ($\Gamma$) of the flow domain. On
both walls, Dirichlet conditions are enforced on the velocity to account for the
no-slip conditions on the fluid–solid interface. In addition, the inflow part of
the fountain flow is considered to be periodic as shown in Fig. 13.7. The
boundary conditions on the free surface will be discussed in the next section.

## 13.4  DOMAIN PERTURBATION TECHNIQUE

The $\theta$ method described in the previous section is discussed for flows defined
on stationary domains. As a consequence, the important class of free surface
flows (and flows with internal fluid/fluid interfaces) has to be considered.
Because of the deformation of the interfaces as a result of the perturbation of
the flow, the computational domain should also be allowed to deform
according to the interface perturbation. Hence, next to the perturbation
variables defined previously, the perturbation of the computational domain $\vec{x}$
should also be considered. However, the nature of linear stability analysis
implies that the disturbances are infinitely small and the domain deformation
can be localized to the interface. Hence, in order to account for these

disturbances it is not necessary to consider the full computational domain but rather the part of the domain near the interface. In the sequel, we will use a domain perturbation technique that was derived by Carvalho and Scriven (19) for the prediction of the (Newtonian) ribbing instability observed in deformable roll coating.

If the base flow domain $(\Omega_0, \Gamma_0)$ is considered, as schematically depicted in Fig. 13.9, and allowed to be perturbed into $(\Omega, \Gamma)$, the governing equations are solved on the perturbed domain. For instance,

$$\int_\Omega (\tilde{a} + a)\, d\Omega = \int_{\xi,\eta} (\tilde{a} + a) J\, d\xi\, d\eta \tag{13.34}$$

in which the tilde denotes the base flow and $J$ the Jacobian of the mapping:

$$J = \frac{\partial(\tilde{x} + x)}{\partial \xi} \frac{\partial(\tilde{y} + y)}{\partial \eta} - \frac{\partial(\tilde{x} + x)}{\partial \eta} \frac{\partial(\tilde{y} + y)}{\partial \xi} \tag{13.35}$$

which equals approximately

$$J \approx J_0 + \left( \frac{\partial x}{\partial \xi} \frac{\partial \tilde{y}}{\partial \eta} - \frac{\partial x}{\partial \eta} \frac{\partial \tilde{y}}{\partial \xi} \right) + \left( \frac{\partial \tilde{x}}{\partial \xi} \frac{\partial y}{\partial \eta} - \frac{\partial \tilde{x}}{\partial \eta} \frac{\partial y}{\partial \xi} \right) \tag{13.36}$$

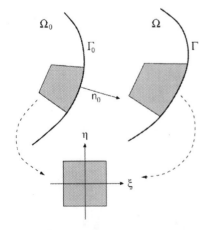

FIGURE 13.9 Mapping of steady computational domain $(\Omega_0)$ with free surface $(\Gamma_0)$ and perturbed domain $(\Omega)$ onto a reference square. The outward normal of the base flow is defined by $n_0$.

when only linear terms are retained. Hence, linearization of the right-hand side of Eq. (13.34) yields

$$\int_{\Omega_0} \tilde{a} \, d\Omega + \int_{\Omega_0} a \, d\Omega + \int_{\xi,\eta} \tilde{a} \left[ \left( \frac{\partial x}{\partial \xi} \frac{\partial \tilde{y}}{\partial \eta} - \frac{\partial x}{\partial \eta} \frac{\partial \tilde{y}}{\partial \xi} \right) \right.$$
$$\left. + \left( \frac{\partial \tilde{x}}{\partial \xi} \frac{\partial y}{\partial \eta} - \frac{\partial \tilde{x}}{\partial \eta} \frac{\partial y}{\partial \xi} \right) \right] d\xi \, d\eta \tag{13.37}$$

The first term of Eq. (13.37) corresponds to the base-state equations, which is zero, whereas the second term is part of the conventional stability equations Eqs. (13.29)–(13.33). The perturbation of the domain is restricted to the interface by the introduction of the function $H^\delta$, which equals 1 on the interface and vanishes smoothly inside the domain (Fig. 13.10). With the definition

$$\vec{x} = hH^0 \vec{n}_0 = hH^0 \left( \frac{\partial \tilde{y}}{\partial \Gamma_0}, -\frac{\partial \tilde{x}}{\partial \Gamma_0} \right) \quad \text{with} \quad H^0 = \lim_{\delta \to 0} H^\delta \tag{13.38}$$

the domain perturbation is governed by the disturbance of the interface in normal direction multiplied by $H^0$. Equation (13.37) can now be simplified significantly since the definition of $H^0$ implies

$$\int_\xi H^0 \, d\xi = 0 \quad \text{and} \quad \int_\xi \frac{\partial H^0}{\partial \xi} \, d\xi = \pm 1 \tag{13.39}$$

Effectively, this means that linearization of Eq. (13.34) now reads:

$$\int_\Omega (\tilde{a} + a) \, d\Omega \approx \int_{\Omega_0} a \, d\Omega + \int_{\Gamma_0} \tilde{a}h \, d\Gamma \tag{13.40}$$

and the last integral has been reduced to a boundary integral on the interface with the addition of the scalar function $h$ (i.e., perturbation in normal direction).

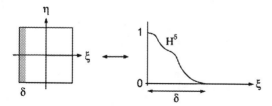

**FIGURE 13.10** Definition of $H^\delta$ which is zero inside the domain and increases smoothly towards unity on the interface. For this example, $\int \partial H^\delta / \partial \xi \, d\xi = -1$ for $\delta \to 0$.

Similar to Eq. (13.34), we can write

$$\int_\Omega \frac{\partial(\tilde{a}+a)}{\partial(\tilde{x}+x)}\, d\Omega = \int_{\xi,\eta}\left[\frac{\partial(\tilde{a}+a)}{\partial\xi}\frac{\partial(\tilde{y}+y)}{\partial\eta} - \frac{\partial(\tilde{a}+a)}{\partial\eta}\frac{\partial(\tilde{y}+y)}{\partial\xi}\right] d\xi\, d\eta$$

$$(13.41)$$

using the definition of spatial derivatives for isoparametric mappings:

$$\frac{\partial a}{\partial x} = \frac{1}{J}\left[\frac{\partial a}{\partial\xi}\frac{\partial y}{\partial\eta} - \frac{\partial a}{\partial\eta}\frac{\partial y}{\partial\xi}\right] \tag{13.42}$$

Retaining only the first-order terms, Eq. (13.41) now reduces to

$$\int_\Omega \frac{\partial(\tilde{a}+a)}{\partial(\tilde{x}+x)}\, d\Omega \approx \int_{\Omega_0}\frac{\partial\tilde{a}}{\partial\tilde{x}}\, d\Omega + \int_{\Omega_0}\frac{\partial a}{\partial\tilde{x}}\, d\Omega + \int_{\Gamma_0} h\frac{\partial\tilde{x}}{\partial\Gamma}\frac{\partial\tilde{a}}{\partial\Gamma}\, d\Gamma \quad (13.43)$$

where the first term of the right-hand side corresponds to the base flow solution and therefore vanishes from the stability equations. The extra contribution due to the perturbation of the computational domain is again reduced to a boundary integral over the stationary interface.

In order to arrive at a complete set of stability equations, the governing equations are expanded using the technique described above. The extra approximation space required for the unknown function $h$ defined on the interface consists of quadratic functions similar to the original grid. The $\theta$ scheme for deformable domains is now defined as follows:

Problem $\theta$—FEM step 1a: Given the base flow variables on $\Omega_0$, the perturbation stress and interface deformation $(\hat{h})$ at $t^n$, find $\vec{u}, p, \overline{G}$, and $h$ at $t^{n+\theta}$such that for all admissible test functions

$$\left\{\text{LHS Eq. } (13.29)\right\} - \int_{\Gamma_0} \nabla\vec{\Phi}_u^T : \mathbf{P}^{-1}$$

$$\cdot\left[\lambda\left(\tilde{u}_x\frac{\partial\tilde{x}}{\partial\Gamma} + \tilde{u}_y\frac{\partial\tilde{y}}{\partial\Gamma}\right)\frac{\partial\tilde{\tau}}{\partial\Gamma} + \tilde{\mathcal{L}}\right]h\, d\Gamma + \int_{\Gamma_0}\left(\frac{\partial\tilde{x}}{\partial\Gamma}\frac{\partial\Phi_u^x}{\partial\Gamma}\vec{e}_x\vec{e}_x\right.$$

$$\left.+ \frac{\partial\tilde{y}}{\partial\Gamma}\frac{\partial\Phi_u^y}{\partial\Gamma}\vec{e}_y\vec{e}_y + \frac{\partial\tilde{y}}{\partial\Gamma}\frac{\partial\Phi_u^x}{\partial\Gamma}\vec{e}_x\vec{e}_y + \frac{\partial\tilde{x}}{\partial\Gamma}\frac{\partial\Phi_u^y}{\partial\Gamma}\vec{e}_y\vec{e}_x\right) : \tilde{\tau}h\, d\Gamma$$

$$(13.44)$$

$$-\int_{\Gamma_0}\left(\frac{\partial\tilde{x}}{\partial\Gamma}\frac{\partial\Phi_u^x}{\partial\Gamma} + \frac{\partial\tilde{y}}{\partial\Gamma}\frac{\partial\Phi_u^y}{\partial\Gamma}\right)\tilde{p}h\, d\Gamma = 0$$

$$\left\{ \text{LHS Eq. (13.30)} \right\} + \int_{\Gamma_0} \Phi_p \left( \frac{\partial \tilde{x}}{\partial \Gamma} \frac{\partial \tilde{u}_x}{\partial \Gamma} + \frac{\partial \tilde{y}}{\partial \Gamma} \frac{\partial \tilde{u}_y}{\partial \Gamma} \right) h \, d\Gamma = 0 \quad (13.45)$$

$$\int_{\Gamma_0} \Phi_h \left[ \frac{h - \hat{h}}{\Theta \Delta t} + \left( \tilde{u}_x \frac{\partial \tilde{x}}{\partial \Gamma} + \tilde{u}_y \frac{\partial \tilde{y}}{\partial \Gamma} \right) \frac{\partial h}{\partial \Gamma} - \vec{n}_0 \cdot \vec{u} \right] d\Gamma = 0 \quad (13.46)$$

with LHS the left-hand sides of the original $\theta$ equations evaluated on the stationary domain $(\Omega_0)$. The Lagrangian residual of the stationary constitutive relation is denoted by $\tilde{\mathcal{L}}$ and defined as

$$\tilde{\mathcal{L}} = -\lambda \left( \bar{\bar{\mathbf{G}}} \cdot \tilde{\tau} + \tilde{\tau} \cdot \bar{\bar{\mathbf{G}}}^T \right) + \tilde{\tau} - \eta \left( \bar{\bar{\mathbf{G}}} + \bar{\bar{\mathbf{G}}}^T \right) \quad (13.47)$$

Equations (13.44)–(13.46) are supplemented by Eq. (13.31), which has no additional terms under the conditions discussed below. Equation (13.46) is the linearized kinematic condition on the interface. The weak form of the constitutive equation in step 1b now reads:

Problem $\theta$—FEM step 1b: Given the base flow variables on $\Omega_0$, the kinematic perturbation variables at $t^{n+\theta}$ (including $h$), find $\tau$ at $t^{n+\theta}$ such that for all admissible test functions

$$\left\{ \text{LHS Eq. (13.32)} \right\} + \int_{\Gamma_0} \left( \Phi_\tau + \frac{h^e}{|\tilde{u}|} \tilde{u} \cdot \nabla \Phi_\tau \right) :$$

$$(13.48)$$

$$\left[ \lambda \left( \tilde{u}_x \frac{\partial \tilde{x}}{\partial \Gamma} + \tilde{u}_y \frac{\partial \tilde{y}}{\partial \Gamma} \right) \frac{\partial \tilde{\tau}}{\partial \Gamma} + \tilde{\mathcal{L}} \right] h \, d\Gamma = 0$$

Notice that in the above formulation, all kinematics including the perturbation of the interface are governed by the operator $\mathcal{A}_1$.

A solution of the transport problem as defined by Eq. (13.33) is obtained from:

Problem $\theta$—FEM step 2: Given the base flow variables on $\Omega_0$ and the perturbation variables at $t^{n+\theta}$, find $\tau$ and $h$ at $t^{n+1-\theta}$ such that for all admissible test functions:

$$\left\{ \text{LHS Eq. (13.33)} \right\} + \int_{\Gamma_0} \left( \Phi_\tau + \frac{h^e}{|\tilde{u}|} \tilde{u} \cdot \nabla \Phi_\tau \right) :$$

$$(13.49)$$

$$\left[ \lambda \left( \tilde{u}_x \frac{\partial \tilde{x}}{\partial \Gamma} + \tilde{u}_y \frac{\partial \tilde{y}}{\partial \Gamma} \right) \frac{\partial \tilde{\tau}}{\partial \Gamma} + \tilde{\mathcal{L}} \right] \hat{h} \, d\Gamma = 0$$

$$\int_{\Gamma_0} \Phi_h \left[ \frac{h - \hat{h}}{(1 - 2\Theta)\Delta t} + \left( \tilde{u}_x \frac{\partial \tilde{x}}{\partial \Gamma} + \tilde{u}_y \frac{\partial \tilde{y}}{\partial \Gamma} \right) \frac{\partial \hat{h}}{\partial \Gamma} - \vec{n}_0 \cdot \hat{\tilde{u}} \right] d\Gamma = 0 \quad (13.50)$$

In the above equations, the line integrals related to the natural boundary conditions have been omitted. For instance, for free surface flows without capillary forces, the perturbation of the normal and shear stress vanishes. Also, notice that the line integrals related to the DEVSS parts of the equations have been omitted from the governing equations. This can be seen if the weak form of the Laplace operator is considered:

$$-(\vec{\Phi}_u, \nabla \cdot \alpha \nabla \vec{u}) = (\nabla \vec{\Phi}_u^T, \alpha \nabla \vec{u}) - \int_\Gamma \vec{\Phi}_u \cdot (\alpha \vec{n} \cdot \nabla \vec{u}) \, d\Gamma \qquad (13.51)$$

and linearization of the boundary integral introduces gradients of the domain perturbation normal to the boundary (i.e., $\partial h/\partial n \neq 0$). However, we cannot compute normal gradients of $h$, which are defined only on the interface, and it is necessary to define $\alpha$ in such a way that the linearized boundary integral resulting from Eq. (13.51) vanishes. Without loss of generality we can set $\alpha = 0$ on the boundary and $\alpha = 1$ inside the domain (with $\alpha$ increasing continuously from 0 to 1). It is easily observed that linearization of Eq. (13.51) with $\alpha = 1 - H^0$ yields a set of equations that only need to be evaluated on the steady-state domain.

## 13.5 RESULTS

The stability of the fountain flow problem is examined using the XPP model. As was already discussed, $\alpha \neq 0$ falls outside the scope of this chapter and hence, we assume $N_2 = 0$ for simple shear flows. The structure of the equivalent Pom-Pom is then fully determined by the nonlinear parameter $r$ ($=$ ratio of relaxation times) and $q$ ($=$ number of arms). The stability of the flow is studied as a function of the relative elastic flow strength, i.e., the dimensionless Weissenberg number We, which is defined as

$$\text{We} = \frac{2\lambda_b Q}{H^2} \qquad (13.52)$$

and is based on the imposed volumetric flow rate $Q$ and a characteristic length scale (half the channel width $H/2$).

It is well-known that most rheological models based on tube theory can show excessive shear thinning behavior in simple shear flows. This holds for the original Pom-Pom equations for which the steady-state shear stress decreases with increasing shear rate when $\lambda_b \dot{\gamma} = O(1)$. However, it also holds for certain combinations of the nonlinear parameters of the XPP model, although, compared with the original Pom-Pom equations, this maximum is shifted several orders to the right depending on the material parameters.

We use a single mode model to investigate the influence of the rheology of the polymer melt on the stability behavior of the injection molding process. Therefore, the nonlinear parameters should be chosen in such a way that the shear stress remains a monotonically growing function for the shear rates that fall within the range of the investigated flow situations. In this work, we will model differences in polymer melt rheology by variations of the parameter $q$. Figure 13.11 shows some of the steady-state viscometric functions for different values of the number of arms attached to the backbone ($q = 5, 9, 13$) and constant ratio of relaxation times ($r = 2$). This choice of parameters has a major impact on the extensional behavior of the model where extensional hardening is increased significantly with increasing number of arms. The influence of the viscometric functions in simple shear is much less severe, although it can be seen that the maximum in the shear stress–shear rate curve shifts to the left when the number of arms is decreased.

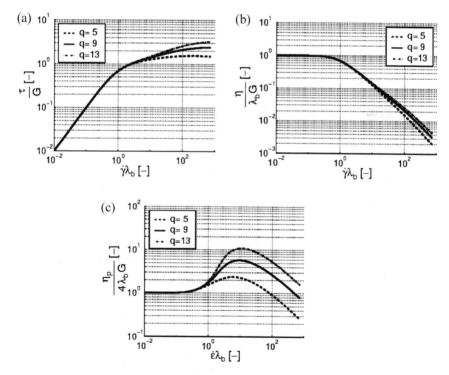

FIGURE 13.11   Steady-state viscometric functions (a) shear stress–shear rate, (b) viscosity–shear rate, and (c) planar elongational viscosity–extension rate, for different numbers of arms $q$ and $r = 2$.

TABLE **13.1**  Characteristics of the Grids Used for the Fountain
Flow Computations

|                              | Mesh 1 | Mesh 2 | Mesh 3 |
|------------------------------|--------|--------|--------|
| No. of element               | 892    | 1468   | 2202   |
| No. of elements on free surface | 20  | 30     | 40     |
| Total length to wetting point | 14H   | 14H    | 14H    |
| Length periodic inflow       | 3.5H   | 3.5H   | 3.5H   |

### 13.5.1 Steady-State Results

The computational results of viscoelastic fountain flows using the previously described XPP fluids are presented. Table 13.1 and Fig. 13.12 show the meshes that were used to analyze both the steady-state results as well as the linear stability characteristics. In the sequel of this section, steady-state results are presented that are computed on the most refined mesh, which is shown in Fig. 13.12.

Figure 13.13 shows the computed base flow solution of the fountain flow surface for various rheologies and Weissenberg numbers. In the left graph, one half of the free surface is shown for $(r = q) = (2, 9)$. Comparing the shape of the free surface as a function of We, one can observe that for small Weissenberg numbers as well as for higher We the shape of the surface is flattened. Moreover, the relative position of the stagnation point on the free surface exhibits a maximum for varying We (Fig. 13.13, right). Due to the differences in the rheologies of the XPP fluids, this maximum shifts considerably to the left for the extensional hardening fluid.

Figure 13.14 shows some base-state variables along the steady free surface. Since the fountain flow surface varies in shape and has therefore

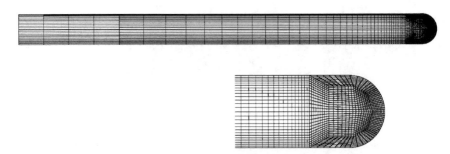

FIGURE **13.12**  Finest mesh used for the computations (mesh 3), full mesh, and detail of the fountain flow region.

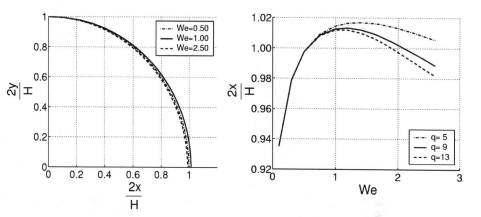

**FIGURE 13.13** Shape of the steady-state free surface (left) for $(r,q) = (2,9)$ and various values of the Weissenberg number. For $(r,q) = (2,9)$ and various values of the Weissenberg number the position of the stagnation point is located at $y = 0$. Position of the stagnation point relative to the intersection with the wall as a function of Weissenberg number (right).

variable length, we have scaled the position on the surface by its total length. The left graph shows the tangential velocity along the surface. We observe that in the stagnation point where the fluid deformation is purely extensional, the extension rate is somewhat higher for the strain hardening material. Also, from the right graph, a strong buildup of the tangential stress $\tau_n$ is seen near the walls of the mold, whereas the tangential stress is relatively constant for a large part of the free surface. Contour plots for the different rheologies at

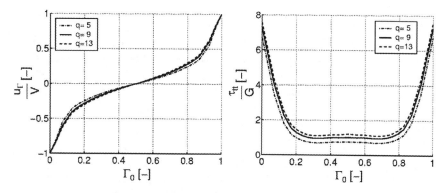

**FIGURE 13.14** Steady-state velocity (left) and the tangential stress $\tau_n$ (right) on the free surface for We = 2.5 and various values of the number of arms $q$. The position on the free surface is scaled with the total surface length.

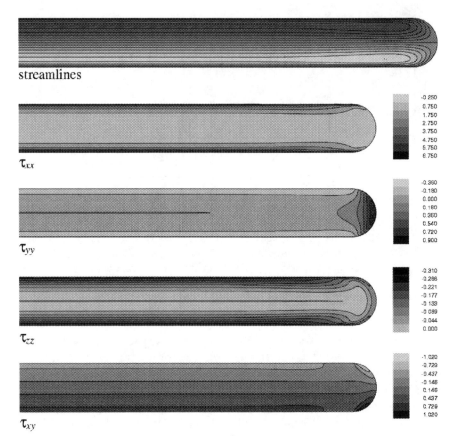

streamlines

$\tau_{xx}$

-0.250
0.750
1.750
2.750
3.750
4.750
5.750
6.750

$\tau_{yy}$

-0.360
-0.180
0.000
0.180
0.360
0.540
0.720
0.900

$\tau_{zz}$

-0.310
-0.266
-0.221
-0.177
-0.133
-0.089
-0.044
0.000

$\tau_{xy}$

-1.020
-0.729
-0.437
-0.148
0.146
0.437
0.729
1.020

FIGURE 13.15   Steady-state result of the XPP fluid for We $= 2.5$ and $(r,q) = (2,5)$.

We $= 2.5$ are presented in Figs. 13.15–13.17. Although there are only minor differences between these graphs, we observe that the viscoelastic stress decay more rapidly for $q = 5$. Due to the increasing shear thinning behavior of the fluid with $q = 5$, the streamlines are somewhat more compressed as compared to the more extensional hardening rheologies.

## 13.5.2   Stability Results

Results of the linear stability analysis of the fountain flows are presented. Starting with an initially random perturbation of the extra stress tensor, we track the $L_2$ norm of the perturbed variables in time. Assuming that the most important eigenmodes are excited in this way, exponential behavior will be obtained when all rapidly decaying modes have died out. Hence, the norm of

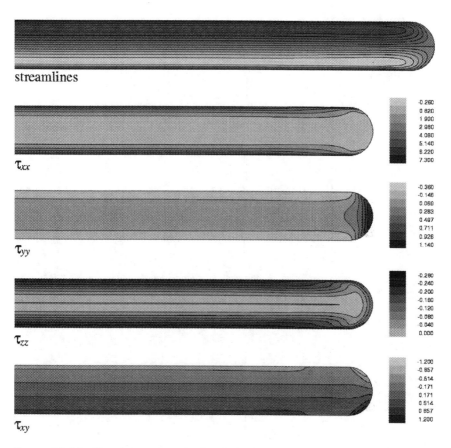

streamlines

$\tau_{xx}$

$\tau_{yy}$

$\tau_{zz}$

$\tau_{xy}$

FIGURE **13.16**    Steady-state result of the XPP fluid for We = 2.5 and $(r,q)$ = (2,9).

the perturbed variables can grow exponentially in time, in which case we will call the flow unstable, or the norm can decay and the flow will remain stable for the limit of small perturbations. From the temporal growth (or decay) of the perturbation norm, the approximate leading growth rate of the flow is determined, which will be larger than zero when the flow is unstable. In addition to an estimate of the most dangerous growth rate, information is obtained about the structure of the leading eigenmode from the solution of the perturbed variables after exponential growth.

Figure 13.18 shows the estimated growth rates for the fountain flow simulations. For the meshes described in Table 13.1 and $(r,q)$ = (2,9) stability results are given in the left graph. We observe that the results converge with decreasing grid size. It should be noted though that for all three meshes the

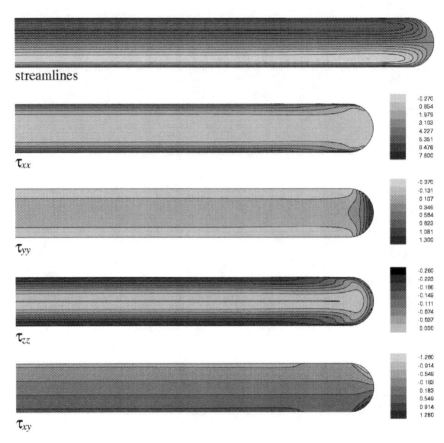

streamlines

$\tau_{xx}$

$\tau_{yy}$

$\tau_{zz}$

$\tau_{xy}$

FIGURE 13.17   Steady-state result of the XPP fluid for We = 2.5 and $(r,q)$ = (2,13).

grids are relatively coarse in the vicinity of the geometrical singularity. This proved necessary because the stability equation for the perturbation of the free surface is very sensitive to disturbances of the wall normal velocity close to the wall. Still, the estimated growth rates converge towards a single curve. The results presented in the sequel are computed on mesh 3 (Table 13.1).

For the XPP fluid with $(r,q)$ = (2,9), which corresponds to the moderate strain hardening material, we observe that the flow loses stability at We ≈ 2.8. Similar trends of the stability curves are plotted in the right graph for the other fluids with $(r,q)$ = (2,5) and $(r,q)$ = (2,13). From this figure it is clear that the point of instability is shifted toward lower Weissenberg numbers when $q$ = 5. Although the steady-state computations failed to converge for We >2.5 and

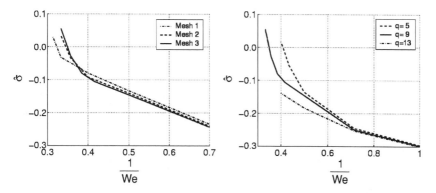

FIGURE **13.18** Comparison of the estimated growth rates for the meshes described in Table 13.1 and $(r,q) = (2,9)$ as a function of the inverse Weissenberg number (left) as well as linear stability results for several fluid rheologies on mesh 3 and $r = 2$ (right).

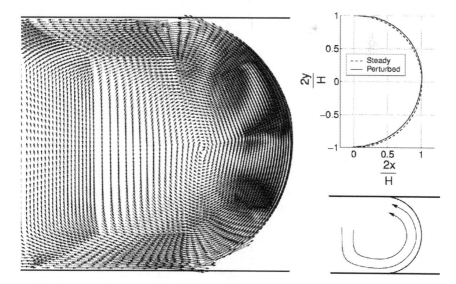

FIGURE **13.19** Results of the linear stability analysis for an XPP fluid for $(r,q) = (2,5)$ and We $= 2.5$. Shown are the perturbation velocities near the free surface (left). Also shown are the linearly perturbed shape of the free surface and a schematic drawing of the swirling flow (right).

$q = 13$, the estimated growth rates are considerably below the other curves, which suggests that the flows may be stabilized when $q$ is increased.

After exponential growth or decay, the characteristic eigenfunction of the flow is obtained. For the unstable flow $(r,q) = (2,5)$ at We $= 2.5$, the perturbation velocity vectors are shown in Fig. 13.19. It can be seen that this eigenfunction corresponds to a swirling motion very similar to the unstable flows that were observed by Bulters and Schepens (16,17). The periodic motion that may be expected from these experiments was not observed during the temporal integration of the disturbance variables. Instead, the perturbation is either clockwise or counterclockwise depending on the initial conditions. If the shape of the perturbed free surface is inspected by adding to the base flow shape an arbitrary constant times the surface perturbation $h$, we obtain the upper right graph of Fig. 13.19. Summarizing the other flow situations (both stable and unstable), the observed characteristic spatial eigenmode was always similar to the swirling flow near the fountain flow surface.

## 13.6 DISCUSSION

A numerical investigation of viscoelastic flow instabilities that may occur during the injection molding of polymer melts was presented. For industrial applications the most important questions to be answered are the following: "What are the critical flow conditions for the onset of instability?" and "How does this relate to the rheology of the polymer itself?" In other words, can the polymer be modified in such a way that the onset of secondary flow is shifted to higher processing rates?

To elucidate these questions and to provide a (numerical) basis for the prediction and analysis of the stability of complex processing flows, some important issues needed to be addressed. This concerns the choice for the constitutive model and the applied numerical algorithm that both play an important role in the analysis of viscoelastic flow instabilities. For instance, it is well-known that most numerical methods can produce approximated solutions that are not solutions of the original continuous problem.

This research focused on the stability problem of the injection molding process. One of the numerically challenging tasks of this flow is the fact that there is a free surface during the filling stage of the mold. This free surface cannot remain stationary due to perturbations of the flow field and, therefore, the deformation of the computational domain was included in the numerical analysis.

A generic injection molding flow has been investigated using a single mode of the eXtended Pom-Pom (XPP) constitutive equations. The influence of the fluid rheology on the stability characteristics of the injection molding flows has been investigated and it was found that the occurrence of the onset of a linear instability can be postponed when the number of arms in the XPP model is increased. Although this has some effect on the shear properties of the different XPP fluids, the major influence of varying the number of arms can be found in the extensional behavior of the fluids. This indicates that flow stabilization can be obtained with increased strain hardening. The structure of the leading eigenmode turns out to be a swirling flow near the fountain flow surface. This is consistent with the experimental observations of Bulters and Schepens (16,17).

This work focuses on the injection molding process but these numerical methods can also be applied to other processing flows that are troubled by viscoelastic instabilities. However, a specific problem of injection molding flows is the presence of geometric singularities where the fountain flow surface intersects with the mold walls. Numerically, this poses problems for both the computation of the steady-state solution as well as the dynamic stability behavior. This is easily seen because the disturbance of the fountain flow surface is defined in the direction parallel to the mold walls. The spatial derivative of the free surface position in the wall normal direction will then approach infinity near the singularities. This results in a formulation where the stability of the flow is sensitive to variations of the wall normal velocity on the free surface near the walls. Overall, this leads to a problem that is not very well posed. This is "resolved" by application of rather coarse grids near the singularities to keep the spatial derivative of the free surface position in the wall normal direction bounded. Further investigation or even reformulation of the stability problem near the singularities may be necessary to fully elucidate the influence of these contact points on the overall stability of the flow.

Variation of the nonlinear parameters of the constitutive model allows for the investigation of the influence of the fluid rheology on the onset of secondary flows. However, for the XPP model (and most other models) the viscometric functions in shear and extension cannot be controlled independently. This means that it is rather difficult to find a general criterion for the occurrence of instability. Such a criterion would be desirable if the susceptibility of polymer melts for the onset of instability needs to be determined beforehand. This drawback could be resolved by application of multimode models as is customary for real polymer melts. Hence, in order to model real viscoelastic melts and to find a general criterion for the onset of unstable flow, a next step would be to incorporate multimode models. Closely related to this is the fact that the physical mechanism of the secondary flows still needs to be elucidated. The analysis of simple shear flows can be performed by a so-called energy analysis where the different contributions of the perturbed variables to the energy perturbation are determined over a period of time. For complex flows, this analysis is somewhat more complicated, and further investigations are necessary to develop the complex variant of this one-dimensional energy analysis.

Not only the experiments of Bulters and Schepens (16,17) but also the preliminary experiments of Rielly and Price (74) indicate that the steady injection molding flow breaks down into a time periodic motion. One of the questions that may be asked is whether this motion is dictated by the linear dynamics (Hopf instability) or by the nonlinear interactions between the perturbed flow variables. The leading eigenmode that is computed using the linear stability analysis described in this work did not show the periodic motion. This means that depending on the initial conditions for the disturbed flow, the swirling flow is either clockwise or counterclockwise. Hence, in order to investigate the time periodic disturbed flow, the full set of nonlinear equations should be taken into account.

It is well-known that comparison between experiments and linear stability analysis often fails. One of the reasons for this is that linear stability analysis provides the critical flow conditions above which no stable flow can exist. Therefore, if the instability observed in the experiments is subcritical, quantitative comparison between the linear stability analysis and experiments will be difficult. Still, with the numerical toolbox that is developed, the linear stability of flows of real viscoelastic melts can be computed and irrespective of the nature of the critical point, experimental validation is not only necessary but essential for the correct modeling and understanding of viscoelastic flow instabilities. In conclusion, one might pose that the stability analysis of a real polymer melt flow is "as good as" the rheological model applied. Therefore, modeling of these secondary flow phenomena is probably one of the most stringent tests of the validity of nonlinear constitutive models.

## REFERENCES

1. Alves, M.A.; Pinho, F.T.; Oliveira, P.J. Study of steady pipe and channel flows of a single-mode Phan-Thien–Tanner fluid. J. Non-Newton. Fluid Mech. 2001, *101*, 55–76.

2. Baaijens, F.P.T. Mixed finite element methods for viscoelastic flow analysis: A review. J. Non-Newton. Fluid Mech. 1998, *79*, 361–386.

3. Beris, A.N.; Avgousti, M.; Souvaliotis, A. Spectral calculations of viscoelastic flows: Evaluations of the Giesekus constitutive equation in model flow problems. J. Non-Newton. Fluid Mech. 1992, *44*, 197–228.

4. Beris, A.N.; Edwards, B.J. *Thermodynamics of Flowing Systems with Internal Microstructure*; Oxford, UK: Oxford University Press, 1994.

5. Bishko, G.B.; Harlen, O.G.; McLeish, T.C.B.; Nicholson, T.M. Numerical simulation of transient flow of branched polymer melts through a planar contraction using the "Pom-Pom" model. J. Non-Newton. Fluid Mech. 1999, *82*, 255–273.

6. Blackwell, R.J.; McLeish, T.C.B.; Harlen, O.G. Molecular drag–strain coupling in branched polymer melts. J. Rheol. 2000, *44* (1), 121–136.

7. Blonce, L. Linear stability of Giesekus fluid in Poiseuille flow. Mech. Res. Commun. 1997, *24* (2), 223–228.

8. Bochev, P.B.; Gunzburger, M.D. Finite element methods of least-squares type. SIAM Rev. 1998, *40* (4), 789–837.

9. Bogaerds, A.C.B.; Grillet, A.M.; Peters, G.W.M.; Baaijens, F.P.T. Stability analysis of the injection molding process using mixed viscoelastic finite element techniques. XIIIth International Congress on Rheology, Cambridge, UK, 2000.

10. Bogaerds, A.C.B.; Grillet, A.M.; Peters, G.W.M.; Baaijens, F.P.T. Stability analysis of polymer shear flows using the extended Pom-Pom constitutive equations. J. Non-Newton. Fluid Mech. 2002, *108*, 187–208.

11. Bogaerds, A.C.B.; Hulsen, M.A.; Peters, G.W.M.; Baaijens, F.P.T. Stability analysis of injection molding flows. J. Rheol. 2002. *submitted.*

12. Bogaerds, A.C.B.; Hulsen, M.A.; Peters, G.W.M.; Baaijens, F.P.T. Time dependent finite element analysis of the linear stability of viscoelastic flows with interfaces. J. Non-Newton. Fluid Mech. *accepted for publication.*

13. Bogaerds, A.C.B.; Verbeeten, W.M.H.; Baaijens, F.P.T. Successes and failures of discontinuous Galerkin methods in viscoelastic fluid analysis. In *Lecture Notes in Computational Science and Engineering. Discontinuous Galerkin Methods*; Springer-Verlag, 2000; 263–270.

14. Brooks, A.N.; Hughes, T.J.R. Streamline Upwind/Petrov Galerkin formulations for convection dominated flows with particular emphasis on the incompressible Navier–Stokes equations. Comput. Methods Sci. Technol. 1982, *32*, 199–259.

15. Brown, R.A.; Szady, M.J.; Northey, P.J.; Armstrong, R.C. On the numerical stability of mixed finite-element methods for viscoelastic flows governed by differential constitutive equations. Theor. Comput. Fluid Dyn. 1993, *5*, 77–106.

16. Bulters, M.; Schepens, A. Flow mark surface defects on injection moulded products: Proposed mechanism. Proceedings of the Annual Meeting of the American Institute of Chemical Engineers, Los Angeles, CA, 2000.

17. Bulters, M.; Schepens, A. The origin of the surface defect "slip–stick" on injection moulded products. Proceedings of the 16th Annual Meeting of the Polymer Processing Society, Shanghai, China, 2000.

18. Canuto, C.; Hussaini, M.Y.; Quarteroni, A.; Zang, T.A. Spectral methods in fluid mechanics. *Springer Series in Computational Physics*; Springer-Verlag, 1998.

19. Carvalho, M.S.; Scriven, L.E. Three-dimensional stability analysis of free surface flows: Application to forward deformable roll coating. J. Comput. Phys. 1999, *151*, 534–562.

20. Chang, M.C.O. On the study of surface defects in the injection molding of rubber-modified thermoplastics. ANTEC '94, 1991, 360–367.

21. Chen, K. Elastic instability of the interface in Couette flow of viscoelastic liquids. J. Non-Newton. Fluid Mech. 1991, *40*, 261–267.

22. Denn, M.M. Extrusion instabilities and wall slip. Annu. Rev. Fluid Mech. 2001, *33*, 265–287.

23. Drazin, P.G.; Reid, W.H. *Hydrodynamic Stability*; Cambridge University Press, 1981.

24. Fan, Y.; Tanner, R.I.; Phan-Thien, N. Galerkin/least-square finite element methods for steady viscoelastic flows. J. Non-Newton. Fluid Mech. 1999, *84*, 233–256.

25. Fortin, M.; Fortin, A. A new approach for the FEM simlation of viscoelastic flows. J. Non-Newton. Fluid Mech. 1989, *32*, 295–310.

26. Fyrillas, M.M.; Georgiou, G.C.; Vlassopoulos, D. Time-dependent plane Poiseuille flow of a Johnson–Segalman fluid. J. Non-Newton. Fluid Mech. 1999, *82*, 105–123.

27. Ganpule, H.K.; Khomami, B. An investigation of interfacial instabilities in the superposed channel flow of viscoelastic fluids. J. Non-Newton. Fluid Mech. 1999, *81*, 27–69.

28. Glowinski, R. Finite element methods for the numerical simulation of incompressible viscous flow: Introduction to the control of the Navier–Stokes equations. In *Vortex Dynamics and Vortex Mechanics*; Lectures in Applied Mathematics; American Mathematical Society, 1992; Vol. 28, 219–301.

29. Glowinski, R.; Pironneau, O. Finite element methods for Navier–Stokes equations. Annu. Rev. Fluid Mech. 1992, *24*, 167–204.

30. Gorodtsov, V.A.; Leonov, A.I. On a linear instability of a plane parallel Couette flow of viscoelastic fluid. J. Appl. Math. Mech. 1967, *31*, 310–319.

31. Gottlieb, D.; Orszag, S.A. Numerical analysis of spectral methods: Theory and applications. *Regional Conference Series in Applied Mathematics 26*; Philadelphia: Society for Industrial and Applied Mathematics, 1977.

32. Graham, M.D. Effect of axial flow on viscoelastic Taylor–Couette instability. J. Fluid Mech. 1998, *360*, 341–374.

33. Graham, R.S.; McLeish, T.C.B.; Harlen, O.G. Using the Pom-Pom equations to analyze polymer melts in exponential shear. J. Rheol. 2001, *45* (1), 275–290.

34. Grillet, A.M.; Bogaerds, A.C.B.; Peters, G.W.M.; Baaijens, F.P.T. Stability analysis of constitutive equations for polymer melts in viscometric flows. J. Non-Newton. Fluid Mech. 2002, *103*, 221–250.

35. Grillet, A.M.; Bogaerds, A.C.B.; Peters, G.W.M.; Bulters, M.; Baaijens, F.P.T. Numerical analysis of flow mark surface defects in injection molding flow. J. Rheol. 2002, *46* (3), 651–670.

36. Grillet, A.M.; Yang, B.; Khomami, B.; Shaqfeh, E.S.G. Modeling of viscoelastic lid driven cavity flow using finite element simulations. J. Non-Newton. Fluid Mech. 1999, *88*, 99–131.

37. Guénette, R.; Fortin, M. A new mixed finite element method for computing viscoelastic flows. J. Non-Newton. Fluid Mech. 1995, *60*, 27–52.

38. Hamada, H.; Tsunasawa, H. Correlation between flow mark and internal structure of thin PC/ABS blend injection moldings. J. Appl. Polym. Sci. 1996, *60*, 353–362.

39. Hinch, E.J. The flow of an Oldroyd fluid around a sharp corner. J. Non-Newton. Fluid Mech. 1993, *50*, 161–171.

40. Ho, T.C.; Denn, M.M. Stability of plane Poiseuille flow of a highly elastic liquid. J. Non-Newton. Fluid Mech. 1977, *3*, 179–195.

41. Hobbs, S.Y. The development of flow instabilities during the injection molding of multicomponent resins. Polym. Eng. Sci. 1996, *32*, 1489–1494.

42. Inkson, N.J.; McLeish, T.C.B.; Harlen, O.G.; Groves, D.J. Predicting low density polyethylene melt rheology in elongational and shear flows with "Pom-Pom" constitutive equations. J. Rheol. 1999, *43* (4), 873–896.

43. Isayev, A.I. *Injection and Compression Molding Fundamentals*; Marcel Dekker: New York, 1987.

44. Joo, Y.L.; Shaqfeh, E.S.G. A purely elastic instability in Dean and Taylor–Dean flow. Phys. Fluids A 1992, *4* (3), 524–543.

45. Kamal, M.R.; Chu, E.; Lafleur, P.G.; Ryan, M.E. Computer simulation of injection mold filling for viscoelastic melts with fountain flow. Polym. Eng. Sci. 1986, *26* (3), 190–196.

46. Kamal, M.R.; Goyal, S.K.; Chu, E. Simulation of injection mold filling of viscoelastic polymer with fountain flow. AIChE J. 1988, *34* (1), 94–106.

47. Keiller, R.A. Numerical instability of time-dependent flows. J. Non-Newton. Fluid Mech. 1992, *43*, 229–246.

48. Khomami, B.; Ranjbaran, M.M. Experimental studies of interfacial instabilities in multilayer pressure-driven flow. Rheol. Acta 1997, *36*, 345–366.

49. King, R.C.; Apelian, M.R.; Armstrong, R.C.; Brown, R.A. Numerically stable finite element techniques for viscoelastic calculations in smooth and singular domains. J. Non-Newton. Fluid Mech. 1988, *29*, 147–216.

50. Larson, R.G. *Constitutive Equations for Polymer Melts and Solutions*; Butterworths: Boston, MA, 1988.

51. Larson, R.G. Instabilities in viscoelastic flows. Rheol. Acta 1992, *31*, 313–363.

52. Larson, R.G.; Muller, S.J.; Shaqfeh, E.S.G. The effect of fluid rheology on the elastic Taylor–Couette instability. J. Non-Newton. Fluid Mech. 1994, *51*, 195–225.

53. Larson, R.G.; Shaqfeh, E.S.G.; Muller, S.J. A purely elastic instability in Taylor–Couette flow. J. Fluid Mech. 1990, *218*, 573–600.

54. Lee, K.-C.; Finlayson, B.A. Stability of plane Poiseuille and Couette flow of a Maxwell fluid. J. Non-Newton. Fluid Mech. 1986, *21*, 65–78.

55. Legrand, F.; Piau, J.-M. Spatially resolved stress birefringence and flow visualization in the flow instabilities of a polydimethylsiloxane extruded through a slit die. J. Non-Newton. Fluid Mech. 1998, *77*, 123–150.

56. Leonov, A.I. Analysis of simple constitutive equations for viscoelastic liquids. J. Non-Newton. Fluid Mech. 1992, *42*, 323–350.

57. Leonov, A.I.; Prokunin, A.N. *Nonlinear Phenomena in Flows of Viscoelastic Polymer Fluids*; Chapman & Hall: London, UK, 1994.

58. Li, J.-M.; Burghardt, W.R.; Yang, B.; Khomami, B. Flow birefringence and computational studies of a shear thinning polymer solution in axisymmetric stagnation flow. J. Non-Newton. Fluid Mech. 1998, *74*, 151–193.

59. Lim, F.J.; Schowalter, W.R. Pseudo-spectral analysis of the stability of pressure-driven flow of a Giesekus fluid between parallel planes. J. Non-Newton. Fluid Mech. 1987, *26*, 135–142.

60. Marchal, J.M.; Crochet, M.J. A new mixed finite element for calculating viscoelastic flow. J. Non-Newton. Fluid Mech. 1987, *26*, 77–114.

61. Mavridis, H.; Hrymak, A.N.; Vlachopoulos, J. Finite element simulation of fountain flow in injection molding. Polym. Eng. Sci. 1986, *26* (7), 449–454.

62. Mavridis, H.; Hrymak, A.N.; Vlachopoulos, J. The effect of fountain flow on molecular orientation in injection molding. J. Rheol. 1988, *32* (6), 639–663.

63. McKinley, G.H.; Pakdel, P.; Öztekin, A. Rheological and geometric scaling of purely elastic flow instabilities. J. Non-Newton. Fluid Mech. 1996, *67*, 19–47.

64. McLeish, T.C.B.; Larson, R.G. Molecular constitutive equations for a class of branched polymers: The Pom-Pom polymer. J. Rheol. 1998, *42* (1), 81–110.

65. Monasse, B.; Mathieu, L.; Vincent, M.; Haudin, J.M.; Gazonnet, J.P.; Durand, V.; Barthez, J.M.; Roux, D.; Charmeau, J.Y. Flow marks in injection molding of polypropylene: Influence of processing conditions and formation in fountain flow. Proceedings of the 15th Annual Meeting of the Polymer Processing Society, 's Hertogenbosch, the Netherlands, 1999.

66. Öztekin, A.; Brown, R.A.; McKinley, G.H. Quantitative prediction of the viscoelastic instability in cone-and-plate flow of a Boger fluid using a multi-mode Giesekus model. J. Non-Newton. Fluid Mech. 1994, *54*, 351–377.

67. Petrie, C.J.S.; Denn, M.M. Instabilities in polymer processing. AIChE J. 1976, *22* (2), 209–225.

68. Rajagopalan, D.; Armstrong, R.C.; Brown, R.A. Finite element methods for calculation of steady viscoelastic flow using constitutive equations with a Newtonian viscosity. J. Non-Newton. Fluid Mech. 1990, *36*, 159–192.

69. Renardy, M. Location of the continuous spectrum in complex flows of the UCM fluid. J. Non-Newton. Fluid Mech. 2000, *94*, 75–85.

70. Renardy, M. Effect of upstream boundary conditions on stability of fiber spinning in the highly elastic limit. J. Rheol. 2002, *46* (4), 1023–1028.

71. Renardy, M.; Renardy, Y. Linear stability of plane Couette flow of an upper convected Maxwell fluid. J. Non-Newton. Fluid Mech. 1986, *22*, 23–33.

72. Renardy, Y. Stability of the interface in two-layer Couette flow of upper convected Maxwell liquids. J. Non-Newton. Fluid Mech. 1988, *28*, 99–115.

73. Renardy, Y.; Olagunju, D.O. Inertial effect on the stability of cone-and-plate flow: Part 2. Non-axisymmetric modes. J. Non-Newton. Fluid Mech. 1998, *78*, 27–45.

74. Rielly, F.J.; Price, W.L. Plastic flow in injection molds. SPE J., 1961; 1097–1101.

75. Rose, W. Fluid–fluid interfaces in steady motion. Nature 1961, *191*, 242–243.

76. Saramito, P. Efficient simulation of nonlinear viscoelastic fluid flows. J. Non-Newton. Fluid Mech. 1995, *60*, 199–223.

77. Sato, T.; Richardson, S.M. Numerical simulation of the fountain flow problem for viscoelastic liquids. Polym. Eng. Sci. 1995, *35* (10), 805–812.

78. Seydel, R. *Practical Bifurcation and Stability Analysis: From Equilibrium to Chaos*; Springer-Verlag: New York, 1994.

79. Shaqfeh, E.S.G. Purely elastic instabilities in viscometric flows. Annu. Rev. Fluid Mech. 1996, *28*, 129–185.

80. Shen, S.-F. Grapplings with the simulation of non-Newtonian flows in polymer processing. Int. J. Numer. Methods Eng. 1992, *34*, 701–723.

81. Smith, M.D.; Armstrong, R.C.; Brown, R.A.; Sureshkumar, R. Finite element analysis of stability of two-dimensional viscoelastic flows to three-dimensional perturbations. J. Non-Newton. Fluid Mech. 2000, *93*, 203–244.

82. Sureshkumar, R. Numerical observations on the continuous spectrum of the

linearized viscoelastic operator in shear dominated complex flows. J. Non-Newton. Fluid Mech. 2000, *94*, 205–211.

83. Sureshkumar, R. Local linear stability characteristics of viscoelastic periodic channel flow. J. Non-Newton. Fluid Mech. 2001, *97*, 125–148.

84. Sureshkumar, R.; Beris, A.N. Linear stability analysis of viscoelastic Poiseuille flow using an Arnoldi-based orthogonalization algorithm. J. Non-Newton. Fluid Mech. 1995, *56*, 151–182.

85. Sureshkumar, R.; Smith, M.D.; Armstrong, R.C.; Brown, R.A. Linear stability and dynamics of viscoelastic flows using time-dependent numerical simulations. J. Non-Newton. Fluid Mech. 1999, *82*, 57–104.

86. Swartjes, F.H.M. Stress Induced Crystallization in Elongational Flow. PhD thesis, Eindhoven University of Technology, November 2001.

87. Szady, M.J.; Salomon, T.R.; Liu, A.W.; Bornside, D.E.; Armstrong, R.C.; Brown, R.A. A new mixed finite element method for viscoelastic flows governed by differential constitutive equations. J. Non-Newton. Fluid Mech. 1995, *59*, 215–243.

88. Tadmor, Z. Molecular orientation in injection molding. J. Appl. Polym. Sci. 1974, *18*, 1753–1772.

89. Talwar, K.K.; Ganpule, H.K.; Khomami, B. A note on the selection of spaces in computation of viscoelastic flows using the hp-finite element method. J. Non-Newton. Fluid Mech. 1994, *52*, 293–307.

90. Talwar, K.K.; Khomami, B. Higher order finite element techniques for viscoelastic flow problems with change of type and singularities. J. Non-Newton. Fluid Mech. 1995, *59*, 49–72.

91. Verbeeten, W.M.H.; Peters, G.W.M.; Baaijens, F.P.T. Differential constitutive equations for polymer melts: The eXtended Pom-Pom model. J. Rheol. 2001, *45* (4), 823–844.

92. Wang, S.-Q. Molecular transitions and dynamics at polymer / wall interfaces: Origins of flow instabilities and wall slip. Adv. Polym. Sci. 1999, *138*, 227–275.

93. Wilson, H.J.; Rallison, J.M. Instability of channel flow of a shear-thinning White–Metzner fluid. J. Non-Newton. Fluid Mech. 1999, *87*, 75–96.

94. Wilson, H.J.; Renardy, M.; Renardy, Y. Structure of the spectrum in zero Reynolds number shear flow of the UCM and Oldroyd-B liquids. J. Non-Newton. Fluid Mech. 1999, *80*, 251–268.

# Index

Printed in the United States
by Baker & Taylor Publisher Services